ESSAYS ON GIORDANO BRUNO

ESSAYS ON GIORDANO BRUNO

HILARY GATTI

PRINCETON UNIVERSITY PRESS

PRINCETON AND OXFORD

Copyright © 2011 by Princeton University Press

Published by Princeton University Press, 41 William Street,
Princeton, New Jersey 08540
In the United Kingdom: Princeton University Press, 6 Oxford Street,
Woodstock, Oxfordshire OX20 1TW

press.princeton.edu

Library of Congress Cataloging-in-Publication Data

Gatti, Hilary.
[Essays. English. Selections]
Essays on Giordano Bruno / Hilary Gatti.
    p.   cm.
Includes bibliographical references and index.
ISBN 978-0-691-14574-7 (cloth : alk. paper) —
ISBN 978-0-691-14839-7 (pbk. : alk. paper)
1. Bruno, Giordano, 1548-1600. I. Title.
 B783.Z7G38 2011
   195--dc22   2010012090

British Library Cataloging-in-Publication Data is available

This book has been composed in Sabon

Printed on acid-free paper. ∞

Printed in the United States of America

10 9 8 7 6 5 4 3 2 1

To my friend Ingrid Rowland

CONTENTS

PREFACE

A good question to start with might be: "Why Giordano Bruno?"

IT ALL BEGAN WITH A COURSE I was giving on Shakespeare's *Hamlet* at the University of Rome "La Sapienza" in Italy. During my reading of a small slice of the daunting amount of critical material on Shakespeare's tragedy (someone somewhere has done a statistical count and arrived at the conclusion that to read all the criticism dedicated to Hamlet alone would require more than a single life span), I came across a reference to a theory that some of Hamlet's characteristics might have been based on the life and tragic death of Giordano Bruno, in the Campo dei Fiori (The Field of Flowers), in Rome. Although I was living in Rome at the time, and aware of where the Campo dei Fiori was, and of the imposing statue that dominates that lovely square, I realized that I had little idea of whom the statue was commemorating or why. I dedicated some hours to it, and the next day went on a visit of discovery to the philosophy library that constitutes one of the areas of excellence of my erstwhile university, where I would spend many hours in the coming years. I brought home with me the dialogue especially mentioned on the page I had been reading, *La cena de le ceneri*, or *The Ash Wednesday Supper*, in the Einaudi edition of 1955 edited by Giovanni Aquilecchia—a fine scholar who would soon become a major point of reference throughout my Bruno studies. I remember reading it through in a few hours, getting to the end almost out of breath. I found it stunning, overwhelming. What was this? Philosophy or cosmology or drama? Or philosophy and cosmology in the form of a drama? And what kind of language was this? Like no Italian I had ever read or heard anywhere before, either ancient or modern. And how did this man from Nola (where was Nola?) know so much about Elizabethan London? Of one thing only I felt quite sure then, and I have never changed my mind: Shakespeare had been through those pages. Somehow, somewhere, he had either read them or had them read to him. For *The Ash Wednesday Supper* is an extraordinary theater of conflicting ideas, as indeed (as I would discover later) are all Bruno's Italian dialogues, and his Latin works as well. But *The Ash Wednesday Supper* in particular, which I would later translate into English, remained and still remains a center of particular interest and concern: a hub around which would coil and develop in the coming years and decades my various and varied studies of Bruno's life and thought.

A large number of those studies are collected together in this volume, arranged in sections according to some major themes, rather than chronologically. It is no coincidence that the volume begins with some of my most recent thoughts on *The Ash Wednesday Supper,* for that is where it all started. It is also a work in which Bruno clearly and explicitly depicts a prophetic image of where it would all end: in his death at the stake as a despised heretic. Shakespeare too understood the drama of minds so innovative, so audacious, that their societies were unable to respond to them except by decreeing their eternal silence.

During the course of my Bruno studies, I have accumulated more debts of gratitude to colleagues, librarians, friends, and family members than I can possibly mention. There are some, however, who cannot remain unnamed. This book was made possible by the constant support and encouragement of the two deans of the newly established Faculty of Philosophy of the University of Rome "La Sapienza," Marco Olivetti, whose premature death was a tragedy for us all, and Marta Fattori, who has so ably carried on his inheritance. Much of my work on Bruno, however, was done in London, beginning with a sabbatical year in 1981–1982 during which I discovered the extraordinary library of the Warburg Institute: an academic institution whose name had already for many years been closely associated with Bruno's own. The then director of the Warburg Institute, J. B. Trapp, became at once, and remained until his death, a constantly courteous and stalwart friend of my Bruno studies, to whom I am grateful in more ways than I can hope to express. It was Professor Trapp who introduced me one memorable morning to Frances Yates, in the Warburg Institute common room, by then an imposing elderly and somewhat forbidding lady, slightly alarmed at the arrival on her territory of an aspiring Bruno scholar not especially concerned with either magic or the occult. She did, nevertheless, offer me generous encouragement to use my knowledge of the Italian language to translate Bruno, insisting on the need for a properly coordinated edition of the philosophical dialogues in English, which only today is slowly getting under way. The CMRS Bruno is to be part of the Lorenzo da Ponte Italian Library (LPIL) published by the University of Toronto Press: a series of English translations of canonical Italian texts. Produced by the Center for Medieval and Renaissance Studies (CMRS) at UCLA under the general editorship of David Marsh, the Bruno project aims to present a complete English version, with facing-page Italian texts, of all six of Bruno's Italian dialogues. My translation of *The Ash Wednesday Supper,* on which I started work shortly after my meeting with Frances Yates, has been incorporated into this series.

A much later stage of my Bruno studies is associated with a term of membership at the School of Historical Studies of the Institute for Advanced Study in Princeton, New Jersey. I am grateful to Jonathan Israel, who coordinates the early modern scholars there, for encouragement to continue my Bruno studies as part of a wider project on ideas of liberty in sixteenth-century Europe. I also owe a debt of gratitude to Quentin Skinner who, in a later meeting in England, advised me to approach the UK office of Princeton University Press. The competence of the humanities editor there, Ian Malcolm, has been essential in getting this book through the press.

Last, in the private sphere, no words can suffice to thank my late husband, Mariano, our children, and their children for the good humor with which they have always accepted the intrusion of Bruno not only into my professional life but also, at times, within our domestic walls as well. Without them, this book would never have been written.

HILARY GATTI
May 2010
Chianni (Italy)

ACKNOWLEDGMENTS

APART FROM THE INTRODUCTORY and final chapters to this volume of collected Bruno essays, which are so far unpublished, the other chapters listed here are all reproduced with kind permission of the publishers and/or editors of the volumes in which they first appeared. Those that were published originally in Italian have been translated into English by the author. In some cases, the essays have been modified to adapt them to the format and purposes of the present volume. Footnotes have been updated, and occasional passages have been rewritten in the light of more recent discussion of Bruno's life and works.

BEGINNING AS NEGATION IN THE ITALIAN DIALOGUES OF GIORDANO BRUNO

This previously unpublished introduction is based on a paper read at the annual conference of the Californian Italianists, held at Stanford University on March 6–7, 2009. The subject of the conference was "Beginnings." I am grateful to Michael Wyatt for his invitation to give the keynote introductory lecture to a particularly lively and attentive public of scholars.

BETWEEN MAGIC AND MAGNETISM: BRUNO'S COSMOLOGY AT OXFORD

An early version of this paper was read, in English, at the "History of Science" seminar of the Modern History Faculty of the University of Oxford directed by Professor Robert Fox. A longer version of the paper, in Italian, was published in *Paradigmi: Rivista di critica filosofica* XVIII (May/August 2000): 237–60.

BRUNO'S COPERNICAN DIAGRAMS

This paper was first published, in English, in a collection of essays by various hands dedicated to the diffusion of the Copernican theory. It appeared in *Filozofski vestnik* (the journal of the Institute of Philosophy at ZRC SAZU, Ljubljana, Slovenia) 25, no. 2 (2004): 25–50.

BRUNO AND THE NEW ATOMISM

A first version of this paper was read at the conference "Late Medieval and Early Modern Corpuscular Matter Theories" held at St. Andrews University in Scotland in August 1996. It was first published, in a slightly different form, in the volume of the same title edited by Christoph Lüthy, John E. Murdoch, and William R. Newman, Leiden, Brill, 2001: 163–80. Its title in that volume was "Giordano Bruno's Soul-Powered Atoms: From Ancient Sources toward Modern Science."

THE MULTIPLE LANGUAGES OF THE NEW SCIENCE

An early Italian version of this paper was read at the conference "Giordano Bruno e la scienza nuova: storia e prospettive" held in Rome on February 16–19, 2000. A longer version titled "I linguaggi molteplici della nuova scienza nel pensiero di Giordano Bruno" was published with the conference papers, edited by Eugenio Canone and Arcangelo Rossi, in *Physis: Rivista internazionale di storia della scienza* XXXVIII (2001): 391–411.

PETRARCH, SIDNEY, BRUNO

The original draft of this paper was read at the British Academy in London in 2004 at the conference on "Petrarch in Britain" celebrating the 700th anniversary of the poet's birth. It was read again as a lecture in the Italian Department of Yale University, New Haven, Connecticut, on February 17, 2005, and has greatly benefited from discussion on that occasion. I am grateful to Professor Giuseppe Mazzotta for the invitation to lecture at Yale. The paper appeared in print in 2007 in the volume *Petrarch in Britain: Interpreters, Imitators, and Translators over 700 Years*, edited by Martin McLaughlin and Letizia Panizza for the British Academy and published by Oxford University Press.

THE SENSE OF AN ENDING IN BRUNO'S *HEROICI FURORI*

This paper was originally read at the American Academy in Rome at a conference dedicated to Bruno's *Heroici furori* on May 9–10, 2003. It was first published in *Nouvelles de la republique des lettres* II (2006): 77–90, and then in the volume *The Alchémy of Extremes: The Laboratory of the* Heroici furori *of Giordano Bruno*, edited by Eugenio Canone and Ingrid D. Rowland, Pisa–Rome, Istituti Editoriali Poligrafici Internazionali, 2007.

BRUNO AND SHAKESPEARE: *HAMLET*

The earliest version of a paper on Bruno and *Hamlet* appeared in my volume *The Renaissance Drama of Knowledge*, London, Routledge, 1989, 114–64. Appendix II of the same volume presented a critical bibliography of the history of Bruno–Shakespeare criticism, with particular attention to the discussion of Bruno and *Hamlet*. An invitation from Agostino Lombardo to translate this material into Italian for his series *Piccola Biblioteca Shakespeariana*, published in Rome by Bulzoni, led to a radical rethinking of the subject and a substantially modified text titled *Il teatro della coscienza: Giordano Bruno e "Amleto,"* published in the same series in 1998. The chapter published here is an English translation of this small volume, shortened and to some extent modified to align it with the format of the present volume.

BRUNO'S *CANDELAIO* AND BEN JONSON'S *THE ALCHEMIST*

This paper was first read, in Italian, at a conference titled "Filosofia e commedia" held at Torre in Pietra (Rome) and was published in the volume *Teatri barocchi: tragedie, commedie, pastorali nella drammaturgia europea fra '500 e '600*, edited by Silvia Carandini, Rome, Bulzoni, 2000: 325–37.

BRUNO AND THE STUART COURT MASQUES

This paper was first published in *Renaissance Quarterly* XLVIII, no. 4 (Winter 1995): 809–42.

ROMANTICISM: BRUNO AND SAMUEL TAYLOR COLERIDGE, AND BRUNO AND THE VICTORIANS

These two papers were first published as a single text, in Italian, titled "Giordano Bruno nella cultura inglese dell'800." They appeared in the volume *Brunus redivivus: Giordano Bruno nell'800 europeo*, edited by

Eugenio Canone, Pisa–Rome, Istituti Editoriali e Poligrafici Internazionali, 1998: 19–66. A slightly different version of the section on "Coleridge's Reading of Giordano Bruno" had previously been published in English in *The Wordsworth Circle* XXVII, no. 3 (Summer 1996): 136–45. The two chapters published here are English translations of the Italian language text, divided into two separate chapters and slightly modified to align them with the format of the present volume.

Bruno's Natural Philosophy

This paper was first published in *Midwest Studies in Philosophy*, in a number dedicated to Renaissance and Early Modern Philosophy, edited by Peter A. French and Howard K. Wettstein, XXVI (2002): 111–23.

Bruno's Use of the Bible in His Italian Philosophical Dialogues

This paper was first read, in Italian, at the conference "La filosofia di Giordano Bruno: Problemi ermeneutici e storiografici," held in Rome on October 23–24, 1998. It was published in the volume of the same title, edited by Eugenio Canone, Florence, Olschki, 2003: 199–216.

Science and Magic: The Resolution of Contraries

Read in an early version, in Italian, at a conference held in Naples in the year 2000, this paper was first published with the title "Scienza e magia nel pensiero di Giordano Bruno" in the volume *La mente di Giordano Bruno*, edited by Fabrizio Meroi, with an introductory essay by Michele Ciliberto, Florence, Olschki, 2004, 307–32.

Bruno and Metaphor

This paper was first read, in English, at a conference on "The History of the Humanities" held at the University of Amsterdam in Holland. I am grateful to Rens Bod for an invitation to speak at this conference. The chapter is also scheduled to appear in R. Bod et al., eds., *The Making of the Humanities. Vol I: The Humanities in Modern Europe*, Amsterdam, Amsterdam University Press, 2010.

Why Bruno's "A Tranquil Universal Philosophy" Finished in a Fire

This so far unpublished paper was first read, in English, in the version presented here, at the Italian Cultural Institute in Toronto, Canada, on May 27, 2004. A shorter version was read some days later at the conference "Ideas under Fire," sponsored by the Jacques Maritain Society of Canada, at the University of Manitoba in Winnipeg on May 30–31, 2004. I am grateful to Jonathan Lavery for his invitation to speak at this conference and to the Italian Cultural Institute of Toronto for sponsoring my journey to Canada.

ESSAYS ON GIORDANO BRUNO

Introduction

BEGINNING AS NEGATION IN THE ITALIAN DIALOGUES OF GIORDANO BRUNO

SUPPOSE THAT AN AUTHOR HAS written a book dedicated to a figure of importance—for example, the French ambassador in London in the year 1584. The book opens with a gesture of almost theatrical physicality: the author represents himself as handing his book respectfully to his patron. It is a present. Without further ado, he starts to outline its contents. It is a supper (the reader knows that from the title, *La cena de le ceneri / The Ash Wednesday Supper*): but what kind of a supper is it? The author of the book, who has already revealed himself on the title page as Giordano Bruno, at this point launches into a scintillating, half ironic description, not of what his book is, but of what it is not.[1] Both classical and Biblical antiquity are eliminated in a trice as the reader learns that the supper in question is *not* a celestial banquet with Jupiter as its host, *nor* a supper in Paradise with our first parents, Adam and Eve. The first sentence, which occupies half the opening page of the book, continues by summarily deleting a string of other celebrated, mythical suppers, thus ushering the reader into the reality of the modern world. The second sentence modulates from straight negatives into a series of contraries or opposites, each of which negates the other. The supper is, at the same time, large and small, sacrilegious and religious, Florentine-lean and Bolognese-fat, comic and tragic, as well as many other contrary adjectives besides. The third sentence elaborates a little further toward a positive stance: the supper is going to be hard on the Aristotelians because the Peripatetics smell; on the other hand, it is possible to eat and drink with the Pythagoreans and the Stoics, respectively.[2] Something can thus be salvaged of classical antiquity. What this leads to for Bruno, however, is *not* the New Testament any more than the Old. For there too we find a negative: the book may be called *The Ash Wednesday Supper*, but it is *not* a supper of ashes. Only at that point does the reader finally start to learn what the supper really was and is: a symposium hosted by Sir Fulke Greville, among whose guests were two absurd Neoaristotelian scarecrows who will be derided mercilessly throughout the text. This remarkable opening gambit is brought to an end with the claim that the satire that animates the narrative is not to be seen as an end in itself. At the heart of the book, Bruno claims, there is serious speculation, rational and moral, metaphysical and mathematical, as well as a natural philosophy.

This beginning of Bruno's text is the beginning of his preface, or his *Proemiale epistola*, only. The beginning of the text proper, or of the first of the five dialogues of which it is composed, is elaborated in terms of a semiserious excursus on the significance of the number two. Before engaging with this further beginning, it needs to be noticed that recent critical consensus tends toward a reading of Bruno's six Italian dialogues written and published in London as a composite whole. In that case, the narrative developed from these beginnings is brought to its end in the equally remarkable final pages of the last of these dialogues, the *Heroici furori*, or *Heroic Frenzies*. There Bruno brings his story to its climax in a moment of ecstatic vision on the part of nine blind philosophers who have left Italy and arrived on the banks of the river Thames. The chief nymph of the place (commonly considered as an image of Queen Elizabeth I, the by then almost mythical Virgin Queen) sprinkles on their eyes a liquid contained in a vase given the philosophers at the beginning of their journey by Circe, who had struck them blind. The powerful but obscure magic of Circe had not been sufficiently strong to allow her to open the vase and restore the philosophers' sight. That only the English nymph is empowered to do.[3]

It should be noticed here that the power of the English nymph remains a double power symbolized by her two shining eyes, images of beauty and truth. The ecstatic moment of illumination that closes this cycle of philosophical texts thus remains in the world of multiplicity, completing what Bruno had already announced in the opening page of the *Supper* as a natural philosophy. Furthermore, the ending recalls the beginning as a modern version of Pythagoreanism insofar as the moment of illumination is presented in terms of music, with all the philosophers playing on their various instruments in an ecstatic vision of a newly infinite universe pervaded by a spirit of rational unity and harmony. Beginnings beget endings, not only of texts but also of the world itself. Bruno remembers this from what was clearly an intense study of the Bible in his early monastic years. Even as he repudiates the theological and spiritual message of both the Old and the New Testaments, he incorporates into his own philosophical vision, and his own texts, the sense of Genesis and Apocalypse. Yet at the same time, he denies them, for his own infinite universe is eternal in time. Although the single bodies in it are born and die, time itself stretches to eternity in a rapturous celebration of eternal life *in this world*.[4]

Back to a beginning, then, that is certainly more than just a beginning of a text, but not a representation of a beginning of time itself. Rather, the first dialogue of *The Ash Wednesday Supper* begins by picking up the contraries of the *Proemiale epistola* and celebrating, in part seriously and in part ironically, the number two: the conceptual or logical starting point of a world of multiplicity.[5] Again what is being stressed is the Pythagorean root of Bruno's thought: not for nothing does *The Ash Wednesday Supper* itself define Bruno's own philosophical "school" of thought as "the Pythagorean school and our own." For Pythagoras is considered to have given us the theory that what gives form to the Unlimited is Limit: an idea that finds its major expression in the discovery of the numerical ratios that determine the intervals of the musical scale. Another field of study in which the ancient Pythagoreans discovered the idea of limits was medicine, given that the body is governed by opposites such as hot and cold, wet and dry. The good physician finds the proper blend between these contraries, obtaining a reconciliation or harmony among opposites, according to an analogy between the human body and a musical instrument. In a famous page of Plato's *Phaedo*, Simmias, the Pythagorean philosopher from Thebes who has come to Athens to comfort Socrates in his last moments, claims that the body is strung like a musical instrument, negatives such as hot and cold, wet and dry, taking the place of high and low in music.[6] Behind this Pythagorean discourse lies the idea that all things are numbers and that this principle applies throughout an infinite and eternal cosmos. The explicit Pythagorean reminiscences that color Bruno's beginnings of both the *Proemiale epistola* and of Dialogue 1 of *The Ash Wednesday Supper* prepare the reader for Bruno's own cosmological discourse. For Bruno in this work, in the course of intense discussion with his scandalized Neoaristotelian opponents, will extend the new Copernican idea of heliocentricity to an infinite, eternal universe inhabited by an infinite number of worlds. Multiplicity, we might say, is taken by Bruno to both its cosmological and its logical extreme.

If *The Ash Wednesday Supper* was immediately considered a "scandalous" text already within the Elizabethan culture of Renaissance England where it was originally written and published, that was largely because it denied the creation story of the Bible. Bruno himself underlines this in the opening of the fourth dialogue of his work, where it is pointed out by the Englishman Smitho that the Divine Scriptures in many places suppose and state exactly the opposite with respect to the cosmological hypotheses put forward by Theophilus, the character who stands here for Bruno himself.[7] Yet it should be remembered that alternative creation myths, alternative stories of the beginning of the world, were not absent from the culture of early modern Europe. Much emphasis has been placed in recent years on the importance of the texts forming the so-called *Hermetica*, such as the *Pimander*: a dialogue in which an alternative creation myth is proposed by Hermes Trismegistus. Marsilio Ficino was so struck with the *Pimander* that, in 1471, he used the name as the title of his entire collection of Latin translations of the Hermetic texts from the original Greek. The work of Frances Yates has established that Bruno was certainly reading the Hermetic texts in Ficino's Latin translation, so it is worth pausing a moment to look briefly at this alternative myth of the beginnings of the world.[8]

At one point in the *Pimander*, Trismegistus sees within himself, in his mind, the light and an innumerable number of Powers, a limitless world and the fire enveloped in an all-powerful force. He asks Pimander: "The elements of nature—whence have they arisen?" Pimander replies, "From the counsel of god, which having taken in the word and having seen the beautiful cosmos, imitated it, having become a cosmos."[9] And so the mind of god, existing as life and light, brought forth a second mind, or craftsman, who being the god of fire and breath, fashioned the seven governors, who envelop with their circles the sensible world. Now the mind, the father of all beings, gave birth to a man similar to himself, whom he loved as his own child. When he saw the creation that the craftsman had fashioned in the fire, the man wished also to produce a work, and permission to do this was given him by the father. Then the man leaned over the cosmic framework and showed to lower nature the fair form of god. When she saw that he had in him the inexhaustible beauty of the form of god, nature smiled and embraced the man, and thus they became lovers.

Such is the Hermetic account of the beginnings of the universe, in which the fires in the firmament above come to coincide, as in an embrace, with the fires in the firmament below. Now that we know, as Bruno did not, that these texts belong to the first centuries of Christianity, and not to the ancient world, we can see how they have absorbed the spirituality of the opening of the Gospel of St. John: "In the beginning was the Word, and the Word was with God, and the Word was God." There are, however, important differences in the Hermetic account that make it into a definitely alternative version of the beginnings of the universe with respect to the Biblical story. For example, there is the idea of limitless elements existing in a preceding state of powerful chaos before the action of a chain of demiurgic creators, who include a demiurgic man himself. Then there is, as Frances Yates pointed out in some recently published notes on her reading of the *Hermetica*, a total absence of any idea of sin and redemption: the demiurgic man is marvelously beautiful, and nature smiles with love at seeing his shadow fall over the world.[10] We are a long way here from the anguished guilt of Adam and Eve, or from a nature that, to use Milton's phrase, "sighing through all her works gave signs of woe."[11] Yet it can, in my opinion, be fairly claimed that the Hermetic texts played a larger role in the final three Italian dialogues, which develop Bruno's moral philosophy, than in the first three concerned with his cosmology. In the final three dialogues, the *Hermetica* are explicitly cited;

in *The Ash Wednesday Supper*, they are not. Although the "embrace" of the world above and the world below becomes an important element in Bruno's cosmology, he seems mostly to have used in *The Ash Wednesday Supper* the Hermetic idea of a universe without limits, particularly the image of a cosmos seen as an infinite circle whose center is everywhere and whose circumference is nowhere. This image actually derives from a twelfth-century Latin text called the *Book of Twenty Four Philosophers*, full of Hermetic reminiscences, but not included in Ficino's volume of canonical Hermetic texts.[12] If we want to find the alternative philosophical tradition most consistently opposed by Bruno to the Biblical account of beginnings, we should look elsewhere.

Within the culture of the early modern world, we find the survival of yet another account of the beginnings of the universe that had never entirely disappeared, even during the long centuries of the so-called Middle Ages. For there was one Platonic text known throughout the Middle Ages: the *Timaeus*, or the text in which Plato outlines his mythological account of the creation of the universe. The best known of the Latin versions of the *Timaeus* that survived into the Middle Ages was by Calcidius, which included an annotated commentary. So it is interesting to note that Plato's alternative creation myth never completely disappeared from view, even during an era dominated by the Christian faith in Biblical narrative. This is largely because Calcidius, possibly a fourth-century Spanish ecclesiastic (although little is known about him), was clearly working in a cultural context of Christianizing Neoplatonism that attempted to incorporate Platonic ideas into the Christian doctrines.

Bruno is likely to have been familiar with Plato's *Timaeus* since his early monastic years in Naples.[13] By the time he reached London, however, he was thinking about cosmology in philosophical rather than theological terms. It was not so much the details of the creation story at the hands of yet another demiurge, told by Timaeus in Plato's text, that interested him. Rather it was the Pythagorean strand in Plato himself that comes out so elegantly and forcefully in the opening gambit of Socrates, who starts off this dialogue from the phenomenon of numbers: "One, two, three, but where, my dear Timaeus, is the fourth of those who were yesterday my guests ... ?"[14] Plato's beginning of the *Timaeus* must rate by any standards as a stroke of literary genius. The passage from unity to multiplicity is lightly balanced by the passage from the indeterminate two to the determinate, empirical, even banal problem of the number of guests present at the time. The apparently casual conversational tone covers with a smile the statement of the philosophical problem underlying the whole text: that of the vexed relationship between an eternal and unchanging unity and the phenomenological multiplicity of the world and the passing of time. Hans Georg Gadamer, one of the major Platonic commentators of the last century, and himself a philosopher of note, has written on these opening pages of the *Timaeus* in a book titled *Dialogue and Dialectic*:

> The doctrine of the indeterminate Two is a doctrine of the primordial discrepancy between essence and phenomenon, a discrepancy which is as inchoately expressed in the *Timaeus* as it is in Parmenides' doctrinal poem, a poem which appends a description of the dual world of oppositions to the Eleactic teaching on unity.[15]

It is surely with a deliberate if ironic glance at this beginning of the *Timaeus* that Bruno begins his own first cosmological work by comparing innumerable opposites within the natural world, followed by his semiserious excursus on the number two—that is, by introducing his reader at once into a world characterized by a sign of negativity. For once you move conceptually beyond *one* to the idea of *two*, you have something that is *not-one*. As Bruno's Theophilus explains, citing explicitly Pythagoras, the first coordinates in the universe are always two, not one, for which reason two has to be considered as a mysterious number.[16] It is perhaps symbolic that Bruno explicitly cites the *Timaeus* twice in *The Ash Wednesday Supper*. Once, Plato in the *Timaeus* appears after a list of Pythagorean philosophers, as someone who, even if timidly and obscurely, upheld the idea of an earth that moves rather than staying still at the center of the universe. Bruno emphasizes the fact that Copernicus himself had quoted Plato's *Timaeus* in this sense. The second citation occurs when Bruno is arguing for the eternity of his universe: an argument he reinforces by claiming that Plato in the *Timaeus* held that the stars are not subject to dissolution. Although disagreeing with the idea of a universe created in time, and with the idea of a demiurgic creator (in that, he was closer to Aristotle than to Plato), Bruno leaves his reader in no doubt as to the importance of Plato's *Timaeus* as a source for his own cosmological discourse.[17]

For example, Bruno also, like Plato, dwells on the importance of the number four. For as in Plato's discussion concerning the natural universe, his dialogue too is composed ideally of four participants. Both Plato and Bruno are undoubtedly remembering the particular significance given to the number four by the Pythagorean philosophy, whose number symbolism considered the square of two as containing all possible contraries and therefore as standing for the entire ocean of multiple being.[18] This great ocean of multiplicity, which Bruno thinks of as reflecting an infinite substance, is impregnated with a principle of negativity that the mind attempts to resolve in a spirit of dialectic and dialogue. Bruno would certainly have had in mind Plato's Socrates, who arrives at his conclusion that he knows nothing only by constant questioning and doubting: the necessary foundation from which he attempts to rise to a perception of certain and true ideas. Plato's pupil Aristotle would, on the contrary, substitute the form of the dialogue with the treatise, despising the Pythagorean opposites that he felt lead only to confusion and uncertainty. In his *De monade, numero et figura* published in Frankfurt in 1591, and based on mystical Pythagorean number symbolism, Bruno himself will write of the number two that it is the first foundation of all numbers according to which there is one thing on this side and another on that, a subject and an object, something subtracted and something added, so that now concord and agreement will no longer be possible, as division has entered between you and me. Aristotle claims in his *Metaphysics* that the original pairs of Pythagorean contraries were ten, and rejects later additions by disciples such as Alcmaeon. The contraries listed by Aristotle are limited and unlimited, odd and even, one and many, right and left, male and female, rest and motion, straight and curved, light and darkness, good and bad, square and oblong.[19] Substituting Platonic dialectic with syllogism as his basic logical tool, Aristotle proposes a philosophy that aims at moving beyond this dualistic uncertainty in order to arrive at undeniable truths.

It is interesting to notice that an analogous divergence regarding the principle of negativity emerges once again within the modern world. In the seventeenth century, Descartes, in the second of his *Rules to Guide the Intelligence*, claims that doubt appertains to any zone in which there is a lack of consensus, or when the opinions of two people concerning the same thing are in contrast with one another.[20] For Descartes, such a situation of doubt equals one of falsehood in which it is not possible to talk of real knowledge. However, such a premise leads Descartes to find certainty only in mathematics, in particular in arithmetic and geometry—the only two arts, Descartes claims, to which his rule is capable of leading. Descartes thus has to sacrifice much in order to maintain his idea of knowledge as something essentially positive, an assertion of certainty. Negativity, pertaining to doubt and falsehood, is banished to the realm of an imperfect world of pragmatics that, for Descartes, has nothing to do with a true philosophy. It is interesting at this point to remember that Descartes, in his only known mention of Bruno, included him with other renaissance *novatores* whose many maxims he found too often contradictory. He claimed that there was no reason to read their works.[21]

Bruno would reappear seriously on the philosophical scene, as indeed on the literary one, only in the post-Kantian culture of European Romanticism, when his was widely considered as a prophetic voice of a new modernism. This is not the place to retell an already established story. Yet it can be useful here to remember that in England at the beginning of the nineteenth century, Samuel Taylor Coleridge, one of the major English poets of the Romantic period and also a philosopher of some note, became a dedicated reader of Bruno. One of the most original aspects of Coleridge's interest in Bruno regarded his dialectic, or what Coleridge called his "polar logic": a triangular progression of thought that moves from the thesis to the antithesis, to resolve itself in a moment of identity or synthesis, or what Bruno called a "resolution of contraries."[22] In a nowadays little read, and sadly undervalued, biographical introduction to Bruno's life and thought, Dorothea Singer cites Coleridge on this subject:

> Every power in nature and in spirit must evolve an opposite, as the sole condition of its manifestation: and all opposition is a tendency to re-union. This is the universal law of polarity or essential dualism, first promulgated by Heraclitus, and two thousand years afterwards republished and made the foundation of Logic, of Physics and of Metaphysics by Giordano Bruno. The principle may be thus expressed. The identity of thesis and antithesis is the substance of all being; their opposition the condition of all existence, or being manifested, and every thing or phenomenon is the exponent of a synthesis as long as the opposite energies are retained in that synthesis.[23]

Coleridge jumps straight from Heraclitus to Bruno, although modern commentators have noticed the intervening contributions to the subject of a polar logic by such medieval figures as Raymond Lull, whose picture logic deeply influenced Bruno, and Nicholas of Cusa, who offered Bruno a logical vocabulary in which to express his theory of the coincidence of opposites.[24] This gradual emergence of an idea of a modern polar logic is a theme that, some years after Coleridge, would be taken up and developed in his *History of Philosophy* by Hegel—a history that collects together material from his lectures in Berlin delivered in 1829–1830. Hegel thought that Bruno had not fully developed the triadic movement of the history of thought, but that he had made a remarkable attempt to do so—making what Hegel calls "a great beginning of the effort to think unity" through the thesis that matter "has life in itself." Hegel revalued Bruno's art of memory for its effort to found the principles according to which the mind develops multiple systems of symbols and images, even if he found the energy and creative force of Bruno's mind more impressive than the results he obtained. The danger, according to Hegel, was that the moments of an infinite and eternal world process are only "collected" or "enumerated," and not developed into a fully logical progression of thought. In spite of such criticisms, however, Hegel, like Coleridge, saw in Bruno a thinker who had laid the foundation stones of a new logic and dialectic that would lead on to the idealism of the modern world: to a new philosophy of the mind.[25] By the time that Hegel gave his lectures on the history of philosophy in Berlin, Bruno's modern reputation was assured, and the foundations were laid for modern editions of his works to appear and for the first full-scale biographies.[26]

Before Hegel, figures in Germany such as Jacobi, Buhle, and Schelling had promoted Bruno as an important figure in the philosophical tradition, and Goethe had done so in a literary context.[27] However, it is Hegel's treatment of Bruno that is of special interest for its insistence on the importance of the negative in the context of a polar logic. So it is interesting here to remember Hegel's own celebration of negativity in his *Phenomenology of the Spirit*.

In his famous preface to that work, Hegel acknowledges what he calls "the tremendous power of the negative." He identifies the negative with "the pure I," or that part of the self that produces the pure energy of thought. The passage goes on to associate the power of the negative— closely associated with the passage of time—with the phenomenon of death, considered by Hegel as "of all things the most dreadful." But, Hegel continues:

> the life of the spirit is not the life that shrinks from death and keeps itself untouched by devastation, but rather the life that endures it and maintains itself in it. It wins its truth only when, in utter dismemberment, it finds itself. It is this power, not as something positive, which closes its eyes to the negative, as when we say of something that it is nothing or is false, and then having done with it, turn away and pass on to something else; on the contrary, Spirit is this power only by looking at the negative in the face, and tarrying with it. This tarrying with the negative is the magical power that converts it into being.[28]

Only when the negative has been consciously engaged with, becoming an essential part of being rather than merely cast aside in fear, does Hegel feel that he can go on to define the process by which the mind rises to a higher form of perception of a fully synthetic unity. A recent commentator on Hegel's *Phenomenology of the Spirit*, J. N. Findlay, has this to say about Hegel's concern with the essential importance of negation:

> On Hegel's basic assumptions negation, in a wide sense that covers difference, opposition, and reflection or relation, is essential to conception and being: we can conceive of nothing and have nothing if we attempt to dispense with it. But negation in this wide sense always operates with a unity, which is not as such divisible into self-sufficient elements, but is totally present in each and all of its aspects, and we conceive nothing and have nothing if we attempt to dispense with this unity.[29]

Let us now return to Bruno—not in the sense of considering him as a "precursor" of Hegel, in the naive and unfashionable nineteenth-century use of that interpretative category, but nevertheless bearing in mind that he was an author known to and commented on by Hegel. One of the works by Bruno that Hegel would certainly have read, because large parts of it had been translated into German by Jacobi at the end of the eighteenth century, was the second Italian dialogue written and published by Bruno in London in 1584: *The Cause, Principle, and One.* In this work, Bruno considers the metaphysical implications of the natural philosophy he had just presented in *The Ash Wednesday Supper.* The cause, particularly the final cause lying behind the world of phenomena, and the unity of a final vision of the whole, or the one, are always present to Bruno as the ultimate goal of the philosophical endeavor. Yet he sees the human mind as living out its destiny, its inevitable fate, within a world of irregularities and inexactitudes, of conflicting realizations, such as Timaeus had described in Plato's dialogue. The first coordinates in the universe are always two, not one, as Bruno's natural philosopher, Theophilus, had maintained at the beginning of *The Ash Wednesday Supper.* Theophilus, however, is also a lover of god, as his name implies, and although Bruno denies divine inspiration, he does believe in a world soul of Neoplatonic derivation, or a rational principle at work within the infinite vicissitudes of the natural world. Bruno thus underlines the necessity for the human mind to search for the illumination of ever higher forms of beauty and truth by progressing from simple to always more sophisticated visions of an infinite whole. This can be done only by progressing dialectically through ever more complex forms of contraries. So, in conclusion, Bruno writes in the final dialogue of *The Cause, Principle, and One,* "he who wants to know the greatest secrets of nature should observe and examine the minima and maxima of contraries and opposites."[30]

Hegel would speak of "tarrying with the negative," and would define such tarrying as the magical power that converts the negative to being. It is difficult to believe that he had not read the page of *The Cause, Principle, and One* on which Bruno declares that the most profound form of magic is that which knows how to perceive contraries within the point of union.[31] The word "magic" here is not used in its more usual sense as a definition of the irrational powers or qualities of an objective world. Rather, it is used to define that leap of the imagination by which the inquiring intellect achieves a vision of truth as unity, only to move beyond that unity in the recognition of a yet higher play of antitheses. Bruno goes on to define the *sommo bene*, or the highest good, as "the highest form of appetite, the highest perfection, the highest beatification consisting in a unity that complicates everything." Here, the word "complicates" derives, according to common sixteenth-century usage, directly from the Latin, meaning not a difficulty or a problem, but rather the inclusion of opposing strands into one whole, or a complex intertwining of contrary forces. Bruno's greatest good thus includes, rather than excludes, the negative, without which, as Hegel would repeat after him, we can conceive nothing.[32]

The importance of the negative for Bruno is underlined precisely by the fact that, as we have seen, he emphasizes that importance at the very beginning of his six Italian dialogues, written and published in London between 1584 and 1585. The beginning of his first text in that sequence corresponds not so much to a cosmic beginning, which was denied by his conviction of the eternity of his infinite universe, as by a logical or conceptual beginning. The passage from the one to multiplicity, or that "primordial discrepancy between essence and phenomenon," as Gadamer called it, becomes the starting point of the philosophical journey narrated in the six ensuing dialogues. The journey is not going to be simple or linear, not a straight path from darkness to light. It leads the reader through a labyrinth of contradictions and negations, of arduous and often frustrated inquiry that only occasionally finds its climax in a moment of genuine illumination, such as the vision of an infinite universe that the nine philosophers achieve together, at the climax of the *Heroici furori*, on the banks of the river Thames. The fact that these characteristics of Bruno's inquiry are underlined in an apparently light-hearted preface, finding their literary expression in an extraordinary piece of virtuoso writing with what are already baroque ramifications, should not deceive the reader as to the seriousness of Bruno's intentions.[33] It is, indeed, Bruno himself who draws attention to the underlying seriousness of his comic vein in *The Ash Wednesday Supper*, asking his reader to consider:

> that this dialogue tells a story. It narrates occasional events, walks, meetings, gestures, sentiments, speeches, proposals, replies, arguments that are both sensible and absurd. Everything is subjected to the rigorous judgment of our four speakers, so that nothing of any importance is left without a comment. Consider also that there is no superfluous word here, for in every part there is a harvest to be gathered of things of no small importance, perhaps more so precisely where it seems least likely.[34]

NOTES

1. For the pages discussed in this paragraph, see Bruno (2002), vol. 2, 431–32.

2. Bruno's anti-Aristotelian polemic found one of its strongest expressions during his second visit to Paris in 1586, and was considered so scandalous that he had to leave France for Germany. See Bruno (2007). For his positive readings of both the Pythagoreans and the Stoics, see "The Pythagorean School and our Own" in Gatti (1999) and the "Nota sobra el uso de Séneca por Giordano Bruno" in Granada (2005).

3. See chapter 6, "The Sense of an Ending in Bruno's *Heroici furori*," in this volume.

4. For the pages of the *Furori* discussed earlier, see Bruno (2002), vol. 2, 738–53.

5. Ibid., vol. 1, 442–44.

6. See Plato, *The Collected Dialogues,* eds. Edith Hamilton and Huntingdon Cairns, Bollingen Series, Princeton, NJ, Princeton University Press, 1961, 68–69.

7. See Bruno (2002), vol. 1, 522–28.

8. See Yates (1964). Ficino's collection of Hermetic texts with the addition of the *Asclepius* can be consulted in English in *Hermetica,* trans. and ed. Brian Copenhaver, Cambridge, UK, Cambridge University Press, 1992.

9. See *Hermetica,* op. cit., 2.

10. See Frances Yates, "Notes on *Camillo and Hermes Trismegistus*," ed. Hilary Gatti, in *Annali della Scuola Normale di Pisa,* Classe di Lettere e Filosofia, series IV, vol. VI, I, 2001, 171–93.

11. See Milton's account of the fall in *Paradise Lost,* bk. IX.

12. See *Hermetica,* op. cit., xlvii.

13. On the Neoplatonism of Bruno's early years in Naples, see Rowland (2008), 38–61.

14. For the *Timaeus,* see Plato, *The Collected Dialogues,* op. cit., 1151–1211.

15. See Hans Georg Gadamer, *Dialogue and Dialectic: Eight Hermeneutical Studies on Plato,* trans. and ed. P. Christopher Smith, New Haven–London, Yale University Press, 1980, 206.

16. See Bruno (2002), vol. 1, 442.

17. Ibid., 494 and 556.

18. For Bruno's thoughts on the number four, see the *De monade,* his work on Pythagorean number symbolism. In particular, Bruno (2000c), 302–28.

19. See Aristotle's *Metaphysics,* bk. I, (A) 986a1, in *The Complete Works* , op. cit., vol. II, 1559.

20. Descartes' *Regulae ad directionem ingenii (Rules to Guide the Intelligence)* appeared posthumously in a Dutch translation in Amsterdam in 1684, and then in Latin in the same town in 1701.

21. See René Descartes, *Tutte le lettere 1619–1650,* ed. Giulia Belgioioso, Milan, Bompiani, 2005, 158.

22. For Coleridge on Bruno, see chapter 10, "Romanticism: Bruno and Samuel Taylor Coleridge," in this volume.

23. This passage was first published in Coleridge's short-lived intellectual review, *The Friend*; see Singer (1950). Singer's discovery of this important passage is mentioned in Calcagno (1998).

24. On Bruno and Lull, see Cambi (2002). On Bruno and Cusanus, see Mancini (2000), 245–74.

25. Antonio Corsano, in an important study of Bruno's mind in its historical context, underlines the "audacity" of his thought, particularly when "he converts impossibility into necessity and negation into an absolute affirmation." See Corsano (1940), 145.

26. See the ample section on Bruno in Hegel's history of philosophy: G.W. F. Hegel, *Vorlesungen über die Geschichte der Philosophie* (1833), in *Werke,* Frankfurt am Maine, Suhrkamp, 1979, vol. 12.

27. For this German revival of Bruno, see Ricci (1991).

28. See G.W.F. Hegel, *Phenomenology of Spirit,* trans. and ed. J. N.Findlay, Oxford, UK, Oxford University Press, 1977, 19.

29. See the foreword to ibid., ix.

30. See Bruno (1998), 100.

31. Ibid.

32. Ibid., 101.

33. For the baroque elements in Bruno's writings, see Barbieri Squarotti (1958).

34. Bruno (2002), vol. 1, 438. My translation.

PART 1

BRUNO AND THE NEW SCIENCE

1

BETWEEN MAGIC AND MAGNETISM

BRUNO'S COSMOLOGY AT OXFORD

IN 1960, ROBERT McNULTY DISCOVERED and published in *Renaissance News* the bitterly satirical page on Giordano Bruno's lectures at Oxford of 1583 that he had found in the book by George Abbott of 1604, *The Reasons Which Doctour Hill Hath Brought, for the Upholding of Papistry*.[1] It was an important discovery, but it also created an interpretative crux for scholars concerned with the development of Bruno's post-Copernican cosmology. This chapter will take as its subject the technical problem of the precise stage that Bruno's cosmological speculation might have reached when he spoke at Oxford in the summer of 1583—"might have" being obligatory, given that the texts of his lectures have not survived. I shall only briefly touch on the question of the historical or cultural factors that affected the knowledge of Copernicus in the Oxford of the period, or on the ways in which this complex and dramatic episode would affect the rest of Bruno's stay in England. Bearing in mind these limits to this inquiry, let us recall briefly the salient characteristics of Abbott's page, and of its author, who was neither mediocre nor obscure.

In the summer and autumn of 1583, Abbott was twenty-one years old and was about to become a fellow of Balliol: one of Oxford's most prestigious colleges. The detailed character of his account of Bruno's lectures suggests that it was based, in all probability, on a personal and active role in the events that he narrates some twenty years after the event.[2] In the meantime, Abbott's academic and ecclesiastical career had been a success and would take him to the peak of the Anglican Church in 1611, when he was appointed archbishop of Canterbury by James I. In spite of his somewhat radical Protestantism, Abbott was a strenuous defender of the episcopal structure and hierarchy of the Anglican Church, of which James was a jealous guardian. Abbott also satisfied James's political and religious needs by arguing energetically against the Catholics. In 1604, at the beginning of the new reign, after the death of Elizabeth I in 1603, a danger for the stability of the throne came from the Catholic camp, hostile to the arrival from Calvinist Scotland of the new king. For James, although he was the son of the unfortunate Mary Stuart, had long since renounced the Catholic religion practiced by his mother. So it is hardly surprising to find Abbott's attack on Bruno in a book of anti-Catholic polemic, with an evident anti-Italian and anti-humanistic slant: "that little Italian, with a name longer than his body, who claimed to be qualified as a Doctor in Theology, etc., and who called himself Philoteus Jordanus Brunus Nolanus."[3]

It is worth remembering that, in 1600, Abbott had been elected vice chancellor of the University of Oxford, a position that he still held in 1604 and from which he would carry out his violent critique of Bruno. Abbott's page on Bruno supplies information about his Oxford experience that was not evident from the brief mentions of it by Bruno himself in his dialogue *The Ash Wednesday Supper*, written and published in London in 1584.[4] For example, it is Abbott who mentions two distinct visits made by Bruno to Oxford. The first of these, which Bruno mentions himself with indignation, sees him as part of the entourage of the Polish Prince Albert Alasco, during whose visit to Oxford Bruno held a public dispute with an opponent who proved to be hostile (a much quoted note in a margin of one of his books by Gabriel Harvey tells us that he was actually John Underwood, a future vice chancellor of the university). It was during another visit, "not a long time afterwards," that Bruno attempted to give the series of lectures that are the subject of the comments by Abbott.[5] It is necessary to add to these two visits the publication, at a date that remains uncertain, of the famous letter addressed to the vice chancellor of the university that Bruno added to some copies of his work on the art of memory, the *Explicatio triginta sigillorum*, railing against the ancient British university for its inability to measure up to new ideas brought in by outsiders.[6] This letter oscillates between moments of eloquent and powerful rhetoric in defense of the right of the modern scholar to think independently, and indignant and at times decidedly offensive criticism of what Bruno considered the obtuse arrogance of the Oxford dons. Abbott, who mentions the *Explicatio* with precision in a marginal note, makes it clear that in this way, Bruno had stimulated a deep and lasting ill humor on the part of the Oxford dons, which was still remembered by Abbott more than twenty years later. So it is in a context of feeling marred by resentment and rancor that Abbott offers a series of indications of the contents of Bruno's lectures, which had taken place many years before he wrote his account.

On a specifically cosmological level, Abbott writes:

1. That Bruno enjoyed "telling us much of *chentrum* and *chirculus* and *circumferenchia*."
2. That "he undertooke among very many other matters to set on foote the opinion of Copernicus, that the earth did goe round, and the heavens did stand still."

3. That his first three lectures were taken "almost *verbatim*" from the works of Marsilio Ficino. (In a marginal note, Abbott specifies: *De vita coelitus comparanda.*)

Once this plagiary had been discovered, the lectures were brought to a close; or at least this has always been the interpretation offered of the relevant passage in Abbott as well as of Bruno's own words in *The Ash Wednesday Supper.*[7] It is, however, possible to query this interpretation and to ask whether the lectures were actually brought to a close or whether Bruno was not simply warned to stop quoting "almost *verbatim*" from Ficino's works. Abbott tells us that after the third lecture, the increasing suspicion of plagiary on the part of the dons was communicated to Bruno. "They caused some to make knowne unto him their former patience, and the pains which he had taken with them, and so with great honesty on the little man's part, there was an end of that matter." On his part, Bruno makes no mention of the suspicion of plagiary, but speaking to his readers, he exhorts them: "To find out about how they made him finish his public lectures (*"gli hanno fatto finire le sue pubbliche letture"*), both those *de immortalitate animae,* and those *de quintuplici sphera"*[8]: a phrase that has always been interpreted as a cry of distress in front of the humiliation of seeing his lectures interrupted and the lectureship that had been temporarily awarded to him taken away. But it is possible to construe that "made him finish his public lectures" as expressing a sense of satisfaction at being allowed to bring them to a regular end—in which case, one has to presume that Bruno admitted "honestly" to having followed too closely the texts of Ficino, while at the same time promising to make amends in the remaining lectures. In that case, his phrase would express recognition of the fact that his Oxford lectures, in spite of the difficulties that occurred due to the plagiary of Ficino, were nevertheless completed. It may indeed have been precisely at that point that Bruno started to speak about Copernicus, given Abbott's insistence that on mentioning Copernicus, Bruno's head "did not stand stil." As far as the accusation of plagiary is concerned, it should be remembered that the Protestant cultures, from Luther onward, regularly accused the Catholic world of superficiality and a lack of originality, due to the mental habit of having to refer any form of discourse to the canonical sources recognized by the tradition. As Giovanni Aquilecchia has already underlined, Abbott's book consisted of an accusation of plagiary against Hill, who was explicitly repeating what had been said by a Catholic author who preceded him.[9]

However we wish to read this aspect of Abbott's page on Bruno, it is clear that the interpretative difficulties derive from the emphasis he also puts on the other aspect of Bruno's lectures—that is, the fact that Bruno attempted to use his temporary Oxford chair to "set on foote" the Copernican astronomy. This was still firmly repudiated by an academic body impregnated by Aristotelian paradigms in the philosophical and cosmological as well as the theological fields. The most recent archival research has established that Copernicus's name had already been mentioned publicly at Oxford by Henry Savile, although Giovanni Aquilecchia, who made a study of the manuscripts of Savile's lectures, established that they mention him only briefly in the context of mathematical hypotheses.[10] It would appear, at least according to the present state of research, that Bruno was the first to propose at Oxford a reading of Copernicus in realist terms, and that by doing so he became the target of the derision of Abbott and his contemporaries, who accused him of being completely mad. But even if this aspect of the question can be considered as established, an enigma remains concerning the terms of Bruno's proposal of Copernicanism in lectures that are said to be taken "almost *verbatim*" from the pre-Copernican works of Marsilio Ficino.

It is curious how few commentators of Bruno have faced up squarely to this problem. One of the first attempts to do so is to be found in the pages of Frances Yates's book of 1964 that interprets Bruno's thought in the light of the Renaissance Hermetic tradition.[11] Writing only a few years after the publication of McNulty's article, Yates ably reinforces her own argument by underlining the largely magical and astrological character of Ficino's *De vita coelitus comparanda*, as well as Abbott's use, in his description of Bruno, of words such as "juggler," which are often found in the anti-magical polemics of this period. Yates claims that such elements in Abbott's narrative not only reinforce her thesis of a Hermetic Bruno but also serve to show how his post-Copernican and infinite cosmological picture was nothing other than a magical talisman of astrological origin, without any real astronomical validity. In this way, Yates appears to resolve the apparent contradiction between Ficino's text and Copernicus's astronomy by dissolving his Copernicanism in the occult currents of Renaissance Hermeticism.

It was only several years later that the problem of Bruno at Oxford was taken up once again in serious terms, when Michele Ciliberto studied it in his two books on Bruno's philosophy published in 1986 and 1990.[12] In Ciliberto's opinion, it is not possible to reduce Bruno's cosmology to a mere astrological emblem—on the contrary, Abbott's page makes it clear that his cosmology was the element in his thought that created the most difficulties for the Oxford dons, who remained faithful to the Aristotelian–Ptolemaic universe. On the other hand, Ciliberto also underlined the importance, with respect to Bruno's Oxford lectures, of the work titled *Sigillus sigillorum*, published by Bruno in London as an addition to the *Explicatio triginta sigillorum*. Ciliberto considers this to be a work of fundamental importance insofar as it represents a "gnoseological chrysalis" within which the difference between the infinite worlds and the infinite universe begins to germinate. It is precisely this concept of an infinite universe that calls into question both the Parmenidian unity as well as the Anaxagorean variety and vicissitude that "postulated everything in everything, because the soul, the spirit or the universal form is in everything; so that from everything, all things can be produced." In Bruno, the individual soul searches for unity with the world soul, through the effort of the various grades of reason, intellect, and sense of which it disposes. The *Sigillus* thus unites "an analysis of the *rectores* of the active intellect" to a preliminary definition of the constitutive elements of Bruno's cosmology, leading Ciliberto to conclude that "it is not difficult to imagine what Bruno might have said at Oxford."

Giovanni Aquilecchia, finding that this conclusion failed to satisfy him, took up the subject again in his volume *Le opere italiane di Giordano Bruno* of 1991, where he underlined in particular the problem of the interruption of Bruno's lectures, which in his opinion is to be considered as certainly having taken place.[13] Aquilecchia points out that the *Sigillus* does not contain a fully fledged Copernican discussion, in astronomical terms, such as the one that can be found in the later *Ash Wednesday Supper*, and that Abbott's page suggests already took place at Oxford. For Aquilecchia, furthermore, any conclusion had to be avoided that led back to Yates's thesis, which he considered a serious oversimplification, for Aquilecchia never accepted that Bruno's defense of the Copernican astronomy was nothing more than a part of his revival of the Egyptian Hermetic religion, becoming (as Yates claimed) a hieroglyph that announces the return of an astral magic.[14]

Another significant treatment of this subject can be found in a much discussed paper read by Rita Sturlese at the conference on "Fonti e motivi dell'opera di Giordano Bruno" held at the University of Cassino in 1993.[15] Sturlese develops Ciliberto's thesis that the *Sigillus* contains the essence of what Bruno said at Oxford, and by following this path (as Aquilecchia had predicted), she arrives at a conclusion similar to that reached by Yates. For the astronomical component of Bruno's Oxford lectures is considered by Sturlese as largely irrelevant, in spite of the fact that Abbott's page seems to testify to the contrary. Sturlese reaches this conclusion by approaching the subject from the angle of the titles of his Oxford lectures that are specified by Bruno himself in *The Ash Wednesday Supper*, where he claims to have spoken on two different subjects: *de immortalitate animae* and *de quintuplici sphera*. Sturlese starts with an analysis of paragraph 31 of the *Sigillus*, titled *de quintuplici et simplici progressionis gradu*, where the soul is considered to be the unique and unifying center of the five senses. Through a purifying ascension, the soul is identified as a spherical monad, although not a transcendental one, as in the Plotinian and Ficinian sources of the *Sigillus*. Sturlese's analysis is thus centered on Bruno's concept of soul, and of the process by which it becomes conscious of itself as a unity and a dynamic entity (that is to say, as a "quintuple sphere"). Only in this way does the soul become aware of the immanent nature of its founding and unifying One. In this concept of immanence, Sturlese might have found, as Ciliberto had attempted to do before her, a connection between the journey of the soul narrated in the *Sigillus sigillorum* and the beginnings of a properly cosmological discourse that would rapidly develop in the coming years. However, like Yates, she also insists on seeing Bruno's astronomy at Oxford as a superfluous irrelevance, basing this conclusion on the rather surprising assertion that "on examination, the expression *quintuplex sphaera* makes no sense either in a Copernican or in a Ptolemaic context."

Developing this varied and variegated discussion in the following years, and particularly between 1993 and 1995, Aquilecchia returns to the subject of *Bruno at Oxford* in terms that place it on a new basis, particularly by introducing into the discussion the discovery on the part of Mordechai Feingold of some important and previously unknown documents. These were the texts of the already mentioned lectures of Henry Savile, which were probably delivered in 1573, or a decade before Bruno's visit to Oxford.[16] According to Feingold, Savile's lectures contained the first consistent public reference at the ancient English university to the new Copernican cosmology. However, Aquilecchia, after a detailed study of these manuscripts, claimed that this is only partially exact. For Savile made only a brief reference to Copernicus's heliocentric cosmology in a context of mathematical calculation—thus reducing it to the role of a pure hypothesis. On the contrary, Bruno in *The Ash Wednesday Supper* proposes a fully realistic interpretation of the Copernican theory and appears to have already anticipated this realist reading at Oxford.

There are other documents that suggest the possibility that a realistic interpretation of Copernicus was already proposed by Bruno at Oxford. In particular, Aquilecchia refers to a document of great historical interest already mentioned by Dorothea Singer—that is, a letter written by Alberigo Gentile, a Protestant refugee in England known for his study of international law, to the French Calvinist Jean Hotman. The letter gives lively expression to the hostility felt generally by the Protestant culture of the time toward the heliocentric theories that were upsetting the traditional Aristotelian cosmology.[17] It was written at Oxford, where Gentile taught, and is dated simply November 8th. According to Aquilecchia, its position in the collection of letters in which it was published in 1700 permits it to be considered as belonging to 1583. This means that it could have been written only a short time after Bruno's lectures that presumably took place in the late summer, and at which Gentile was almost certainly present. In the letter, Gentile refers to: "*absurdas, & fatuas assertiones maximorum Virorum audivimus, Coelum lapideum, Solem bipedalem, Lunam multarum urbium atque montium orbem, Terram moveri*" [a fatuous assertion that we have heard made by a well-known man that the sky is made of stone, the sun is only two feet wide, that the moon contains many cities and mountains, and that the earth moves]. Although Gentile is probably exaggerating to gain a ridiculous effect, such statements, as Aquilecchia points out, could only have been made by Bruno at Oxford at that time. So that, even if Bruno is not explicitly mentioned by Gentile in this letter, Aquilecchia maintains nevertheless that it is possible to include it among the growing number of documents that refer to Bruno's Oxford lectures. Undoubtedly it is a document of great interest that appears to sweep away many uncertainties over the contents of Bruno's lectures at Oxford. They clearly contained a strong cosmological component, presented in overtly realist terms that even an advanced and cosmopolitan spirit such as Gentile considered as expressions of pure folly, just as Abbott would judge them to be many years later.

As far as Abbott is concerned, it is evident that he wished to present Bruno as quite literally a madman. Precisely this desire appears to lie behind his claim that "he undertooke among very many other matters to set on foote the opinion of Copernicus, that the earth did goe round, and the heavens did stand still; whereas in truth it was his owne head which rather did run round." Carefully considered, this sentence could give rise to the suspicion that what Bruno proposed at Oxford was a transitory solution in which the earth revolved around its own axis, but not around the sun.[18] Strictly speaking, this solution could not properly be called Copernican given that the *De revolutionibus* proposes a fully heliocentric astronomy based on multiple movements of the earth. Nevertheless, whatever Abbot and Gentile heard him say, it is the fully Copernican heliocentric proposal that Bruno will endorse, in his own realist terms, in *The Ash Wednesday Supper*, written and published early in 1584, only a few months after his lectures at Oxford.

A careful study of Gentile's interesting letter actually suggests some confusion and ambivalence on his part as to what had really been said in the lectures he claims to have heard. The first two cosmological arguments he mentions—that is, the immobile, unchanging and stone-like sky, together with a sun only two feet wide—are certainly to be found among the subjects that Bruno discusses later in *The Ash Wednesday Supper*. However, he discusses them only to criticize them sharply as elements of an ancient cosmology that are to be considered as seriously erroneous. The immobile sky clearly derives from Aristotle's *De caelo* and forms the most negative and outdated part of the dualistic Aristotelian–Ptolemaic cosmology that Bruno attacks, and that had in any case been proved mistaken by the recent appearances of new comets that were being much discussed in scientific circles at that time.[19] The idea of a small sun, not much larger than it appears to be when seen from the earth, derives instead from Epicurus, who is explicitly criticized for this doctrine by Bruno in the third dialogue of *The Ash Wednesday Supper*, where Theophilus demonstrates that the sun must necessarily be larger than the earth in order to give rise to the nocturnal shadow.[20] On the other hand, the other two cosmological arguments mentioned by Gentile—that is, a moon that cannot be considered perfectly round or smooth because it is formed of the same substance as the earth, as well as the idea of an earth in movement—are to be numbered among the positive arguments proposed by Bruno in *The Ash Wednesday Supper* as essential components of his own infinite and heliocentric cosmology. It would thus seem that Gentile, in this letter (if he was referring to Bruno), was unable to distinguish the arguments that Bruno criticized from those that he was proposing in positive terms.

This reading of Gentile's letter is in line with the role he would later play in one of Bruno's major cosmological dialogues, *De l'infinito, universo et mondi*. Aquilecchia, once again, has identified the figure of Albertino, who appears as one of the most important characters in dialogue V of *De l'infinito*, as representing Alberigo Gentile, who is sometimes called Albertus in documents of the period.[21] This Albertino is a character of great interest in the structure of the final dialogue of Bruno's work, for he passes from an initial attitude of skeptical disapproval with respect to a post-Copernican cosmology, which closely echoes the sentiments expressed in Gentile's letter, to a final attitude of consent that confers seriousness and conviction on the cosmological arguments that he had previously covered with ridicule and scorn. In this final spirit of consent, Albertino, at the end of the five dialogues that make up the *De l'infinito, universo et mondi*, exhorts Philoteus, who is the mouthpiece of Bruno's own ideas in this dialogue, to continue unperturbed his work of diffusion of a new image of the universe:

> Continue to make known to others what the sky is really composed of, as well as the planets and the stars; how the infinite worlds are all distinguished one from the other; how it is not only possible but necessary that there should be an infinite space; how such an infinite effect must have an infinite cause; what is really the substance, the matter and the efficient cause of everything; from what principles and elements everything in the sensible world is formed.[22]

This overview of the discussion that has developed concerning the subject of Bruno at Oxford during the past half century allows us to assert that, largely due to the work of Giovanni Aquilecchia, a number of conclusive results have now been reached. By and large, Abbott's page has proved to be accurate. It seems certain at this point that Bruno spoke about the new Copernican cosmology at Oxford, anticipating the realist terms of his reading of Copernicus that he would develop in more detail in the first three Italian dialogues written and published in London in 1584 and that would remain from then on at the basis of his philosophy. Nevertheless, some important points remain to be clarified. Above all, two problems still need to be posed: What was the meaning in all this of the reference to Ficino's pre-Copernican *De vita coelitus comparanda* that seems to have given rise to the interruption of the lectures, or at least to a warning to Bruno on the part of the academic authorities not to continue to plagiarize this text? And how exactly are we to interpret the title *de quintuplici sphera* that Bruno himself indicates in *The Ash Wednesday Supper* as the title of a part of his lectures?

The recent hypothesis formulated by Aquilecchia that the title *de quintuplici sphera* referred to by Bruno himself could call into question the cosmology of Tycho Brahe, which Bruno might have wanted to discuss at Oxford, is undoubtedly of great interest, even if Aquilecchia presents it with an eloquent question mark. He reminds us that Brahe had already formulated his "compromise" system of the universe, partly geocentric and partly heliocentric, in 1583, even if he would not make it public until 1588. It was, however, already circulating and being discussed by those in the know. In Brahe's system, the earth remains at the center of the universe, with the sun, the moon, and the planets Mars, Jupiter, and Saturn revolving around it, while Mercury and Venus revolve around the sun. Brahe's cosmology could thus be considered as a *quintuplici sphera*.[23] Furthermore, it is well known that Bruno knew of and admired the Danish astronomer's observational skill and that in 1588 he sent him a copy of his *Camoeracensis acrotismus*—a gift that, it seems, was not appreciated.[24] This is not surprising, as Bruno's and Brahe's cosmologies were clearly opposed to one another, both in the way they developed the heliocentric aspect of the Copernican revolution, which Bruno accepted without reserve, and concerning the infinity of the universe proposed by Bruno, which Tycho Brahe never accepted. If Bruno did mention Brahe's cosmology at Oxford, it might have been in critical terms. In any case, neither Abbott nor Gentile mention Brahe's name.

One rather curious fact remains, which it is worth underlining, and that is that none of the studies of Bruno at Oxford has been based on a close reading of the Ficinian text mentioned by Abbott as the immediate cause of the accusation of plagiary—that is, the *De vita coelitus comparanda*. It is true that Frances Yates underlined emphatically Abbott's indication of Bruno's use of this text, so full of talismanic magic and Renaissance astrological doctrine. Even so, she did not develop a precise analysis of the text itself, which might have led to suggestions concerning the terms of that reference on the part of Bruno. Sturlese, for her part, carries out a comparison between Bruno's *Sigillus* and Ficino's *Theologia platonica*, which Abbott fails to mention, and produces only one parallel between Bruno's text and the *De vita coelitus comparanda*. Yet Abbott's page is characterized by a somewhat pedantic attention to detail, referring specifically to the *Explicatio triginta sigillorum* as the text that contained Bruno's famous letter to the vice chancellor of the University of Oxford. There seems to be no reason to doubt his claim that it was Ficino's *De vita coelitus comparanda* that a person of authority in the English university (described by Abbott as a "grave man") went to consult in his study in order to verify the suspicion of plagiary. This makes it seem advisable, and even necessary, to ask oneself what it might have been in this text of Ficino's that interested Bruno when he decided, during his lectures at Oxford, to endorse the Copernican astronomy. Perhaps he was already inspired with the intention of expanding Copernicus's universe to infinite dimensions—or so Gentile's letter suggests, when it mentions a moon made of the same substance as the earth.

It is well to remember the date of Bruno's visits to Oxford, which took place only a few weeks after his arrival in London from Paris.[25] In the French capital, in 1582, Bruno had published the first of his works to have survived, the *De umbris idearum*, the *Cantus circaeus*, the *De compendiosa architectura*, and the comedy in Italian, *Candelaio*, while the works published immediately after his arrival in London, the *Ars reminiscendi* and the *Explicatio triginta sigillorum* (of which the *Sigillus sigillorum* is a part), had probably been already written in Paris. Frances Yates, and those commentators who follow her Hermetic interpretation, have often underlined the importance, in the formation of Bruno's thought in these early works, of the cultural atmosphere that surrounded the French court of Henri III. An early interest in the new cosmology is testified to there by the translation in 1552 of the first book of Copernicus's *De revolutionibus* by Pontus de Tyard, published with the title *Discours des parties de la nature du monde*. This early Copernican work was impregnated by a deep interest in Hermetic and Neoplatonic texts.[26] It is a mixture that can be found in Bruno's early works as well, and in particular in the *De umbris idearum*, where an already central sun is invoked with words and alchemical images that remind us more of Hermes Trismegistus and of Ficino in the *De vita coelitus comparanda* than of Copernicus himself. And in the *De umbris*, as Yates continually emphasizes, we find a significant reference to Hermes, who appears as the hero and teacher of the new philosopher Philothimus (a mouthpiece of Bruno himself, who will reappear in the Italian dialogues written and published in London as Theophilus or Philoteus), while in the *Cantus circaeus*, we find a clear reference to the *De vita coelitus comparanda* of Ficino in the invocation, in its opening pages, to the magical powers of the sun.

To underline the importance of this mixture of solar mysticism and heliocentric cosmology in sixteenth-century France does not necessarily imply a return to Frances Yates's thesis of a thoroughly Hermetic Bruno. Hermeticism appears indeed to be something that, to some extent at least, Bruno left behind him once he arrived on the banks of the river Thames. He himself suggested this in the clearly autobiographical pages that close the last of his Italian dialogues written and published in London, the *Heroici furori (Heroic Frenzies)*. There what Bruno describes is a qualitative intellectual leap that takes the form of a reawakening from a previous state of melancholy and blindness: an illumination that turns him into a newly natural philosopher.[27] One may suppose, however, that this illumination was actually somewhat less sudden than Bruno himself described it as. Clearly the joyful discovery of an infinite universe as the "greatest good on earth," which he celebrates in the song of the *Illuminati* that closes the *Heroic Frenzies*, symbolizes the moment in which Circe, who had imprisoned his spirits in the labyrinth of base matter, is repudiated as the bearer of a shadowy blindness. But this could have been a more gradual process than the *Frenzies* suggest. That could explain why, in his lectures at Oxford, Bruno continued to introduce elements of Ficinian Neoplatonism into his treatment of the new astronomy, according to the claim made by Abbott, using them to describe the terms of the reawakening of a clearer and purified intellect. At the same time, however, Bruno was proposing a more technical discourse centered on the new astronomy, which would become the primary theme of the first three Italian dialogues written and published in London in 1584.[28]

At a more general level, as Frances Yates claimed, such a mixture of discourses, lacking in homogeneity from a specifically scientific point of view, undoubtedly played an important role in the development of a new natural philosophy at the end of the sixteenth century. Bruno's French experience can be correctly claimed in this sense to assume particular value as a point of departure for his philosophical speculation.[29] For it is precisely in a French context that the work that Abbott specifically refers to—that is, Ficino's *De vita coelitus comparanda*—had been translated, by Guy Le Fèvre de la Boderie, in 1582. This is the same year as Bruno's first Parisian publications. It is also the only sixteenth-century translation of Ficino's *De vita* that contains the third book—that is, the *De vita coelitus comparanda*.[30] Bruno himself would have been reading Ficino in Latin, but he would almost certainly have been aware of the publication of such an important translation of Ficino's work in the city he was living and lecturing in at the time.

The habit of translating the first two books only of the *De vita* can be explained by the fact, underlined by Ficino himself in the *Proemio* of the *De vita coelitus comparanda*, that this text had been conceived originally in 1489 as an autonomous comment on a passage from the *Enneads* of

Plotinus, and had only been joined to the other two books of the *De vita* in August 1489, when Ficino dedicated the whole book to Lorenzo dei Medici.[31] Nevertheless, the reticence of the translators in front of the third book probably also depended on the strong element of magical and occult doctrine that it contains. As Paul Oscar Kristeller wrote: "Ficino's occult doctrines appear mainly in the third book of the *De vita* that was composed in 1489 as a part of his commentary on Plotinus, Jamblichus and Proclus, and, as we now know, of the Arabian *Picatrix*."[32]

The importance of Ficino's multiple and significant references to the medieval *Picatrix*, with its doctrines of astral magic, has been understood only recently by his commentators. The crucial event was the find in 1976 of an unpublished letter dictated by Ficino to Michele Acciari and addressed to Filippo Valori.[33] From this letter, it is clear that Ficino not only had a text of the *Picatrix* available to him, but that he also had it on the table in front of him during the composition of the *De vita*. The text of the *De vita coelitus comparanda*, full as it was of doctrines traditionally considered as suspect by the Christian churches, would understandably have appeared as forbidden territory to the sixteenth-century translators, and it is hardly surprising to discover that the Oxford dons were worried by Bruno's continual references to it. Linked to the proposal of a realist reading of the new and still suspect Copernican astronomy, Bruno evidently was expounding to the academic culture of his times an explosive mixture of magical and astronomical doctrines that could easily have appeared as a provocation. It was clearly perceived as a danger to the young minds being trained by the Oxford dons, among which has to be counted that of George Abbott, future Archbishop of Canterbury.

The text of Plotinus that Ficino commented on in the *De vita coelitus comparanda* has been identified by Kristeller and Eugenio Garin as *Enneads, IV, III, 11*, where it is shown that all beings are governed by a unifying principle. In an essay of great interest for our subject, published in 1960, Garin observes that the Plotinian text that interested Ficino "illustrates the mediating function of the soul, which creates things according to rational forms hidden within it." Garin underlines the importance of Ficino's reference in this context to the *De sacrificio et magia* of Proclus, which "defines the chain linking beings one to another, according to a universal sympathy." From these texts, and particularly from Plotinus as well as Proclus, there emerges a concept of the world soul as the "interpreter" of what the sensible world derives from the intelligible one, while it is precisely the universal principle of this work of "interpretation" that assures that "there is no distance or separation between things." "Every idea," continues Plotinus, "is of and by itself, although not in a spatial sense, and, although joined to matter, is separated from it." According to this conception of the world, in the opinion of Plotinus,

the heavenly bodies are divine, because they are never separate from Intelligible principles but are linked to the original Soul: "so that their souls
look in no other direction except toward what is above them."[34]

The way in which Ficino develops these themes in chapter III of the *De
vita coelitus comparanda* is of particular interest here because of the need
that he feels to introduce the concept of *spiritus* in the play between the
anima mundi, or universal idea, and material things. According to Ficino,
such a spirit is a necessary requirement as a mediating factor between a
soul of divine origin, which is too different and removed from the dense
body of the sensible world to be able to communicate with it. Help is thus
required from a particularly excellent kind of body, which Ficino calls
spirit and which he defines as "nearly without body, and nearly a soul"
(*quasi non corpus, & quasi aim anima*).[35] But the most interesting aspect
of this introduction of "spirit" on Ficino's part, in the context of a possible plagiary of his text by Bruno while he was proposing his realist reading of Copernicus, is undoubtedly the near-identity that Ficino operates
between this mediating universal spirit and the heavenly quintessence.
Spirit, in this sense, according to Ficino, may well be considered heavenly,
or a kind of fifth essence. Nor would it be wise to undervalue the fact
that there was a more recent philosophical school of thought, with which
Bruno was undoubtedly familiar, where such ideas were considered important. This was the Telesian circle, or the group of philosophers who
became followers of Bernardino Telesio of Cosenza (1509–1588). This
same page of Ficino's in the *De vita coelitus comparanda* had already
been used by one of Telesio's closest followers, Antonio Persio, to underline the ubiquity of "pure spirit, of a universal and subtle kind." There
would be little difficulty in applying to Persio as well the accusation of
plagiary, for he literally translates long passages of Ficino's text, without
any explicit recognition of his source.[36]

It is not necessary here to rehearse the long discussion about the nature
of the Aristotelian quintessence that had developed in the course of the
Middle Ages and that would assume a particular importance in Bruno's
interpretation of the Copernican astronomy in infinite terms. For within
his new infinite cosmology, all essential differences between the matter
that makes up our world and the matter of the heavens disappear, allowing Bruno to propose a universe populated by infinite worlds, all made of
the same primary substance as our own.[37] It seems that Bruno was already
suggesting something of this kind at Oxford—or so it would appear from
Gentile's letter, with its references to the "absurd" idea that the moon or
other planets might be populated by living beings, just as our world is.

It is well known that it was Aristotle who divided the universe into two
spheres of being in the *De caelo*, distinguishing them clearly one from the
other. Underneath was the sphere of sublunar being, composed of the

four elements derived from Empedocles, while above was the sphere of the quintessence composed of a pure, extremely subtle, unalterable, and eternal substance.[38] The medieval commentators, and above all the Arabs, had debated at length on the physical and metaphysical implications of this Aristotelian dualism. Avicenna had written a work on the heavenly bodies that dealt with what he called "the fifth substance," and Averroes had replied to this work with his *De substantia orbis*, which was well known to Bruno.[39] According to Averroes, every heavenly body possesses a heavenly soul with two functions: on the one hand, the heavenly soul moves its body in a perfect circle, while on the other, it looks toward the prime mover as the object of its desire. The nature of the prime mover, according to Averroes, is that of an immaterial intellect that the heavenly soul contemplates from its heavenly body as the object of its thought, while it moves in a perfect and eternal circle.[40] Averroes thus operates a clear distinction between the heavenly bodies in circular movement and earthly bodies that move in straight lines, and by doing this, he maintains Aristotle's distinction between the two spheres of being. Nevertheless, his definition of the heavenly soul as an intellect clearly extends it throughout the universe, as a unifying and regulating principle that involves also the sublunar sphere, thus tending to challenge the Aristotelian dualism that had dominated the *De caelo*. In Ficino, that challenge becomes even more evident with the introduction of the concept of *spiritus*, above all when the spirit is identified with the quintessence itself. For if the heavens above the moon, in Ficino's view, are composed of the quintessence, he does not want to limit the quintessence to the heavens but extends it, as a substrate of the *anima mundi*, to all things in the universe. Even the most material things imaginable, according to Ficino, such as the huge stones we see around us, are actually pervaded by a certain quantity of the quintessence, which is what renders them part of a unique and unified world. This ubiquity of the quintessence seems to have been derived by Ficino from the *Picatrix*, where it is called "elisir," and it is on that ubiquity that he founds the possibility of "comparing" the heavens and the earth.[41] Consequently, if in the heavens the perfect movement is the circular one, then the same thing must correspond on earth—for that reason, Ficino advises the wise philosopher to "walk in a circular sense for as long as you can, trying to avoid giddiness, and cast your eyes on heavenly things, contemplating them also with your mind" (*Si ipse quoque leniter & ferme similiter movearis, quosdam pro viribus giros agens, vertigine devitata, celestia lustrans oculis, mente versans*).[42]

Garin has written acutely of Ficino's thought that it is not possible to find in it the coherence of a linear development, because "on the one hand Ficino insists to exasperation on the differentiated unity of all things, on the graduality of the different levels of being, on the delicately articulated mediation between things, but on the other, he insists forcefully on the distinctions between them."[43] What Bruno does to Ficino's thought is to defend rigorously a new coherence founded on the concept of a differentiated unity of all things, thus abolishing the concept of distinctions between levels of being: an idea that Bruno translates into cosmological terms through his reading of the new Copernican astronomy in the light of a new concept of cosmological infinity. Thus we arrive at the pages of *La cena de le ceneri* where, against the objections of the two scandalized Oxford doctors who had been invited by Fulke Greville to debate with the Nolan philosopher about the paradoxes of his philosophy, Bruno, in the figure of Theophilus, proposes once again his infinite universe, in which all distinctions between levels of elementary and quintessential being have completely disappeared:

> And where the tail of the Bear lies is no more worthy of being called the eighth sphere than where the earth is, on which we live: for all the heavenly bodies are placed distinctly in one and the same ethereal region, which may be considered a single great space or field, and they move nearer or further from each other with certain regular intervals.[44]

It is hardly surprising, after reading Abbott's page on Bruno, not to find in *The Ash Wednesday Supper* any reference to Ficino or to the *De vita coelitus comparanda*: the accusations of plagiary by the Oxford dons had clearly had an effect. Nevertheless, it was logically possible for Bruno to argue the terms of his new cosmological vision on the basis of the Ficinian concept of a quintessence that runs through all things like a divine elixir. Bruno's lectures titled *De quintuplici sphera* could well have been based on the idea of a quintessence running through all things, as we find it in chapter III of the *De vita coelitus comparanda*, while the priority of circular movement throughout the infinite spaces of the newly united universe (evoked by Abbott with his references to Bruno's insistence on *chentrum & chirculus & circumferenchia*) may well have included a celebration of circular movement such as we find in chapter 11 of the *De vita coelitus comparanda*. For his part, Aquilecchia has already postulated a use on Bruno's part at Oxford of the Hermetic formula of the infinite circle whose center is everywhere and whose circumference is nowhere—a formula amply evoked by Bruno in his cosmological dialogues when he needs to define the spatial shape of his infinite universe.[45] Nor should it be forgotten that the circular movement of heavenly bodies was still accepted as a dogma by Copernicus himself, whose book, according to Abbott, was being read by Bruno together with Ficino's.[46] Thus the apparent dilemma evoked by this reading of the *De vita coelitus comparanda* together with the *De revolutionibus* of Copernicus tends to disappear in the light of a detailed reading of Ficino's text.

Although he never goes so far as to abolish the Aristotelian distinction between two spheres of being, Ficino insists that there is a quintessence that resides beyond the sublunar sphere of the four elements of earth, water, fire, and air, but that from its original state of a heavenly substance turns into a divine elixir that runs through all things in the universe. Surely it is this idea that Bruno takes from Ficino's text, bending it to his own use by postulating a *unique quintuple substance* to be identified with his newly infinite universe. That universe is seen by Bruno as populated by an infinity of solar systems, defined by a heliocentric principle mediated by Bruno from the book of Copernicus. The "quintuplici sphera" could thus well be nothing more nor less than Bruno's infinite universe itself (whose center is everywhere and whose circumference is nowhere) composed of the four canonical elements with the addition of a *spiritus*, or *elisir*, which, by running through all things, joins them to the *anima mundi*, or world soul, of which it is the subtle voice or breath. Such ideas may find some of their sources in the magical tradition, but they are transposed, first more timidly by Ficino and then more boldly by Bruno himself, into the sphere of a properly astronomical and physical discourse.

It has already been stressed that the texts of Bruno's Oxford lectures have not survived, or at least are not known to have survived except in the form of their titles given by Bruno himself in *La cena de le ceneri*. This means that any comment on them necessarily remains in the field of conjecture. Nevertheless, the most recent documents presented and discussed here clearly confirm Abbott's declaration that there was a definite astronomical component to them: "he undertooke among very many other matters to set on foote the opinion of Copernicus."[47] It is clear that there was a constant attention on Bruno's part, in the context of his cosmological speculation, for the workings of the soul within his new infinite universe, and in particular for the possibility, or impossibility, of maintaining a traditional conception of the immortality of the individual soul. We still find this attention in the late *De triplici minimo*, the first work of the Frankfurt trilogy of 1591, in which Bruno develops his atomistic doctrine fully for the first time.[48] In the third chapter of the first book of this work, immediately after the explanation of the conceptual bases of his atomism, Bruno discusses at length the question of death "which does not involve the corporeal substance, and even less the soul." So it is not surprising if already in the *Sigillus sigillorum*, which is a work close in time to the Oxford lectures, we find Bruno underlining the importance of what Ciliberto calls "the *rectores* of intellectual activity" and what Sturlese identifies as the five grades of ascension of the soul toward knowledge of the divine: *de quintuplici et simplici progressionis gradu*. It is furthermore probable that the lectures titled *de immortalitate animae*, which Bruno says he delivered together with those *de quintuplici sphera*, discussed the workings of the soul within the newly infinite universe, possibly using as a source, as Sturlese suggests, the *Teologia platonica* or the *de immortalitate animae* of Ficino. What seems much more questionable, on the other hand, is the claim that the phrase *de quintuplici sphera* should necessarily be considered as without astronomical implications, for Abbott unquestionably asserted that Bruno attempted to introduce the Copernican astronomy at Oxford.

Earlier in this chapter, I have attempted to demonstrate that the title given by Bruno himself in *The Ash Wednesday Supper* for a part of his Oxford lectures—that is, *de quintuplici sphera*—could quite plausibly have included an attempt, surprising only at first sight, to merge together, in properly astronomical terms, a reading of Ficino's *De vita coelitus comparanda* with the *De revolutionibus* of Copernicus, precisely as Abbott claims. The final part of this inquiry into Bruno's Oxford lectures will underline the relevance for this claim of a page of the *De vita coelitus comparanda* that discusses magnetism. Ficino's concern with this subject tends, in traditional terms, to underline the occult powers of the magnet, as Agrippa of Nettesheim had already done in his *De occulta philosophia* as well as Della Porta in his *Magia naturalis*.[49] A particular characteristic of Ficino's treatment of this subject, however, was his insistence on the nature of the attraction of the magnet for the polestar: a phenomenon that tended to undermine any emphatic dualism between the changing sublunar world of the four elements and that of the eternal heavens. Ficino writes of the action of the magnet that "in a unique series of linked events, that which is above attracts that which is below and draws it towards itself."[50] It is precisely because of his insistence on an "infused [magnetic] virtue" that physically links the constellation of the Bear to the earth in a unity that defies any kind of dualism that Ficino would be attacked more than a century later by the Jesuit Niccolò Cabeo. This attack is developed in indignant tones that are remarkably similar to those used by Abbott to denounce exactly the same "madness" in Bruno at Oxford in 1583.[51]

It has to be noted that Bruno would never make magnetism the cause of the movements of heavenly bodies, as William Gilbert and his circle of magnetic philosophers would do in London some years later, soliciting the explicit approval of Galileo in his *Dialogue on the Two Major World Systems*. For Bruno always thought of the heavenly bodies as being moved by internal stimuli corresponding to biological and natural necessities, expressions of the instinctive intelligence with which all things are endowed.[52] Even so, it is probable that Bruno noticed the daring use on Ficino's part of the magnetic argument to sustain his "comparison" between earthly and heavenly things. For it was precisely by following that path that Ficino would open the door—although not himself prepared to cross the threshold—that would later lead toward that full unification of the entire universe proposed by Bruno in his Italian cosmological dialogues written and published in London in 1584. Such a development, in its turn, would point the way toward the absolute space of Newton.

There is one more aspect of Bruno's evident use as one of his sources of Ficino's *De vita coelitus comparanda*, according to the testimony of Abbott when referring to his Oxford lectures, which needs to be underlined, and that is its character as a text of medicine and alchemy. The *De vita* was designed as a therapeutic text that would lead to the purification of the weary and melancholy soul of the wise philosopher. According to Kristeller, there was nothing new in Ficino's attempt to introduce astrological themes into a medicinal tract.[53] What was more likely to have struck Bruno as interesting was the marked insistence with which Ficino postulates a substantial unity between celestial harmonies and the desired harmonies on earth, for it was precisely this unity that would become the dominating theme of Bruno's new infinite universe. To be precise, it is more correct to talk about his "ancient" universe, which is how Bruno himself would define it in *The Ash Wednesday Supper*, indicating the ancient Pythagorean cosmology as his principal source and inspiration.[54] It is precisely in the context of this typically Renaissance theme of the superior wisdom of the ancients that Bruno, in the first dialogue of the *Supper*, would celebrate the temperate habits, the expert medicine, and the long lives of those who already knew, in ancient times, how to live comfortably within the confines of an infinite universe. What was destined to upset this original harmony was the great "error" of a heaven divided from the earth by an eternal and immobile substance located above the moon: the quintessence introduced by Aristotle.[55]

It would seem that this mixture of cosmological and medical themes, which appears to have characterized Bruno's lectures at Oxford, remained as a persistent memory within the walls of the ancient British university. For the year 1632 saw the publication of the celebrated and much read work by Robert Burton, *The Anatomy of Melancholy*. Burton's lifelong link with Oxford is well known. He lived in Oxford for most of his life as a tranquil scholar and an Anglican churchman. Yet surprisingly Giordano Bruno is remembered in the very first page of Burton's work. There the melancholy author presents himself humorously to his reader as a new Democritus who, like Epicurus and his master Leucippus, "believed in the Paradox of the Earth's motion, of infinite worlds *in infinito vacuo, ex fortuita atomorum collisione*"—beliefs in an infinite space inhabited by colliding atoms that had, according to Burton, been lately revived by Copernicus, Bruno, and others.[56]

It is not easy to understand the exact position of Burton himself within the complex play of paradoxes that surrounds the theme of melancholy and its treatment in his erudite work. There can, however, be no doubt of the importance for Burton of the *De vita* of Ficino, which is continually referred to in marginal notes and which clearly constitutes an essential point of reference. Furthermore, Burton too, like Bruno before him, adds to his therapeutic reflections a section on the heavens: "having ever beene especially delighted with the study of *Cosmogrophy*."[57] It is to be found in the second part of his book, which is entirely dedicated to the treatment of melancholy, the causes and symptoms of which had already been the subject of the first part. The second section of the second part is titled *Ayre Rectified. With a Digression of the Ayre*, and it follows a section dedicated to the problems of diet. From a medical point of view, the second section insists on the necessity of avoiding polluted air and celebrates life in the country and the beneficent effects of quiet walks under the open sky. Nevertheless, the "digression" has a properly cosmological character and traces a rapid panorama of the most recent astronomical theories, where Bruno is referred to constantly as the proponent of a new infinite universe composed of a single substance: "one matter throughout, saving that the higher, still the purer it is, and more subtle."[58]

It is clearly significant that Burton develops his cosmological argument in a *Digression of the Ayre*, inserted into a medical treatise, with a specific reference to the infinite cosmology of Bruno. For his part, Smitho, the English counterpart of Theophilus in Bruno's *The Ash Wednesday Supper*, had asked, in the fifth dialogue of that work, for a final exposition of the movements of the earth, claiming that they cannot be considered proper matter for a mere "digression":

> as for myself I am of the opinion that the earth must necessarily move rather than that spherical system of fixed lamps: and in order to convince those who are still uncomprehending, it is better to announce it as a principal subject rather than to confine it into the space of a digression. So that if you wish to please me, I pray you to specify now the movements of our globe.[59]

It is a passage that could be taken to suggest that, at Oxford, Bruno too had spoken only in a digression on his astronomical theme, which becomes his principal subject in the *The Ash Wednesday Supper*—that is, "the movements made by this globe." By doing so too hastily at Oxford, he may have left many who were still uncomprehending—such as, for example, George Abbott and Alberigo Gentile.

Burton's later work is not founded, as Ficino's was, on an astrological conception of medicine. It is rather part of a new era influenced by Francis Bacon that had become more worldly and more deeply rooted in a search for natural causes and effects. Nevertheless, Burton has a clear if at times ironic perception of the trauma and confusion caused by the new cosmological theories. He evokes, with a gusto that seems to oscillate between melancholy and enjoyment, the sensation of living in a world that has gone quite mad, just as Abbott had more bitterly claimed before him. For the new universe is involved in a process of constant mutability, and by turning round and round, the newly moving earth has destroyed all Burton's previous certainties: "The whole world belike should be new moulded ... and turned inside out, as we do haycocks in Harvest, toppe to bottome, or bottome to top: or as we turn apples to the fire, move the world upon his Center."[60] So the center fails to hold, and the melancholy, which is Burton's subject, appears in part to derive precisely from this radical and terrifying upheaval. In the England of 1632, this had become not only a question of cosmology, but also one that was fraught with social and political dangers. Only eight years later, the new middle classes, inspired by Parliamentary sentiments of an increasingly radical nature, would declare civil war on a monarchy that had become actively anti-Parliamentarian and was pervaded by absolutist tendencies. It is well to remember that Oxford, at that point, would become the meeting place and center for the vacillating monarchical forces of Charles I. And so Burton's book seems to close a circle, opened by Bruno's Oxford lectures, that had mixed Ficino's astral medicine dangerously with the new Copernican astronomy. From there, Bruno would go on to develop his idea of an infinite, post-Copernican universe in his London dialogues, giving rise to the later protest of George Abbott, future Archbishop of Canterbury. Robert Burton would complete the circle with a book designed to maintain a physical and mental balance within a world run mad.

Bruno, however, saw things differently. Whereas Burton considered melancholy as the pervading humor of an unstable modern world, Bruno, in the pages of *The Ash Wednesday Supper* where he remembers his visit to Oxford, attempts to present himself as the true doctor of the modern soul. He does it by launching a "prophecy": that it will be his new infinite universe, composed of a single homogeneous substance that is both material and spiritual and that links all things within it into a harmonious whole, that will save a new era from despair. .

NOTES

1. See McNulty (1960).

2. The entry regarding George Abbott in the *Dictionary of National Biography* specifies that he became a fellow of Balliol at the beginning of the academic year 1583–1584—that is, very shortly after the probable date of Bruno's lectures that Abbott would criticize so sharply in 1604.

3. Abbott (1604). Hill's page on Bruno is quoted in full in English and commented on in Italian by Aquilecchia (1995a).

4. The text of Bruno's Italian dialogues used in this paper is Bruno (2002). *La cena de le ceneri (The Ash Wednesday Supper)* is in vol. I, pp. 427–590.

5. Some of the details of Bruno's first visit to Oxford can be deduced from the marginal note to a book owned by Gabriel Harvey (I. Ramus, *Oikonomia, sev DISPOSITIO REGULARUM VTRIVSQUE IVRIS IN LOCOS COMMUNES*, 1570, 192), which suggests that Harvey too was present at the lectures. For a comment on this note, and on the speculations to which it has given rise, see Aquilecchia (1995a).

6. The importance of this letter was first underlined by Limentani (1933), and more recently by Ciliberto (1986), 95–99. It is considered by most commentators to have been written and published between Bruno's two visits to Oxford, although Bassi (2004), 25, has argued for an alternative hypothesis of publication following Bruno's final return to London after his second visit, in the wake of the confusion caused by his lectures.

7. See Bruno (2002), vol. I, 535.

8. For the bibliographical references to these passages, see notes 3 and 6.

9. See Aquilecchia (1995a), 34.

10. See Aquilecchia (1993a), 376–93.

11. Yates (1964), 206–11.

12. See Ciliberto (1986), 91–101, and Ciliberto (1990), 31–45.

13. Aquilecchia (1991), 77–85.

14. On the precise terms of the difference of opinion between Aquilecchia and Yates concerning the nature of Bruno's thought, see my forthcoming paper on "Giovanni Aquilecchia's Contribution to the History of Renaissance Science," to be published in a volume of conference essays scheduled to appear in the Warburg Institute Studies series.

15. Sturlese (1994), 89–167.

16. See Feingold (1984) and Aquilecchia (1993a) and (1995). This subject is revisited in a later and decidedly hostile comment on Bruno at Oxford by Feingold (2004).

17. See Aquilecchia (1993a) and (1995), and Singer (1950).

18. A reading of Bruno's cosmology at Oxford in line with this possibility can be found in Knox (1999).

19. For Bruno's contribution to this subject, see Ingegno (1978), 1–25.

20. See Bruno (2002), 498–502.

21. See the second of the *Tre schede su Bruno ad Oxford* in Aquilecchia (1993a).

22. In Bruno (2002), 165–66. My translation.

23. See Aquilecchia (1995a), 36.

24. For the history of the Bruno–Brahe relationship, see Sturlese (1985).

25. See Rowland (2008), 139–48.

26. See Yates (1964), 197–200.

27. For the importance of these pages of the *Furori* in the context of Bruno's intellectual biography, see chapter 6, "The Sense of an Ending in Bruno's *Heroici furori*," in this volume.

28. The complex and at times ambiguous role played in the scientific revolution by the Platonic tradition has been discussed by Luciano Albanese in chapter III of his *La tradizione platonica: aspetti del platonismo in occidente*, Rome, Bulzoni, 1993, 125–214.

29. The importance of Bruno's French connection has been underlined by Nuccio Ordine. See Ordine (2007).

30. See Paul Oscar Kristeller, *Marsilio Ficino and His Work after Five Hundred Years*, Florence, Olschki, 1987, appendix VII, 149, and also Patrizia Castelli, "Per una storia degli scritti di Ficino tra '500 e 600. Note preliminari sugli scritti medici e astrologici," in *Il lume del sole: Marsilio Ficino medico dell'anima*, Florence, Opus libri, 1984, 65–69.

31. For the history of the composition of the *De vita*, see A. Tarabocchia Canavero, "Il *De triplici vita* di Marsilio Ficino: una strana vicenda ermeneutica," in *Rivista di filosofia neo-scolastica* LXIX (1977): 697–717.

32. See Kristeller, *Ficino and His Work after Five Hundred Years,* op. cit., 9. Also Kristeller, *Supplementum ficinianum*, Florence, Sansoni, 1937, vol. I, lxiv–lxvi.

33. The letter was first published in D. Del Corno Branca, "Un discepolo di Poliziano: Michele Acciari," in *Lettere italiane*, XXVIII (1976): 470–71.

34. See Eugenio Garin, "Le 'elezioni' e il problema dell'astrologia," in *Umanesimo e esoterismo. Atti del V Convegno internazionale di studi umanistici*, ed. Enrico Castelli, Padua, Cedan, 1960, 17–37.

35. M. Ficino, *De triplici vita libri tres,* Florence, 1489, fol. Miiii r. For a modern English translation of this work, see Ficino, *Three Books on Life*, eds. Carol V. Kaske and John R. Clark, Binghamton, NY, Medieval and Renaissance Texts and Studies, 1989. (For the traditional concept of the world soul, see Tullio Gregory, "Anima mundi: la filosofia di Guglielmo di Conches e la scuola di Chartres," in *Mundana sapientia: Forme di conoscenza nella cultura medievale*, Rome, Edizioni di Storia e Letteratura, 1992, where, in particular at p.134, it is underlined how frequent it is in the Middle Ages to find quotations from the Biblical verse in Genesis, *Spiritus Dei ferebatur super aquas,* together with quotations from the passage in the Hermetic *Asclepius* that talks of a spirit that is infused throughout the world, and similar passages to be found in Plato's *Timeus* 34b and Virgil's *Aneid*, VI, 723–26—all of them pages that were well known to Bruno. For the importance for Ficino of the idea of *spiritus*, see Walker (1958). For the development of the idea of *spiritus* between Ficino and Bruno, in authors such as Cardano, Scaliger, and Pico della Mirandola, see the introduction by Garin (1984), 3–14.

36. See Antonio Persio, *Trattato dell'ingegno dell'huomo*, ed. Luciano Artese, Pisa–Rome, Istituti editoriali e poligrafici internazionali, 1999, 40, and also Canone (1998b).

37. Developed in depth for the first time in *La cena de le ceneri (The Ash Wednesday Supper)*, Bruno's doctrine of a unique, homogeneous substance becomes a central doctrine in both *De l'infinito, universo et mondi* and *De immenso et innumerabilibus, seu de universo et mundis*, the third and last work of the Frankfurt trilogy published in that town in 1591.

38. See in particular the first two books of Aristotle's *De coelo*, where, at 270a 12–15, the quintessence is defined as untouched by generation or corruption, and as unsusceptible to change.

39. For the relationship between the thought of Bruno and that of Averroes, see Sturlese (1992), where Bruno's familiarity with the *De substantia orbis* is underlined. See also Granada (1999).

40. See Averroes, *De substantia orbis,* ed. and translated by. A. Hyman, Cambridge, MA–Jerusalem, The Medieval Academy of America, 1986.

41. Ficino writes of the quintessence at fol. F.vii.r of the *De vita coelitus comparanda*: "elixir arabes astrologi nominant."

42. *Three Books on Life,* op. cit., p.221, and *De triplici vita libri tres,* op. cit., fol. g.viii.r.

43. See Garin (1976), 77.

44. See Bruno (2002), vol. I, 544.

45. See Aquilecchia (1991), 84. This Hermetic definition of the infinite universe does not appear in the canonical Hermetic texts translated by Ficino with the title of *Pimander*. According to Brian Copenhaver, it goes back to *The Book of the Twenty-Four Philosophers*, a Hermetic compilation of the twelfth century, where it appears as the twenty-fourth proposition; see *Hermetica*, ed. and trans. Brian Copenhaver, Oxford, UK, Oxford University Press, 1992, xlvii. This Hermetic formula for the infinite universe is repeated by Bruno in dialogue V of *De la causa*, in dialogue III of *De l'infinito*, and in book IV, chapter 7 of the *De immenso*.

46. See in particular chapter IV, book I of Copernicus's *De revolutionibus* titled "Why the movement of the heavenly bodies is uniform and perpetually circular or composed of circular movements," in Copernicus (1978).

47. See note 3 earlier.

48. The *De triplici minimo* is in Bruno (2000c), 1–227. For a comment on this aspect of Bruno's atomism, see my essay "Giordano Bruno's Soul-Powered Atoms: From Ancient Sources towards Modern Science," republished as chapter 3, "Bruno and the New Atomism."

49. See on this subject Gatti (1999), 45–46.

50. See the fifteenth chapter of the *De vita coelitus comparanda,* op. cit. Ficino deals with the magnet also in chapters 2, 8, and 26.

51. See Niccolò Cabeo, *Philosophia magnetica,* Ferrara, 1629, 25–28.

52. For Bruno's relationship with the circle around William Gilbert, see chapter 5, "Bruno and the Gilbert Circle," in Gatti (1999). For Galileo's reference to Gilbert's magnetic philosophy, see Gatti (1997).

53. See Kristeller, *Marsilio Ficino and His Work,* op. cit., 9.

54. See chapter 1, "'The Pythagorean School and Our Own," in Gatti (1999).

55. See Bruno (2002), vol. I, 462.

56. See Robert Burton, *The Anatomy of Melancholy,* 2 vols., eds. Thomas C. Faulkner et al., Oxford, UK, Clarendon Press, 1989, vol. I, 1.

57. Ibid., 4.

58. Ibid., vol. II, 33–67.

59. See Bruno (2002), vol. I, 563.

60. Burton, *The Anatomy of Melancholy,* op. cit., vol. II, 39.

2

BRUNO'S COPERNICAN DIAGRAMS

THE STUDY OF GIORDANO BRUNO'S COPERNICANISM has a long and distinguished history, going back to the nineteenth century and continuing until the present day. It has involved a number of prestigious scholars, both historians of science and historians of philosophy, such as Paul-Henri Michel, Alexandre Koyré, Hélène Vedrine, Thomas Kuhn, and Robert Westman, among many others.[1] This notable body of comment on Bruno as one of the major Copernican philosophers of the sixteenth century will be taken as given, and mention will be made of the details of his reading of the *De revolutionibus* only when necessary to the development of the subject of this chapter. This intends to be a comment on the way in which Bruno attempted to pilot a recalcitrant sixteenth-century public, convinced of the falsity of the Copernican hypothesis except within a strictly mathematical formulation of it, toward a realist acceptance of the heliocentric principle, together with much else (such as, for example, the infinity of a universe filled with an infinite number of worlds) that Copernicus himself would not have been prepared to accept. It was precisely this realist heliocentric stand, however, shared by only a small handful of his contemporaries, that involved Bruno in the attempt to visualize a new world picture—for he left to others the task of calculating more precisely the movements of the heavenly bodies. At the same time as he praised Copernicus publicly as one of the most audacious and innovative minds of all times, he also chided him for being "too much of a mathematician, and not enough of a natural philosopher."[2]

Bruno did not make the mistake of identifying Copernicus himself with the famous anonymous preface to the *De revolutionibus*, actually written by Andreas Osiander, which advised use of the astronomical system proposed in the volume only in terms of a mathematical hypothesis. Indeed Bruno was the first to declare publicly that Copernicus himself could not possibly have written that preface, although he seems not to have known who the true author was. But Bruno did think that Copernicus himself had not stood out strongly enough in defense of the realist nature of his own proposal. Bruno saw himself as assuming Copernicus's mantle insofar as he accepted the difficult challenge of making people see the world in its new shape, not just mathematically but physically. For

Bruno, who was a philosopher, not an astronomer, the new universe was the place we have to live in, and he hoped that it would be possible to live better there than in the world people had thought they were living in before. This was made all the more difficult by the fact that Bruno also extended the Copernican hypothesis to infinite dimensions, proposing not a unique universe with a single sun at its center but an infinite world inhabited by an infinite number of solar systems. For, as Michel-Pierre Lerner has recently once again underlined, Bruno was among the first to develop a radical criticism of the finite cosmology delimited by the so-called planetary spheres. These were supposed to carry the planets round in their harmonious circles in a crystalline quintessence of Aristotelian origin: for Bruno, they were pure fictions with no physical basis at all.[3] Bruno's own cosmology derives from Epicurus and Lucretius rather than Aristotle. Space becomes an infinite envelope filled by a tenuous ether that pervades it in all its parts. Visualizing our own solar system in Copernican terms thus meant for Bruno not visualizing the universe as such, but visualizing only a small speck of it floating within an immense and infinitely populated whole. Although today we have become used to seeing the earth as a minute, hardly visible point within immense vistas of space and time, such an idea at the end of the sixteenth and beginning of the seventeenth centuries appeared overwhelmingly unfamiliar and strange. Even those who had made the effort to accommodate their minds to the new Copernican system, such as Johannes Kepler, found Bruno's overall cosmological picture totally unacceptable. Kepler referred to it as Bruno's "innumerabilities," expressing concern for his friend Johann Matthäus Wacker von Wackenfels's "deep admiration for that dreadful philosophy."[4] On the other hand, it was precisely Bruno's conceptual leap toward the idea of an infinite universe that lead Alexandre Koyré to exclaim, four hundred years later:

> On reste confondu devant la hardiesse, et le radicalisme de la pensée de Bruno, qui opère una transformation—révolution véritable—de l'image traditionelle du monde et de la réalité physique. [The audacity and radicalism of Bruno's thought are stunning. He operates a transformation—truly a revolution—of the traditional image of the world and of physical reality.][5]

THE PHYSICALLY REAL

To be sure, the criterion of scientific realism that inspired Koyré's outburst of praise for Bruno's conceptual leap into infinite space appears now as part of the "traditional" view of the so-called scientific revolution. Proponents of the more recent historiographical criteria of contingency and scientific sociology, or social constructivism, would be quick to brand it as suspect "for want of a right reason constituted by nature."[6] It would overrun the bounds of this paper to enter into our contemporary debate concerning the respective claims of a logical system of reasoning based on a coherent concept of scientific objectivity, and the idea of science as "a form of intellectual ecology rather than of inductive logic."[7] It is worth pointing out, however, that Bruno himself, placed at the very beginning of what still continues to be called "the scientific revolution," was aware of precisely this problem, and discussed it openly in his cosmological dialogues. In the remarkable second dialogue of his major cosmological work in Italian, *La cena de le ceneri*, or *The Ash Wednesday Supper*, written and published in London in 1584, Bruno pictures himself as "the Nolan philosopher" (he was born in Nola, near Naples) and sees himself as undertaking a nighttime journey that will eventually lead him to the rooms of Sir Fulke Greville, where the supper and the cosmological discussion were held. Traveling in an ancient creaking boat down the Thames, followed by an adventurous walk through the muddy streets of the still crowded city—metaphors of a world still enclosed within the gradually disintegrating structure of the traditional Aristotelian–Ptolemaic universe—Bruno notes how on the way he cannot avoid meeting with "a princely palace here, there a wooded plain with a glimpse of the sky lit by the morning sun."[8] The dialogue continues by offering a wealth of further information about the London of the day: how the unfriendly English servants dress and behave, the affectations and at times the arrogant behavior of Bruno's aristocratic hosts, how in England wine at table was often drunk out of a communal cup (complete with only half-hidden references to the Protestant transformations of the rituals of the Catholic mass). Such was the social context in which a cosmological discussion based on Bruno's reading of Copernicus's *De revolutionibus* was held on the evening of Ash Wednesday, 1584, in the rooms of Sir Fulke Greville, friend and future biographer of Sir Philip Sidney, whom Bruno praises in his work as one of the most brilliant minds of his time. Bruno is aware that all this cannot but affect the way in which Copernicus's book was being read and discussed in London on that momentous evening.

Nevertheless, having dealt with such "preliminaries" in the first two dialogues of the *Supper*, in the third dialogue, where the cosmological discussion properly begins, Bruno does call upon a criterion of physical objectivity in his defense of the Copernican astronomy. He does this in the first place by mounting a bitterly ironic attack on the writer of the anonymous preface, whom he brands as an unfaithful doorkeeper of Copernicus's new edifice. This in itself is clearly a metaphor pregnant with important meanings, for an edifice must have its mathematical coordinates, but it is clearly in the first place a physical construction. Although "set" within a definable social, geographical, and historical landscape, nevertheless an edifice constitutes an autonomous architectonic structure within which its inhabitants live, move, and create their world. There is clearly a sense in which a physical edifice is more "real" and lasting than the mathematical calculations that have served to create it or than the social and historical context within which it has been built. Bruno's choice of metaphor, at the very beginning of the discussion of Copernicus's book that the final three dialogues of *The Ash Wednesday Supper* narrate, is thus a conceptually appropriate one with which to define the complex but nevertheless "realist" terms in which, as the Nolan philosopher, he intends to conduct the debate.

Robert Westman, in what he has called the "Wittenberg interpretation" of Copernicanism in the sixteenth century, has demonstrated how rare were the early attempts to read the new astronomy in realist terms, in the Protestant parts of Europe as well as in the Catholic ones. He includes Bruno among the very few Copernican realists active in sixteenth-century Europe.[9] Undoubtedly, given the fact that the discussion narrated by Bruno in the *Supper* took place in London and that he wrote about it and published his work in that city, the most important precedent to Bruno's realist stand was that of Thomas Digges. First published in 1576, and presented somewhat slyly as a mere addition to his father's completely traditional work on astrology, in particular in its practical application to weather forecasting, *A Prognostication Everlastinge*, Digges's few Copernican pages are partly direct translation from book I of *De revolutionibus* and partly stringent comment on their implications. Unlike Bruno, Digges does all he can to avoid underlining the "revolutionary" nature of the Copernican proposal. Insofar as he also sees it as opening out the universe to possibly infinite dimensions, he proclaims his entirely traditional acceptance of the four elemental spheres reaching as far as the moon, surrounded by a crystalline semidivine substance identifiable as Aristotle's quintessence. Thus, for Digges there is only one solar system, not an infinite number, as Bruno would proclaim. So Digges saw no need for his readers to be alarmed by the new astronomy, and he precedes his Copernican pages with the picture of a ship sailing in calm waters—presumably a tranquilizing message to Sir Edward Fines, the Lord High Admiral, to whom the book, in his father's name, is dedicated. Within this overall strategy of underplaying the innovative aspects of his own pages, it is entirely characteristic of Digges that he should give his key punch for a realist reading of the heliocentric proposal almost in a throwaway aside. It is not clear how many of his English readers (for Digges was writing in English rather than in Latin, as his father had done before him) understood the literally world-shattering implications of his claim:

Copernicus mente not as some have fondly excused him to deliver these grounds of the Earthes mobility onely as Mathematicall principles, fayned and not as Philosophicall truly averred.[10]

Bruno himself, on the other hand, had already discovered that even in England the waters of Copernican discussion tended to be remarkably agitated, and not tranquil at all. By the time Sir Fulke Greville invited him to supper to discuss his reading of Copernicus as well as other "paradoxes" of his new philosophy, Bruno had already been publicly derided by the Oxford dons after his attempts to explain the Copernican astronomy in lectures at the university given during the summer of 1583.[11] His own ship diagram in *The Ash Wednesday Supper* depicts stormy waters, in the course of being stirred up to further tempests by a chubby-cheeked north wind. Nevertheless, Bruno's ship image may be, and frequently has been, compared with Digges's ship insofar as both authors are concerned to argue that the impetus of a ship's movement would be "impressed" on a weight dropped from the mast, which would therefore fall vertically to the foot of the mast and not be left behind by the moving ship. This argument was already known and discussed in the Middle Ages, although in an Aristotelian–Ptolemaic context. It was repeatedly used in early Copernican discussion, up to and including Galileo, to contradict the anti-Copernican objection that a moving earth would leave all the clouds and the birds behind.[12] Bruno never mentions Digges in his work (an example followed by Galileo, who never mentions Bruno, to Kepler's surprise and concern), but it seems more than likely that Bruno at least knew of Digges's work. For Digges was a pupil of John Dee, who also taught mathematics to Sir Philip Sidney, and whose remarkable library, which contained Copernicus's *De revolutionibus*, was the occasion of a meeting with Sidney and his entourage after a state visit to Oxford in which Bruno is known to have participated.[13] Although Bruno, unless aided by a friend, would not have been able to read Digges's English text, he could certainly have contemplated his well-known Copernican picture of the universe, and may have had it in mind when preparing his own rather different Copernican picture to illustrate the text of the fourth dialogue of *The Ash Wednesday Supper*.

Copernican realism, already a characteristic (if constantly underplayed) of Copernicus himself and of Digges, and a defining one of Bruno's readings of his astronomy, caused problems of visualization from the very beginning. It decreed the sudden superfluity of a centuries-long tradition of illustrations of the Aristotelian–Ptolemaic universe, which had assumed a notable aesthetic as well as scientific dimension (see figure 2.1).

Figure 2.1 From Peter Apian, *Cosmographia*, Antwerp, 1524.

The task of drawing a new and unfamiliar image of a now heliocentric cosmology was by no means simple, and Edward Rosen has drawn attention to the fact that difficulties arose at once with relation to the illustration to be included in first editions of the *De revolutionibus*. Copernicus's own diagram was rejected, and another diagram, possibly by Rheticus, was included. This diagram was to be the cause of perplexities and misunderstandings throughout the sixteenth century (see figure 2.2).[14] Digges's Copernican diagram is virtually the same as that in the *De revolutionibus*, except for the suggestion of an infinite number of stars stretching out beyond a unique astronomical system of a heliocentric kind (see figure 2.3).

In dialogue 4 of *The Ash Wednesday Supper*, the published diagram in *De revolutionibus* appears at the center of the heated Copernican discussion between Theophilus, the mouthpiece of Bruno himself, and Torquato, one of the two bejeweled and conservative Oxford dons called in by Sir Fulke Greville to defend the traditional cosmology at his sup-

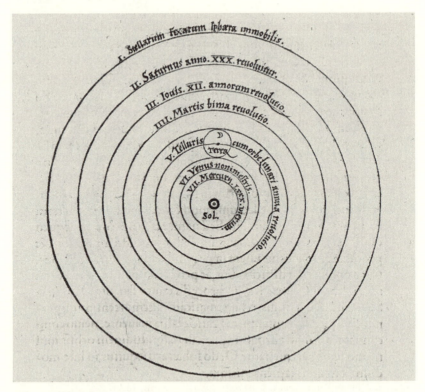

Figure 2.2 The image of a heliocentric model from book I of Copernicus's *De revolutionibus orbium coelestium*, Prague, 1566.

per party. The problem raised by Bruno has been often considered both puerile and mistaken by commentators, especially by those anxious to further Frances Yates's Hermetical and magical reading of Bruno's works, which denies any scientific value to his Copernicanism at all.[15] In fact, Bruno's argument is both justified and not altogether incorrect. Torquato, as Bruno points out, bases his anti-Copernican comments on Rheticus's diagram rather than on a serious reading of Copernicus's text, thus failing to understand that if the orbit of the earth around the sun is seen as perfectly circular, then the sun has to be slightly off-center for the system to save the phenomena. Otherwise, as Bruno puts it, the diameter of the sun would appear constant throughout the year. Another solution to this problem, put forward by Copernicus himself only in book III of *De revolutionibus,* is to keep the sun at the geometrical center of the system and put the earth on an epicycle, which is the solution adopted by Bruno in his own Copernican diagram in *The Ash Wednesday Supper* (see figure 2.4).

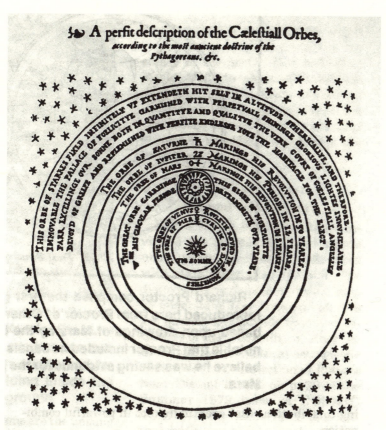

Figure 2.3 Thomas Digges's version of the heliocentric model from book I of Copernicus's *De revolutionibus orbium coelestium*, from *A Perfit Description of the Celestial Orbes*, an Addition to the book by his father Leonard Digges, *A Prognostication Everlastinge*, London, 1576.

Bruno's visualization of the new sun–earth relationship, although very schematic, is thus quite correct—more correct than that suggested by the *De revolutionibus* diagram, and indeed by that of Digges.[16] It is interesting to note, however, that Digges, in a previous Latin work of 1573, *Alae seu scalae mathematicae*, written together with John Dee, had already made a number of references to Copernicanism in Latin. This work could well have been read by Bruno, as in it Digges raises the same questions that Bruno is discussing here—that is, the necessity of introducing either epicycles or eccentrics to guarantee the apparent changes in the sun's diameter.[17] Bruno, furthermore, goes on to make a mistake himself, by putting the moon on the same epicycle as the earth, whereas Copernicus

PTOLEMAEVS.

COPERNICVS.

Figure 2.4 Diagram representing the Ptolemaic world system (upper half) and the Copernican world system (lower half) discussed by Bruno in *La cena de le ceneri* (*The Ash Wednesday Supper*), London, 1584. Courtesy of the Biblioteca Nazionale, Rome. (Unauthorized reproduction of this image is prohibited.)

puts it on a second epicycle centered on the revolving earth. These were still early Copernican times, and mistakes in reading the new cosmology were many. Both Kepler and Galileo made their own seriously mistaken conjectures, raising the whole question of "Copernican mistakes" that are themselves an interesting, and ultimately not unfruitful, aspect of his reception. Where Bruno leaves Digges far behind, although in written text rather than in illustration, is in his attempt to visualize an entirely homogeneous and infinite universe, no longer characterized by those elemental spheres that are so clearly depicted by Digges in his diagram (see figure 2.3) as still dominant in the earth–moon orbit of his newly Copernican world.

WAITING FOR THE TELESCOPE

> Advances in engraving techniques, and in particular the detail made possible
> by copper-plate, meant that illustrations could match the most disparate sub-
> jects. Maps, plans, structural and logical diagrams, mathematical figures,
> drawings of machines and cog wheels, reproductions of animal or plant spe-
> cies, and synoptic tables invaded the printed page, clarifying, qualifying and
> completing it. . . . The image acquired a philosophical role, and the ensuing
> redefinition in figures and signs of the totality of knowledge would play its part
> in the development of a new conception of man and the cosmos.

This eloquent passage written by Luce Giard on illustrations in scientific
texts of the early modern period defines the context in which discussion of
Bruno's illustrations, cosmological and otherwise, should be examined.[18]
Much recent discussion of the problem of visualization of astronomical
objects, however, has concentrated on the hiatus between the pre- and the
post-telescopic age. The advent of telescopic observation with Galileo,
it is argued, raised a whole series of new optical issues, including those
relating to the degree of accuracy of scientific instruments themselves. A
systematic program of observations of the moon, for example, was not
carried out until well after Galileo's death, and even then not without
numerous problems of interference relating to sightings of disks created
by the telescope itself.[19]

It is known that telescopes were already being made and discussed
in Bruno's time. Bruno himself would undoubtedly have known about
them from the work on natural magic of his fellow Neapolitan, Giovan
Battista della Porta, which was also known to Kepler, and possibly also
from the works of Leonard and Thomas Digges.[20] Both Della Porta and
the Diggeses, however, discuss in their works the use of telescopes only
for terrestrial observation, particularly in the field of navigation. Modern
commentators have tended to deduce from this that visualization of the
new astronomy started only with Galileo. The pre-telescopic age appears
relegated by this discussion to a kind of meaningless limbo, as if from
Copernicus himself the reception of his theory jumped to the momentous
event expressed by Galileo's succinct comment of 1610: "But forsaking
terrestrial observations, I turned to celestial ones."[21]

Nobody was more critical of such an approach to the new astronomy
than Kepler himself. For Kepler formulated his theory of the elliptical
orbit of Mars on the basis of observations made with the naked eye. Furthermore he wrote his famous *Dissertatio* or *Conversation* on Galileo's
discovery of the moons of Jupiter, shortly after the discovery had been
published in the *Sidereus nuncius*, before having obtained a telescope
with which to observe the moons for himself. There is a curious note of
disdain in Kepler's disparagement of Galileo's ability to make his own

telescope. Kepler himself is not able, he assures his public, to work with his hands, but soon someone will lend him a telescope, and then he will see Galileo's new moons himself.[22] To his credit, Kepler never doubts the authenticity of Galileo's discovery, as Galileo's ecclesiastical enemies went on doing until well after his trial and house imprisonment. Kepler's instinctive trust in Galileo's observational skill throws a deep shadow over Galileo's own mistrust, indeed total silence, with respect to Kepler's momentous discovery of elliptical orbits. It was Galileo himself who was largely responsible for the assumption, made by so many scholars today, that serious visualization of the Copernican theory began only with telescopic observation of the new pattern in the skies.

A major claim made by Kepler in his *Conversation* is that a number of post-Copernican theories and discoveries formulated before Galileo's observations of the moons of Jupiter made that discovery conceptually possible. He thinks that Galileo should have recognized their importance in his text. And if Kepler's main concern is to insist on the importance of his own theories and discoveries, he also includes Bruno in this context. For Bruno had formulated a clear distinction between bodies such as suns and stars that generate their light from within and moons or earths that are illuminated from without. Kepler agrees with Bruno that it is necessary to move beyond the purely visual outlook of the new system provided by Copernicus himself and to pass from the facts to the causes.[23] This had become imperative to the natural philosopher of the time, as the new system virtually banished from the cosmological picture the traditional Aristotelian "prime mover," which had set the Ptolemaic celestial system in motion in the first place (see figure 2.1). Copernicus himself, as well as an early Copernican such as Thomas Digges, had fleetingly referred to the Neoplatonic concept of elemental motion put forward, in an Aristotelian cosmological context, by Marsilio Ficino. Recently studied by Dilwyn Knox, this doctrine sees gravity and levity as causes of celestial motion, within a conceptual context still founded on the theory of the four elemental spheres as the primary constituents of matter up to the planetary sphere of the moon.[24] However, Bruno repudiated the elemental spheres just as he repudiated the planetary spheres of Aristotelian fame. Serious speculation about the universal causes of the heavenly motions within the new cosmology thus may be seen as starting with Bruno—even if Kepler prefers his own unique world based on his more mathematical idea of a universe defined by the five Platonic solids. Galileo, for his part, had little time to spare for Kepler's mystical Neoplatonism and, in his later *Dialogue concerning the two chief world systems*, preferred to refer to William Gilbert's magnetic explanation of the causes of celestial motions.[25] Kepler knew and admired Gilbert's *De magnete*, which had been published in 1600, the year of Bruno's death. Nevertheless, in his *Conversation* with Galileo, it is through multiple references to Bruno's natural philosophy that Kepler establishes the principle that a new, universally valid cause of the celestial motions was necessary to make sense of Copernicus's theory at all.

Bruno's own solution, already put forward in *The Ash Wednesday Supper* and never abandoned, was based on a thermodynamic concept of the play between the contrary forces of cold and heat. Its root lay in the anti-Aristotelian natural philosophy of Bernardino Telesio, whom Bruno greatly admired.[26] Telesio saw the whole universe as moved throughout by the active principles of heat and cold, even if he himself never abandoned the Aristotelian, finite cosmology. Telesio's thermodynamic doctrine of planetary movement, however, did tend to defy the traditional idea of elemental spheres, for the contrary forces of heat and cold were seen as dominant throughout his still finite and geocentric universe. Kepler was probably thinking of Bruno's enthusiastic adaption of this concept to his infinite universe when he criticized Bruno for "talking in generalities." However, a careful reading of Bruno's *De immenso et innumerabilibus* of 1591 shows that he did attempt to specify his thermodynamic theory of planetary motion by supplying it with a precise mathematical formulation. He does this through the use of a diagram whose importance seems to have escaped the notice of his commentators (see figure 2.5).

Bruno's text claims that in the infinite universe, if considered infinitely, nothing can be said either to act or to be acted upon. But if considered in terms of the finite bodies within it, then they do act and are acted upon. He goes on to consider how, in a general sense, action of one body on another decreases with respect to increase in the distance between them. For example, in figure 2.5, the fire e heats point f according to the distance e–f. If the fire at d is four times as hot as the fire at e, it will heat e according to the distance d–e four times as much as e heats f, but it will heat f only twice as much because it needs to travel twice the distance to reach it. Thus Bruno is introducing a mathematical idea of the ratio of distance to intensity to measure the amounts of heat by which the hot bodies (stars or suns) attract the cold ones (earths or moons) into their orbit. The argument goes on to consider Aristotle's (puerile) claim that if the universe were infinite and the heat of an ethereal fire were of infinite intensity, then there would be no chance of the earth withstanding such heat—therefore, all bodies must be contained within a finite world. Bruno's final claim is that Aristotle would have been right if the elements were confined, as Aristotle thought, to separate spheres, and therefore, fire, in its own sphere, were pure. As we have seen, however, for Bruno, there are no elemental spheres, just as there are no planetary spheres, but only an infinite universe filled with a universal ether within which a homogeneous substance assumes proto-atomistic form. In this universe, in all its parts,

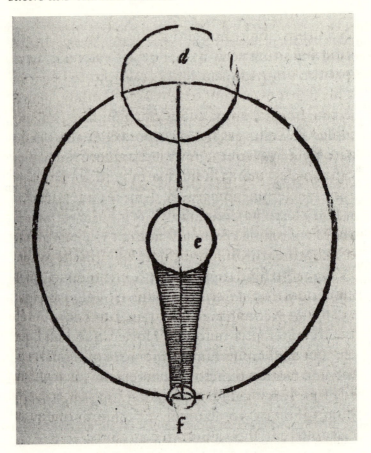

Figure 2.5 Diagram representing the diffusion of heat in Bruno's *De immenso et innumerabilibus*, published together with his *De monade*, Frankfurt, 1591.
© The British Library Board. Shelfmark 532.b.29, p. 239.

Bruno claimed that fire is always united in some degree to humidity, creating an atmosphere in which all the celestial bodies, including the so-called fixed stars, can move and survive.[27] It is true that Bruno's thermodynamic theory of celestial motion got him into difficulties when he had to consider the movements of moons about cold planets. For the moment, however, it is enough to notice that he is already thinking in terms of a universally valid cause of the movements of stars and planets within heliocentric systems, which can be expressed by a mathematical formulation. Kepler was surely right to note that Bruno's published discussions of the heliocentric astronomy constitute a development in the reception of the Copernican revolution that Galileo should not have ignored.[28]

Bruno's diagram also shows that the visualization of the new celestial problems was an important moment of pre-telescopic thought about the new astronomy. Two more of his Copernican diagrams may be mentioned here, although these have already attracted the attention of commentators. Both come from the Copernican discussion in *The Ash Wednesday Supper*. In that work, Bruno makes a considerable use of optics to justify the new astronomy. He makes no mention of his sources, but it has been supposed by his commentators that he had been reading the work of Jean Pena. Before arriving in London, Bruno had been living and lecturing in Paris, where Pena's optical writings, which already apply optics to a discussion of the Copernican theory, were well known.[29] Bruno's reasoning in *The Ash Wednesday Supper* may also have been influenced by the *Optics* of Ibn Al-Haytham (Alhazen), an Arabic mathematician and astronomer who originated from Iraq and was active in Cairo in the first half of the eleventh century. A Latin translation of his work, known as the *Perspectiva*, was published in 1572 by Freidrich Risner in Basle and widely used by the natural philosophers of the period. The ninth earl of Northumberland, who owned one of the most important contemporary collections of Bruno's texts, attributed the change of his life from a frivolous courtier to a dedicated natural philosopher to a reading of this work of Alhazen.[30] In book III, chapter 7, Alhazen considers "The Ways in which Sight Errs in Inference" and writes that "by looking at a fixed star and a planet at the same time sight will not perceive the difference between their distances, but rather perceive them both in the same plane despite the great difference between their distances." These, and similar optical arguments, were used by Bruno to justify not only the astronomy of heliocentric systems but also his theory of an infinite universe. Two of his best known Copernican diagrams in *The Ash Wednesday Supper* (see figures 2.6 and 2.7) are of some importance in his discussion of his new picture of the universe.

In figure 2.6, Bruno is concerned to show that a smaller opaque body placed between the eye and a larger luminous body becomes invisible to the eye at great distances. This simple diagram thus supplies him with a conceptual instrument for challenging the Aristotelian doctrine that the sky contains only those bodies that are visible to the eye. Bruno's frequently expressed conviction that the sky could and undoubtedly did contain numerous bodies that had so far never been seen was probably what Kepler was thinking about when he told Galileo that Bruno was one of those who had helped to prepare the conceptual grounds for his telescopic discovery of the moons of Jupiter.[31]

In figure 2.7, the last of the diagrams in *The Ash Wednesday Supper*, Bruno attempts to visualize the multiple movements of an earth in motion according to the Copernican hypothesis by using the example of a

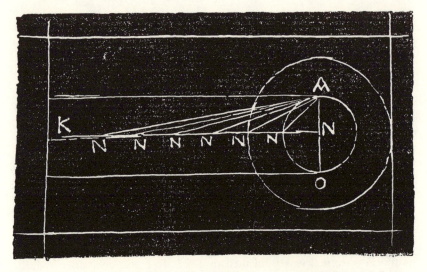

Figure 2.6 Diagram representing vision of a body at increasing distances from the eye. From Bruno's *La cena de la ceneri* (*The Ash Wednesday Supper*), London, 1584. Courtesy of the Biblioteca Nazionale, Rome. (Unauthorized reproduction of this image is prohibited.)

ball thrown into the air. Bruno thinks of the ball as having four different motions, all of them part of one single complex motion. The first and principal one is along the trajectory A–E, the second around its own axis I–K. The third movement consists of an oscillation in the revolution of the moving ball along parts of the circumference that Bruno visualizes in his text by dividing it into eight segments. These segments are not indicated in the diagram, and it is not altogether clear what circumference he is referring to. In a recent edition of this text, it has been assumed to refer to a slipping back of the traveling ball along the circumference of the orbit A–E, which would make it correspond to Copernicus's account of the movement known as the precession of the equinoxes. This, however, presupposed an earth still fixed onto precisely those celestial spheres that Bruno, earlier on in this work, had already denied. Alternatively, Bruno's third movement may have corresponded to what was known as axial precession, composed of an oscillation that traced a figure eight around the two poles of the earth itself. This movement of axial precession, however, could be considered as integrated into Bruno's fourth movement of the ball, visualized as an oblique spin that eventually inverts the positions of O–V. Undoubtedly some obscurity remains in Bruno's account of the third and fourth movements of the ball in the air, largely due to the incomplete nature of his diagram. The important point to be made, however, is that Bruno has understood the principal novelty constituted

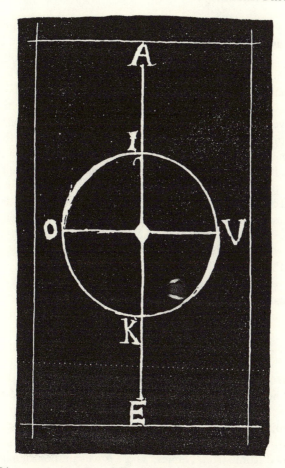

Figure 2.7 Diagram representing differing movements of the earth according to the metaphor of a spinning ball. From Bruno's *La cena de la ceneri* (*The Ash Wednesday Supper*), London, 1584. Courtesy of the Biblioteca Nazionale, Rome. (Unauthorized reproduction of this image is prohibited.)

by the Copernican account of precession of the equinoxes and its accompanying anomalies—that is, that it should be seen as a complex of very slight, long-term variations in the movements of the earth itself, and not of the zodiac or a sphere of fixed stars, as was the case in the traditional astronomy. Bruno thinks of the four movements of the ball in his figure as roughly corresponding to the Copernican annual movement of the earth around the sun, its daily revolutions around its own axis, added to two of the complex set of long-term anomalies associated in Copernicus's still circular astronomy with the precession of the equinoxes, although Bruno never uses that term. Precession remained extremely complicated in Copernicus's system, as it had been in Ptolemy's, and it was giving rise to heated discussion among more technical experts than Bruno. In any case, Bruno thought that the astronomers were not capable of offering more than mathematical approximations of the movements of the earth and the other planets. His main purpose with the ball image and its accompanying diagram was to catapult his readers into a new adventure in outer space, forever ousting them from their once comfortably central and immobile earth. In *The Ash Wednesday Supper*, Bruno insists that the multiple motions of a now moving earth are regular and constant and must be respected as such. If he thought that astronomical calculations were inevitably approximate, that was because of his mistrust of mathematics as the perfect instrument of human prediction, rather than lack of faith in the infinitely complex but ordered regularity of the natural world.[32]

WORK IN PROGRESS

Owen Gingerich's *Annotated Census of Copernicus's "De revolutionibus" (Nuremberg, 1543 and Basel, 1566)* contains a description of a copy in the Biblioteca Casanatense in Rome of the 1566 edition with a signature "Brunus Fr[ater] D[ominicanus]," but no annotations by Bruno.[33] This is claimed by Gingerich as "the bold Giordano Bruno signature from the fly-leaf," although Bruno scholars tend to be more cautious. There are, however, some interesting points to be made about this volume. First, it is almost impossible either to attribute it to Bruno, or not to do so, on the basis of the handwriting of what is not strictly speaking a signature but rather a florid and highly stylized design. Second, the book reached Rome from Naples, where it was in the original nucleus of the library belonging to the Spaniard Matias de Casanate (c. 1580–1651), father of the Cardinal Casanatense who brought the collection to Rome. Matias was a high-ranking judicial official and might have obtained the book during the agitation caused by Bruno's trial and execution in Rome in 1600, when the official investigations into Bruno's previous heresies that had got him into trouble with the authorities of the Dominican monastery in Naples became a subject of attention by the Inquisition. Third, it has been convincingly shown by Miguel Granada that Bruno must have been reading the 1566 edition of *De revolutionibus*, of which this volume is a copy. The 1566 edition also contained the *Narratio prima* of Rheticus, passages of which Bruno often transcribes.[34] Fourth, if this really is Bruno's copy of the *De revolutionibus*, which would not be put on the Index of forbidden books until much later, in 1616, then he was presumably reading Copernicus at a considerably earlier age than commentators have usually supposed. Bruno entered the Dominican monastery in

Naples in 1565, at the age of seventeen, and fled north in 1576, at the age of twenty-eight.

Gingerich's *Census* also contains a description of Kepler's annotations to his 1543 copy of the *De revolutionibus*, at present held by the Universitätsbibliothek at Leipzig.[35] These clearly show how sixteenth-century and early-seventeenth-century readings of the Copernican astronomy were in the form of "works in progress" rather than constituting a definitely acquired body of new astronomical knowledge. They also emphasize how a major problem in the ongoing understanding of Copernicus's system concerned the question of where to situate the center of the new universe. This is the problem raised by Bruno in *The Ash Wednesday Supper*. Also in his case, it is correct to speak of "work in progress"—in fact, it is Bruno himself who, in the fourth dialogue of that work, gives his readers an account of his progressive reactions to the Copernican astronomy. Bruno claims that he had passed through the following stages of growing Copernican conviction: First, he considered the new cosmology a mere joke put forward in debate by those who amuse themselves by trying to demonstrate that black is white. Second, he began wondering why Aristotle had spent so much time in his *De caelo*, book II, criticizing the heliocentric theory of Pythagoras and his followers. Third, in a more mature period of his youth, he began to think of Copernicus's theory as a possibility. Only later (at an unspecified date) came the growing conviction of its certain truth.[36]

In a page of book III, chapter 5 of his later *De immenso*, Bruno harks back to what seems to be the third stage of this story—that is, his growing conviction of the truth of the new theory (see figures 2.8 and 2.9).[37] Referring to a time "when he was younger," he describes a picture he had formulated in his mind of the following cosmological hypothesis: the sun together with the fixed stars orbits annually around the earth through AF; the earth revolves around its center at C along the axis HI in its diurnal rotation; the earth does, however, move from the geometrical center, traveling annually away from the equator of the universe, at times toward the tropical pole E, at times toward the antarctic pole G. The traditional long-term movements of trepidation and oscillation are assured by additional spiraling movements of the earth that expose its surface to the heat or the cold of the poles according to the long-term necessities of its evolution. Bruno illustrates this very schematic cosmological picture with a diagram that he insists represents "the philosophy of the masses," and not his own mature convictions. The question it poses is whether it was possible to maintain a central earth within a compromise solution that took at least some minimal account of the Copernican theory. By 1591, when the *De immenso* was published, such a system had been worked out in much finer technical detail by Tycho Brahe, who had published

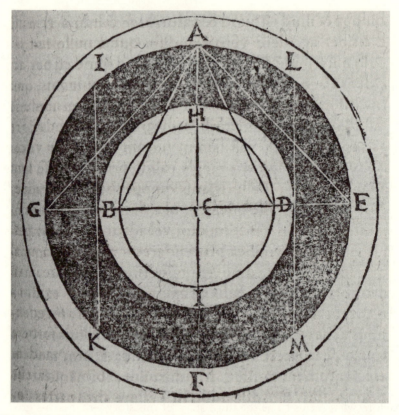

Figure 2.8 Diagram representing a juvenile hypothesis of movement of the earth at the center of the universe. In Bruno's *De immenso et innumerabilibus*, published together with the *De monade*, Frankfurt, 1591. © The British Library Board. Shelfmark 532.b.29, p. 301.

an account of his own partly Copernican cosmology in 1588.[38] Brahe, although not explicitly mentioned, is probably being criticized here as overprudent and "immature" insofar as he failed to step into a fully heliocentric world. Interestingly William Gilbert was aware of this cosmological model of Bruno's "when he was younger" (*cum esset junior*). He commented on it in his posthumously published *De mundo*, adding a diagram of his own. Gilbert criticizes the hypothesis for making the earth move in a straight line, "which is not normally attributed to celestial bodies," although it is probable that Bruno's diagram was not intended to indicate movement in a straight line but rather a small orbit of the earth around the geometrical center, through BD in Bruno's diagram and ae in Gilbert's.[39] It is not clear whether Gilbert was aware of the ironic stance

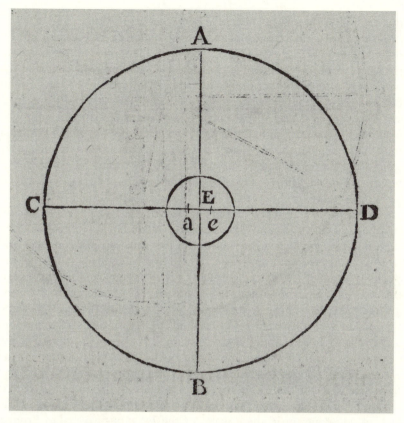

Figure 2.9 Diagram representing the same juvenile hypothesis attributed to Bruno by William Gilbert in *De mundo nostro sublunari*, Amsterdam, 1651. © The British Library Board. Shelfmark 8704.d.12, p. 199.

assumed by Bruno in these pages. Although Gilbert himself was sympathetic to Bruno's cosmological theses, his circle of magnetic philosophers either remained stubbornly Aristotelian in their cosmology, or referred to Tycho Brahe's compromise solution, which Bruno could not accept.[40]

Gilbert's interest in Bruno's cosmological theories did not stop with this diagram. On the very next page, he presents another (by now more fully Copernican) way of visualizing the cosmos in terms of Bruno's ideas (*Alius modus iuxta Nol.*; see figure 2.10).[41] Gilbert found this new theory in the *De immenso*, book III, chapter X.[42] In the pages that interested Gilbert, Bruno appears to be referring to *De revolutionibus*, III, 25, where Copernicus supposes an anomolous heliocentric model in which "the center of the annual revolution be fixed, as though it were the center of the world, but the sun be moveable by two motions similar and equal to those which we have demonstrated for the center of the eccentric, everything will appear just as before. . . . For then the motion of the center of the earth would be a perfect and simple motion about the center of the world, since the two other motions have been granted to the sun." Bruno begins by criticizing Copernicus because he does not normally make the sun orbit at the center of the solar system. Further criticism addresses Copernicus's account of the precession of the equinoxes, which posited a third movement of the earth as if it was carried around on its planetary sphere and therefore had to slip back gradually on its orbit in order to remain constant.[43] Bruno himself had long maintained that there are no planetary spheres and that the earth and other planets hang freely in the universal ether. He now sees it as a principle of rotatory planetary motion that the axis remains parallel to itself and in equilibrium, thus rendering superfluous Copernicus's third motion of the earth—a principle that will later be confirmed both by Gilbert himself and by Galileo. As for the sun, Bruno in these pages, like Copernicus in the passage cited earlier, visualizes it as moving in an oblique orbit with respect to an earth that travels around the center of the system on an axis parallel to the equator of the world. The sun must also rotate around itself with a spiraling motion, according to Bruno, as otherwise it would always seem to rise in the same place. Further oscillations of the earth's poles with respect to the zodiac, Bruno notes with admiration, had been introduced by Copernicus to compensate for the traditional slipping back of the zodiac itself that explained, in the Ptolemaic system, the precession of the equinoxes. The lack of any diagram in these pages of the *De immenso* makes Bruno's text arduous reading. Such must have been the impression of Gilbert, whose second Bruno diagram (figure 2.10) in his *De mundo* illustrates this anomalous heliocentric system described in words by Bruno himself.

In Gilbert's diagram, which correctly illustrates this page of Bruno's, DFCG represents the colure, or limits, of the solstitial points, and C and D the poles of the solar system. AB is the equator of the system around which the earth moves with an annual motion. The earth's equator, ab, also moves daily around its own axis. The sun describes a small circle limited by the equinoctial parallels egi and fhk. If its poles are G and F, the orbit of the sun will pass through g and h, or its two tropical limiting points, although other angulations of the orbit of what seems also here to be a spiraling sun are posited by Gilbert as possible. Bruno himself had further justified this principle as necessary to guarantee the revolution of the planets by supplying them with ever varying quantities of heat and cold. Later, in Galileo, the idea of a sun that revolves around its own axis would become important to explain the sighting of sunspots. Surely Bruno was right to consider early Copernicanism as a slow acquisition of

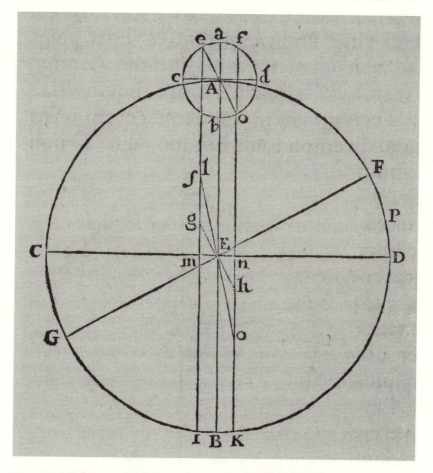

Figure 2.10 Diagram representing a more articulate hypothesis of the movements of the earth and sun, again attributed to Bruno by William Gilbert in *De mundo nostro sublunari*, Amsterdam, 1651. © The British Library Board. Shelfmark 8704.d.12, p. 200.

new astronomical concepts according to various approaches and reached by traveling along many different paths.

WHAT IS RIGHT AND WHAT IS WRONG?

Let us now come back to Kepler's mistaken distrust of Bruno's "innumerabilities," expressed to Galileo in his replies to the *Sidereus nuncius*. Kepler points out that Galileo's discovery does not support it because

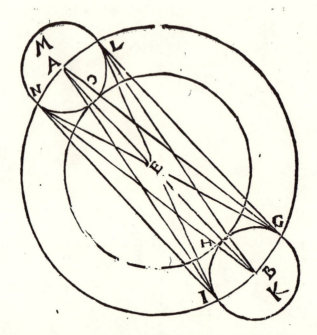

Figure 2.11 Bruno's final cosmological diagram from *De immenso et innu-merabilibus*, published together with *De monade*, Frankfurt, 1591. Courtesy of the Biblioteca Nazionale, Rome. (Unauthorized reproduction of this image is prohibited.)

Bruno thought of earths as circling around suns, while Jupiter is a planet, and yet the new moons circle around it.[44] For Kepler, this suggested that our own solar system constitutes a unique universe—thus saving him from Bruno's "horrible" idea of a plurality of suns. Kepler's observation, however, carries other implications. It highlights the terms of Bruno's "lunar" mistake in *The Ash Wednesday Supper* (see figure 2.4), although Kepler does not mention this specifically. Yet Bruno himself had already realized that his thermodynamic theory of planetary motion did not permit him to put the moon on a further epicycle centered on earth (or an epipicycle), as Copernicus had done to save the phenomena, because this would have meant visualizing a cold moon as circling around the center of a cold earth. Why should it do that? For Bruno, the moon too must circle around the sun as its center: the sun becoming thus the fountain of heat and light for the moon in the same degree as for the earth. This precedent in *The Ash Wednesday Supper* should be remembered when considering Bruno's final cosmological diagram in the *De immenso*, book 3, chapter X (see figure 2.11).

Bruno says of this diagram that it derives from his conviction that the orbits of Mercury and Venus around the sun cannot really be smaller than those of the earth and the moon, as the astronomers claim.[45] He proposes a system in which the earth A, with the moon now on its epicycle NMLO, revolves around the sun E in direct opposition to Mercury B, which carries Venus on its epicycle IHGK. Although in flagrant disregard of astronomical observation as well as of Copernicus's mathematics, this diagram occurs in a part of Bruno's text devoted to praise of Copernicus as the true hero of the modern world. It reflects Copernicus's conviction, eloquently expressed in his own dedicatory letter of the *De revolutionibus* to Pope Paul III, that a well-ordered universe implies uniformity and harmony of the spheres. Undoubtedly the Pythagorean bases of both Copernicus's and Bruno's cosmologies need to be underlined here, as much recent commentary has been doing.[46] Bruno himself refers to both Pythagoras and Plato just before describing this diagram in his text. Nevertheless it was Kepler who understood most clearly the specific technical difficulty that Bruno's thermodynamic theory of planetary movement had led to: if cold planets like the earth fulfill their purpose in the universe by varying on their surface the intensity of heat and light, cold and shadow, through which life evolves on their surface, why should cold moons circle around them at all? Bruno recalls his thermodynamic theory of planetary motion in the opening pages of *De immenso*, book 3, chapter X. His attempt to visualize a rudimentary planetary system in this diagram tries to solve the problem that would later be raised by Kepler. It shows how cold planets and the cold moons that revolve around them, by clinging together in epicycles, all orbit at harmonious distances around the sun, from which their life-giving energies arise.

Two considerations are in order here. First, Bruno is not addressing in these pages the Hermetic magicians or the Neoplatonic magi (although he does do that in other parts of his work). Here, he is explicitly addressing the astronomers. Translated into modern vocabulary, with respect to this rudimentary planetary diagram, Bruno himself admits defeat. He is quite aware that his picture fails to save the phenomena, and therefore that some kind of extension is required to his thermodynamic theory of planetary motion. He tries to turn this into a qualified defeat by pointing out that he at least has a physical theory of planetary motion that postulates a universally valid cause. The empirical problem of saving the phenomena is something that Bruno thinks cannot be solved by simply calculating quantities from the basic observables of time and position. It must be solved within a theoretically acceptable physical framework—a necessity that, in his opinion, Copernicus himself and most of the early post-Copernicans continued to ignore. His own comment on his planetary diagram ends with an appeal to the astronomers to integrate their mathematical skills into a theoretical physics—that, he claims, is all he asks of them in order to be satisfied.

A full consideration of this aspect of Bruno's thought would require a more detailed attention to his semiotics in its relation to the development of modern science.[47] Here, however, it is sufficient to point out that the frequent use of this diagram by those commentators who are concerned to enclose Bruno's thought entirely within a magical and Hermetic tradition that has nothing to do with a scientific logic is questionable, particularly if it implies (as it frequently does) that serious mistakes in reading Copernicus oust the culprit from any valid tradition of properly scientific thought. Such a premise would clearly present problems with Kepler, given his mistaken attempt to construct a new heliocentric cosmology on the basis of the five Platonic solids; with Galileo, who thought he had "proved" the Copernican hypothesis with a mistaken theory of the movements of the tides; as well as with Tycho Brahe, who constructed a short-lived compromise cosmology whose conceptual basis was clearly to a considerable extent religious—an attempt to respect the Biblical cosmology as well as the scientific phenomena. Furthermore, it is worth reflecting on the fact that one of the earliest formulations of an entirely negative judgment on Bruno's Copernicanism derives from the nineteenth-century astronomer Giovanni Virginio Schiaparelli. Appealed to by Felice Tocco, a prestigious nineteenth-century Italian philosopher who was presenting a positive reading of Bruno's Copernicanism as a prelude to Galileo's in a volume that remains essential reading today, Tocco found himself in difficulty when the internationally renowned Schiaparelli replied that Bruno's cosmological arguments were obscure, puerile, and of no validity at all. Tocco found a clever solution to his problem by continuing to develop in his text a fundamentally positive appreciation of Bruno's cosmological speculation, while relegating to a series of much discussed notes the impatient criticisms of Schiaparelli.[48] Those commentators who are still today using Schiaparelli to eliminate Bruno from the scientific arena may, however, wish to reflect on the fact that Schiaparelli himself perpetrated one of the most colossal and colorful of scientific "mistakes" when he claimed that his telescopic sightings had revealed a regular network of canals on the surface of Mars, which it was "not impossible" to conceive of as constructed by intelligent beings. Schiaparelli's sightings gave rise to more than half a century of fervid Martian speculation. This included the lifelong work of Percival Lowell, who built an observatory in California and dedicated his life to what became ultimately a desperate attempt to prove Schiaparelli right. Of course, he may have been, but the Martian probes at present are not pointing in Schiaparelli's direction.[49] Ironically, Schiaparelli may have been thinking about life on Mars because he had been reading the work of the "confused and imprecise" Bruno, whose concept of an infinite universe was based on the postulate that it was a "living" universe in all its parts. It is, in any case, unfortunate that the most recent enthusiast of Schiaparelli's criticisms of Bruno's cosmological speculation is the editor of the important volume recently dedicated to a comment on all Bruno's illustrations and diagrams. Following in Schiaparelli's footsteps has led their editor to take into little or no serious consideration the many diagrams that Bruno uses to illustrate both his atomism and his Copernicanism—two of the most daring scientific speculations of his day.[50]

What is "right" and what is "wrong" is surely not the point that needs to be labored in studying the early readings of the Copernican astronomy. The historian's task is to address those original minds that responded positively to the overwhelmingly unfamiliar implications of a new theory destined to become the foundation stone of modern cosmological thought. Bruno was among the first to understand that this would be the case—that the centuries-old Aristotelian–Ptolemaic cosmos had suddenly become a thing of the past and that a new world picture had to be formulated of a radically different kind. His limited grasp of the mathematics of Copernicus's *De revolutionibus* is more than compensated for by his remarkably subtle and daring speculation into its physical and philosophical implications. His extension of the much-enlarged but still finite Copernican universe to infinite dimensions, conceived of as a new atomistic physics, and not only (or even primarily) as a religious intuition, added, less than half a century after the publication of the *De revolutionibus*, another stone to the foundation of the modern world. Furthermore, Bruno's infinite universe incorporated a Copernican heliocentric principle in a "realist" sense: he thought of his infinite number of finite astronomical systems as all centered on suns, seen as the source both of their revolutions and of their life. Bruno knew that his philosophical achievement in his cosmological works depended on the original "revolution" proposed by Copernicus himself. More than once, he attributed generous public recognition to Copernicus as the genius whose "light" had ushered in a new era:

> For he had a profound, subtle, keen and mature mind. He was a man not inferior to any of the astronomers who preceded him, unless they are considered in their own time and place. His natural judgment was far superior to that of Ptolemy, Hipparchus, Eudoxus, and all the others who followed them, and this allowed him to free himself from many false axioms of the common philosophy, which—although I hesitate to say so—had made us blind.[51]

NOTES

1. Serious consideration of Bruno's Copernicanism starts with Berti (1876). For the twentieth-century comments mentioned here, see Koyré (1957); Kuhn (1957); Michel [1962] 1973; Vedrine (1967); and Westman (1977). More recently, Bruno's Copernicanism has been reconsidered by, among others, McMullin (1987); Seidengart (1992); Mendoza (1995); Gatti (1999); and Tessicini (2007). See also Gatti (2008a).

2. The quotation is from dialogue 1 of *La cena de le ceneri (The Ash Wednesday Supper)*. All references to the Italian dialogues in this chapter are to Bruno (2002). For the preceding quotation, see vol. 1, 449. All translations from Bruno's works are mine.

3. See Lerner (1996–1997), vol. 2, 157–66.

4. Kepler (1965), 37.

5. Koyré (1939–1940), vol. 1, 141.

6. The quotation, which is from Hobbes, is used by S. Shapin and S. Schaffer to question the whole concept of scientific realism. See their much discussed volume, *Leviathan and the Air-Pump*, Princeton, NJ, Princeton University Press, 1985. Previous philosophical discussion concerning the problem of scientific realism had included B. C. Van Fraasen, *The Scientific Image*, Oxford, UK, Oxford University Press, 1980, and Ian Hacking, *Representing and Intervening: Introductory Topics in the Philosophy of Natural Science*, Cambridge, UK, Cambridge University Press, 1983. For a synthesis of this discussion, see W. H. Newton-Smith, "Realism," in *The Routledge Companion to the History of Modern Science*, eds. R. C. Olby et al., London–New York, Routledge, 1990, 181–95.

7. See Stephen Toulmin, "From Logical Systems to Conceptual Populations," *Boston Studies in the Philosophy of Science*, eds. R. C. Buck and R. S. Cohen (eds.), 8 (1971): 552–64. For a balanced discussion of the recent debate concerning scientific realism, see H. Floris Cohen, *The Scientific Revolution: A Historiographical Inquiry*, Chicago, University of Chicago Press, 1994, 230–36.

8. See the *Argomento del secondo dialogo* of *La cena de le ceneri,* in Bruno (2002), vol. I, 434.

9. See Westman (1975).

10. Leonard Digges, *A Prognostication everlastinge . . . lately corrected and augmented by Thomas Digges, his sonne*, London, Thomas Marsh, 1576, fol. Mi, r. and v.

11. This episode, although much discussed and the subject of many conjectures, remains obscure insofar as the texts of Bruno's lectures have not survived. For the known documents, see Aquilecchia (1993a) and chapter 1 in this book.

12. Massa (1973); Westman (1977); and Aquilecchia (1995b) , 485–96.

13. See *John Dee's Library Catalogue*, eds. J. Roberts and A. G. Watson, London, Bibliographical Society, 1990, note 220.

14. See Copernicus, *De revolutionibus,* op. cit., 359, note 21.

15. See Yates (1964).

16. For Bruno's use of book III of *De revolutionibus* in this context, see my chapter on "Reading Copernicus" in Gatti (1999), 43–77.

17. See Thomas Digges, *Alae seu* scalae *mathematicae, quibus visibilium re-motissima Caelorum Theatra conoscendi*, London, Thomas Marsh, 1573, fols. Aiir–Aiiiv.

18. See Luce Giard, "Remapping knowledge, reshaping institutions," in *Science, Culture and Popular Belief in Renaissance Europe*, eds. Stephen Pumphrey, Paolo L. Rossi, and Maurice Slawinski, Manchester, UK, Manchester University Press, 1991, 19–47; 29–31.

19. See Mary G. Winkler and Albert Van Helden, "Representing the Heavens: Galileo and Visual Astronomy," in *Isis* 83 (1992,): 195–217; Mary G. Winkler and Albert Van Helden, "Johannes Hevelius and the Visual Language of Astronomy," in *Renaissance and Revolution: Humanists, Scholars, Craftsmen and Natural Philosophers in Early Modern Europe*, eds. J. V. Field and Frank A.J.L. James, Cambridge, UK, Cambridge University Press, 1993, 97–116; and Isabelle Pantin, "L'illustration des livres d'astronomie à la renaissanceRenaissance: l'evolution d'une discipline à travers ses images," in *Immagini per conoscere: dal Rinascimento alla Rivoluzione Scientifica*, eds. Fabrizio Meroi and Claudio Pogliano, Florence, Olschki, 2001.

20. See Albert Van Helden, "The Invention of the Telescope," in *Transactions of the American Philosophical Society*, 67, pt. 4 (1977); and for a claim that the telescope was invented in England by Leonard Digges, see C. A. Ronan, "The Origins of the Reflecting Telescope," in the *Journal of the British Astronomical Association*, 101 (1991): 335–42.

21. Galileo Galilei, *The Starry Messenger*, ed. and trans. Stillman Drake, New York, Doubleday, 1957, 28.

22. See Kepler (1965).

23. For a comment on Kepler's multiple references to Bruno in this text, see Varanini (2003), 207–15.

24. See Knox (1999).

25. Galileo Galilei, *Dialogue Concerning the Two Chief World Systems*, ed. and trans. Stillman Drake, Berkeley–Los Angeles, University of California Press, 1953, 400–411.

26. On Bruno and Telesius, see the two relevant papers in Aquilecchia (1993a) and Gatti (1994).

27. An anastatic reprint of the first edition of Bruno's *De immenso* [1591] may be consulted in Bruno (2000c), 399–907. For this diagram and its textual explication, 490–92 or, in the Italian translation by Monti, 493–94.

28. For a recent, detailed comparison of Bruno's cosmology with that of Galileo, see Rossi (2001), 283–303.

29. For the importance of Pena's application of optics to the new Copernican astronomy, see W.G.L. Randles, *The Unmaking of the Medieval Christian Cosmos*, Aldershot, UK, Ashgate, 1999, chapter 3, "The Challenge of Applied Optics," 58–79. Bruno's possible use of Pena's optical theories has been considered by Sturlese (1985).

30. For an English translation of this text, see Alhazen [Ibn Al-Haytham], *Optics*, ed. I. Sabra, London, The Warburg Institute, 1989. For Northumberland's claim, see his essay on "Love," first published by Frances Yates in *A Study of*

'*Love's Labour's Lost,*' Cambridge, UK, Cambridge University Press, 1936, 206–11. On the importance of Alhazen's optics in Renaissance thought, see J. V. Field, *The Invention of Infinity,* Oxford, UK, Blackwell, 1997. For Northumberland as a reader of Bruno, see Gatti (1989), 35–73.

31. For Bruno's discussion of this diagram, see Bruno (2002), vol. I, 504.

32. For Bruno's discussion of the earth as a moving ball, see ibid., 566–68. For Bruno's concept of mathematics as an approximate rather than a certain art, see De Bernart (2002).

33. Owen Gingerich, *An Annotated Census of Copernicus's 'De revolutionibus' (Nuremberg, 1543, and Basle, 1566),* Leiden, Brill, 2002, 115.

34. For the history of the Casanatense collection, see Marina Panetta, *La 'Libraria' di Mattia Casanate,* Rome, Bulzoni, 1988, where *De revolutionibus* is listed as no. 1263. For Bruno and Rheticus, see Granada (1990).

35. For Kepler's annotations, see Gingerich, *An Annotated Census,* op. cit., 76–80.

36. See Bruno (2002), vol. I, 535–36, and Gatti (2008a).

37. See, *De immenso,* in Bruno (2002c), 553.

38. For the importance of 1588 in the development of the new cosmology, see Granada (1996).

39. Gilbert (1651), 199–200.

40. See my chapter on "Bruno and the Gilbert Circle," in Gatti (1999), 86–98.

41. See Gilbert (1651), 200–201.

42. See Bruno (2000c), 592–93.

43. For what they consider Copernicus's "interesting" alternative model to his usual solar theory, see N. M. Swerdlow and O. Neugebaur, *Mathematical Astronomy in Copernicus's 'De revolutionibus,'* 2 vols., New York–Berlin, Springer-Verlag, 1984, 159. For Copernicus's account of precession, see N. M. Swerdlow, "On Copernicus' Theory of Precession," in *The Copernican Achievement,* ed. Robert Westman, Los Angeles, University of California Press, 1975.

44. Kepler (1965), 11 and 34.

45. Bruno (2000c), 596–98.

46. For an extended comment on Copernicus's own dedicatory letter, see Robert Westman, "Proof, Poetics and Patronage: Copernicus's Preface to *De revolutionibus,*" in *Reappraisals of the Scientific Revolution,* eds. D. C. Lindberg and R. S. Westman, Cambridge University Press, 1990, 167–205. See also P. L. Rose, "Universal Harmony in Regiomontanus and Copernicus," in *Avant, avec, après Copernic,* Centre National de la Recherche Scientifique, Paris, Blanchard, 1975, 153–63. The Pythagorean and Neoplatonic sources of the new sun-centered cosmology have been discussed by Eugenio Garin, "La rivoluzione copernicana e il mito solare," in *Rinascite e rivoluzioni: movimenti culturali dal XIV al XVIII secolo,*" Bari, Laterza, 1975, 257–95, and Paolo Casini, "Il mito pitagorico e la rivoluzione astronomica," in *Rivista di filosofia* 85, no. 1 (1994), 7–33. On Bruno's Pythagoreanism, see Tessicini (2001), 159–88.

47. Pioneering work on this subject is being done above all by German scholars such as Wolfgang Wildgen. See Wildgen (2001), 473–86.

48. See Tocco (1889). For the letter from Schiaparelli, see the notes at 313–17. For a positive discussion of Bruno's Copernicanism based on a critique of Schiaparelli's remarks, see Ingegno (1978), 63–70.

49. This story is told by F. I. Ordway, "The Legacy of Schiaparelli and Lowell," *Journal of the British Interplanetary Society* 39 (1986): 19–27.

50. See Bruno (2001). For the novelty represented by Bruno's attempt to visualize his atomistic theory of matter, see Lüthy (1998).

51. See Bruno (2002), 448–49.

3

BRUNO AND THE NEW ATOMISM

IN 1417, POGGIO BRACCIOLINI REDISCOVERED the lost *De rerum natura* by Lucretius, the Roman disciple of Epicurus. A largely forgotten and, in religious terms, severely condemned philosophical discourse was reintroduced into western culture. Categories of explanation became available for questions concerning the nature of matter, the mortality or immortality of the soul, and above all, generation and corruption, which the few atomists of the Middle Ages, such as Nicholas of Autrecourt or Nicole Oresme, had had to glean indirectly from the numerical Pythagoreanism of Plato's *Timaeus*; the critical commentary of Aristotle, Cicero, or Lanctatius; or the poetry of Virgil.[1]

Research into fifteenth-century culture in Italy has shown that the early impact of Renaissance Epicureanism was largely limited to moral philosophy and a discussion of the Epicurean *voluptas*.[2] Even after the first Latin translation of the *Lives of the Philosophers* by Diogenes Laertius in 1470 (the last two books of which were dedicated to the atomism of Leucippus, Democritus, and Epicurus), and the *editio princeps* of the *De rerum natura* in 1473, it is rare to find any consistent reference to Epicurean natural philosophy right up to, and including, the Neoepicurean poem by Palingenius, *Zodiacus vitae*, published in 1553. The use by Girolamo Fracostoro, in the middle years of the sixteenth century, of a corpuscular hypothesis to explain the spread of diseases may be seen as the explanation of a local phenomenon relating to medicine rather than an attempt to propose a general atomistic hypothesis in the context of a new theory of matter.[3] The influential and provocatively titled *De rerum natura* by Bernardino Telesio, which began to appear in Naples in 1563, under strict ecclesiastical control, attempted to mount a systematic attack on Aristotle's natural philosophy, but only vaguely adumbrated an atomistic conception of matter.[4] It is not until the publication of the six philosophical dialogues in Italian written and published in London by Giordano Bruno between 1583 and 1585 that it is possible to find a consistent series of references to a Neoepicurean atomism that will become a dominating topic of the natural philosophy of the seventeenth century.

Giordano Bruno's philosophy was published and probably composed within a brief but intense ten-year period from 1582 until 1592, when his work was interrupted by his arrest on the grounds of heresy by the

Venetian Inquisition and the beginning of the long trial that would lead to his execution in Rome in 1600. After the philosophical dialogues of his London period, his atomism receives its next major expression in a Latin work published in Prague, and dedicated to the Emperor Rudolph II, in 1588: *Articuli centum et sexaginta adversus huius tempestatis mathematicos atque philosophos*. Here among the *Axiomata*, we find *Individuum est minimum*, followed by a long series of *Theoremata minimi* that include references to the atom. However, it is only in the first work of his Frankfurt trilogy, *De triplici minimo et mensura*, written in Latin and published in that town in 1591, that Bruno's concept of matter, based on the ancient idea of discrete, indestructible, indivisible atoms, is subjected to a major and in-depth discussion.[5]

One of the problems discussed by commentators of Bruno's atomism is whether he can already be considered an atomist when he wrote his early Italian dialogues in London. Nineteenth-century commentators, such as Felice Tocco, tended to emphasize the development of Bruno's thought between the London and the Frankfurt periods.[6] Modern commentators, on the other hand, largely following the example of P. H. Michel, author of a brief but seminal contribution on Bruno's atomism published in 1957, have tended to stress the pages of the Italian dialogues, particularly *De la causa, principio et uno*, which already contain a clear definition of an atomistic theory of matter.[7] However, the terms in which atomism is introduced in the Italian dialogues are very different from those found in the Frankfurt trilogy. In the first book of the trilogy, *De triplici minimo*, Bruno founds his whole natural philosophy on the idea of the minimum, which in the physical sphere, he defines as the atom: on the basis of that foundation, he will reach the cosmological conclusion of his trilogy with the *De innumerabilibus, immenso et infigurabili*, which presents the final formulation of his infinite and eternal universe. In the earlier Italian dialogues, on the other hand, atomism appears only as a corollary to Bruno's cosmological theses, which are the proper and primary subject of these dialogues.

The importance of this difference cannot be overstressed. Bruno's atomism appears in the Italian dialogues after he has already argued in *The Ash Wednesday Supper* against the Aristotelian–Ptolemaic cosmology, still accepted by both Catholic and reformed Christianity. Bruno proposes, instead, a post-Copernican, infinite universe inhabited by an infinite number of solar systems similar to our own. This revolutionary cosmological thesis, which led to the interruption of Bruno's lectures in Oxford and, by his own account, caused consternation in the intellectual circles of Elizabethan London, was explicitly based on ancient Pythagorean and Epicurean sources. It was developed through a frontal attack on the Aristotelian idea of a hierarchical, finite universe filled with two entirely different types of matter: elemental matter under the lunar sphere, and an immutable, celestial quintessence above. Bruno's newly unified, infinite cosmology had thus already paved the way for a subsequent, logically unimpeachable reference to pre-Aristotelian and pre-Socratic atomism. However, Bruno, in his comments on prime matter in the Italian dialogues, insists on the virtuality of the atoms, which coincide with being only insofar as they represent the possibility of being.[8] That is to say, precisely because they are the foundation of what Bruno calls the "absolute all," they represent by definition extreme purity, simplicity, indivisibility, and unity: if they possessed weight, mass, or other positive properties, they would not be the bases of all things. In the Frankfurt trilogy, this matter theory becomes more substantial and appears as the foundation of the infinite universe celebrated in the final work, the *De immenso et innumerabilibus, seu de universo et mundis.*

The *De triplici minimo* is a work about the minimum. The atom itself is to be understood as one aspect only of what Bruno defines as a triple minimum. The three minimi correspond to the three dimensions of Euclidean geometry. The primary, or one-dimensional, minimum is the monad: the first principle of quantity and as such the basis of metaphysics. The two-dimensional minimum is the mathematical point: the first principle of extension and as such the basis of geometry. The atom is the minimum body, or three-dimensional minimum: and as such the basis of physics. The close relationship between Bruno's three minimi is stressed by Carlo Monti in his introduction to the Italian translation of the trilogy.[9] As Monti points out, it is important to notice that for Bruno, the mathematical idea of the minimum as minimum extension or point and the physical idea of the minimum as minimum body or atom are not considered incompatible but as two different aspects of the concept of the minimum. Pierre Duhem had pointed out the importance of these two points of view when approaching the subject of body in the *Summa theologica* of Saint Thomas Aquinas (part I, question VII, article III): an authority whose work, as an ex-Dominican, Bruno knew well. Duhem writes:

> *Or il faut observer que le corps, qui est la grandeur parfaite, peut etre pris de deux manières. On peut le considérer du point de vue matémathique et ne porter son attention que sur la seule grandeur de ce corps. On peut aussi le considérer du point de vue physique ou naturel, en le regardant comme un composé de matière et de forme.* [It must be observed that body, which corresponds to a perfect size, can be considered in two ways. It can be considered from a mathematical point of view, in which case attention is centered only on the size of the body. Or it can be considered from the physical or natural point of view, in which case it is composed of matter and form.][10]

Book I, chapter 2, of Bruno's *De triplici minimo* specifies that the minimum "is atom in the strict sense of the term in those material entities that constitute the primary parts themselves, and less strictly in those entities that are all in all and in every single part."[11] Such latter entities are the voice, the soul, and so forth. Bruno's atomically structured primal matter is thus made up of a plurality of what he calls entities. Furthermore, he is making a distinction here between different kinds of entities—that is, between material and more spiritual entities. This distinction maintains to some extent the Aristotelian distinction between matter and form. As Michel notes, Bruno remains robustly Aristotelian in his insistence that matter and form, although distinct entities, are strictly related to each other in all their manifestations.[12] However, Bruno is anti-Aristotelian in his rejection of substantial forms, as well as in his insistence that the soul is to the body as a pilot in a ship. This definition of the relationship between body and soul has given rise to much discussion of what is to be considered a complex and perhaps never fully resolved aspect of Bruno's thought.[13] What is certain is that Bruno is using the Platonic and Neoplatonic concept of a world soul when he offers a definition of the primal matter as that within which the voice (logos) or soul (i.e., the formal principle) is "all in all and in every single part." It is never made clear in what sense this vital force, active throughout the infinite universe, is atomistic in a "less strict sense." Later, Bruno will specify that it is "indivisible" like an ether or a vacuum (not made up of indivisibles), and therefore presumably continuous, if, indeed, it is material at all.

Already in the still introductory chapter 2 of book I of *De triplici minimo*, Bruno writes that it is not enough to affirm the existence of the vacuum and of the atoms—it is necessary to postulate also the existence of an element that unites them.[14] For the moment, the nature of this complementary principle that makes up the primal matter is not specified, and the subject of the exact composition of the primal matter is not fully explicated until book I, chapter 9. There Bruno, probably thinking of the alchemists, expresses his approval of those philosophers who distinguish the principles that make up primary matter, as such matter is not constituted by a single or even by a double principle. Bruno, like the alchemists themselves, here defines his own material principles as three: a humid element constituting the underlying substratum, a dry element made up of the atoms, and light, which unites them. Bruno thus follows Aristotle in his negation of a vacuum. The humid principle is not a vacuum filling up the interstices between individual atoms but an infinite substratum in which the atoms exist and move. This humid principle is given various names, but most often identified as ether.[15] The light that Bruno introduces in order to "unite" these two entities is one of the basic principles of his atomic substance intimately joined to both the humid and the arid principles throughout infinite space. It constitutes an essential aspect of his philosophical theory of atomism that has not been sufficiently appreciated by the commentators. Bruno's insistence on the element of light in the primal matter harks back to his first published work *De umbris idearum*, where the whole universe was visualized as made up of more or less densely compact shadows.[16] Within this universe, now composed of agglomerations of atoms moving in the humid principle illuminated by light, there is no absolute material light, for if there were, it would be blinding and consume everything. Equally there is no absolute material darkness, which would also deny the threefold composition of the primal matter and therefore signify nonbeing. What seems to us blinding light or intense darkness is therefore always only relative.[17]

At this point of book I, chapter IX of *De triplici minimo*, a definition is offered of a fourth principle inherent in the primal matter of Bruno's infinite universe, which is called a harmony, or a special form of light, and may be identified as the universal intellect, or a faculty of the world soul—a principle that, as we have seen, had actually been somewhat surreptitiously introduced earlier, and further developed in the all-important third chapter of book I.[18] The rather tortuous development of Bruno's argument here appears clearly related to his anxiety to assure his sixteenth-century reader, from the outset, that he is not proposing an entirely mechanical or random universe, but on the contrary uniting animistic explanations of phenomena to his return to ancient atomism. Michel pointed out already in 1957 that atomistic and animistic theories were not considered incompatible by sixteenth- and seventeenth-century atomists, and the work of Tullio Gregory, Ugo Baldini, and John Henry, among others, has confirmed this to be the case.[19] Bruno's pioneering attempt to create this kind of synthesis of what at first sight appear to be radically contradictory types of explanation is of particular interest, linked as it is to the intense Renaissance debate on the nature of the soul. Bruno fails to comment explicitly on the opinions of Pomponazzi or his major opponents such as Agostino Nifo whose early sixteenth-century debate on the mortality or immortality of the Aristotelian soul had made this a central problem of Renaissance philosophy.[20] Nevertheless the nature of soul within Bruno's own ontology becomes one of the most complex and original aspects of his atomistic philosophy.

In the first place, however, it is necessary to specify clearly the definition that Bruno supplies of the atom itself in book I of *De triplici minimo*. In spite of a close and repeated reference to the atomists of Greek antiquity, together with a clear act of homage to Lucretius expressed in the formal choice of a scientific poem in Latin as the appropriate expression of his atomistic theory, Bruno's atoms are different in many respects from those of both Democritus and Epicurus. It is true that insofar as they are the hard, dry, primal components of matter, they are indivisible and impenetrable, as the ancient Greek atoms were, and also they are infinite in number. However, Bruno's atoms do not come in various shapes, as Democritean and Epicurean atoms did. It is well known from ancient sources that the variety of the original Greek atoms was considered necessary to ensure the infinite variety of structures perceived in the phenomenological world: the atoms of both Democritus and Epicurus formed agglomerations on collision within the void due to the overlappings and interlockings between them. However, this explanation of the infinite variety of phenomena was no longer necessary to Bruno once he had abolished the idea of a vacuum and conceived of the dry atomic minimi as being united throughout the infinite whole by the twin principles of humidity and light. He could thus apply to his atomic minimum a logical argument ignored by the ancient atomists that has been well expressed by Jonathan Barnes in the chapter "The Corpuscularian Hypothesis" in his book *The Presocratic Philosophers*: "if there are, literally, infinitely many differences in the phenomena, that at most requires that there are infinitely many different atomic structures underlying the phenomena. It does not require that the atomic *shapes* be infinitely various; indeed, it does not require that there be more than one atomic shape. How could the [ancient] Atomists have failed to see that?"[21]

Bruno's unique atomic shape is that of the minimum sphere: a choice again argued on logical grounds. The minimum circle or point is the smallest element of extension from which three-dimensional space derives, and the atom is the physical equivalent of the mathematical point. Through the spherical shape of his atoms, Bruno thus operates a close link between two of his three minimi, and further links them both to the primary number or monad, conceived of metaphysically as the infinite sphere whose circumference is everywhere and whose center is nowhere. The importance of this last argument for the idea of soul as an intimate component of Bruno's atom will need to be stressed later. Here the essential point to be made is that Bruno's spherical atoms, although indivisible and impenetrable, have no weight. There is thus no reason why they should fly off into a vortex as Democritus's atoms do, and even less why they should all fall vertically downward toward a center that Epicurus (already under the powerful influence of Aristotle) had failed to see was everywhere and nowhere within his infinite space.[22] Deprived, in their passive material principle, of characteristics that can put them into motion, Bruno's atoms thus require soul as an essential component. It is the soul, or the form-making principle active throughout the infinite universe, that generates the whole system as system, putting it into motion from within, according to an intelligible order that Bruno identifies with a bioethical necessity of preserving to a maximum degree the vital principle of life itself.

There is clearly a reference here to the Platonic *Phaedrus*, where the soul is defined as that which possesses self-motion and its powers likened to "a team of winged steeds and their winged charioteer." In atomistic terms, this self-moving principle seems to correspond to the idea of a "natural force" acting on or within the atom, already proposed in the ancient world, according to Cicero, by Carneades as an alternative to the more mechanical Epicurean "swerve." In the Christian Middle Ages, the idea was developed in more specifically spiritual terms by St. Augustine, who in the *De genesis ad litteram*, claimed that God had deposited in matter a hidden treasure of active forces: the *rationes seminales*, whose successive germination in the womb of matter produce the different species of corporeal beings. The necessary activity of the primary substance was probably mediated by Bruno also through Raymond Lull. Bruno wrote several explicitly Lullian works and recognized him throughout as a major source for his philosophy. Like Lull, he thought that being and activity both belong to the substance of things and are identical. Activity *ad intra* thus becomes a necessary component of all things. This self-moving force in matter may be seen also as corresponding to the Neoplatonic concept of love, which will later be identified by Bacon as Cupid: "whose principal and peculiar power is effective in uniting bodies."[23]

Bruno's insistence on the essentially vital rather than mechanical nature of his atomically structured universe lies behind the claim put forward in book 1, chapter III of *De triplici minimo* that death cannot really be said to affect either the corporeal substance or the soul. Bruno's treatment of his subject is clearly developed in conscious reference to Lucretius's arguments on the same theme in book III of *De rerum natura*. Nevertheless Bruno argues explicitly against the Epicurean ideas on death, accusing them of crass materialism and impiety. Referring to the vertical fall of Epicurus's atoms, which Bruno denies, he informs his reader that he has no intention of leading him into the Epicurean abyss, deprived of divine light. His reader is to be saved by the element of light or soul hidden within the infinite extension of matter according to divine decree. Although there are both Hebrew and Christian theological sources, as well as Neoplatonic and Hermetic ones, behind Bruno's treatment of the element of soul in matter, his own explicit reference in this chapter is pre-Socratic. Those who accept an atomic theory of matter, claims Bruno, will fear death only if they fail to listen to the saintly words of Pythagoras, the philosopher from Samos. The idea of immortality that Bruno puts forward in this chapter is clearly one of metempsychosis: a subject to which he referred often, sometimes affectionately but satirically, for example, in the *Cabala del cavallo pegaseo*, at other times in agnostic terms, as when, during his trial, he told his judges that he thought it might correspond to the truth.[24] Here, however, in this chapter of *De triplici minimo*, metempsychosis becomes part of a physical, mathematical, and metaphysical theory of the atom: death, as Pythagoras affirms, is only a moment of passage. If the material composition of a body dissolves at the moment of death, there is one individual component that remains. This is soul, indestructible and eternal, which searches for new occasions within the infinite whole, continuing its journey through eternal time.

It is not my purpose here to investigate whether the Pythagorean numerology can be seriously considered as a proto-atomistic theory, or a unit-point atomism, as some commentators have claimed and as Bruno appears to be intimating.[25] What needs to be underlined is rather that his reference to Pythagorean doctrine is not being made in the light of the Neoplatonic interpretation of that doctrine, so popular in Renaissance culture and confirmed in our own century by the reading proposed by Burkert.[26] Bruno is approaching Pythagoras through Aristotle, whose critical reading of Pythagorean doctrine distinguished it from the transcendental theories of Plato: an immanent reading, it should be stressed, that corresponds to the most recent treatment of Pythagoras by scholars such as Jonathan Barnes in his book *The Presocratic Philosophers*, as well as by Carl Huffman in his invaluable recent book on the major disciple of Pythagoras in the ancient world, Philolaus.[27] That is to say, the reference to Pythagoras is not made by Bruno in the light of a claim that the souls ultimately transcend the material world after their various reincarnations. Rather, the souls that animate individual agglomerations of atoms represent within the infinite and eternal spaces of Bruno's universe principles of permanence and continuity in search of ever purer forms of perfection: their immortality means eternal life in the sense of seeking eternally renewed expression within a fragmented material infinity composed of indestructible atoms that Bruno considers to be itself impregnated with divine goodness and light.

Such a treatment of the subject of death inevitably poses the problem of the nature of individual agglomerations of atoms, and of the attempt made by Bruno in this chapter to define the sense of a conscious individuality, considered in terms of a soul capable of surviving the moment of breakdown of the physical body. A number of factors need to be born in mind in order to understand the development of Bruno's argument in this crucial moment of his work. One is that in his atomistic universe, there are no soul-atoms like those envisaged by both Democritus and Epicurus—that is, soul made up of some specially fine and tenuous but always material atomistic formation. Soul for Bruno is a purely spiritual substance hidden deep within all atomic minimi as well as being extended throughout the infinite ether. In its all-pervading nature, it corresponds to the Platonic and even more to the Neoplatonic *anima mundi*—although in Bruno's scheme of things, it finds its total explication within the infinite universe itself, rather than emanating from a transcendental sphere. A consequence of this immanent formulation of the traditional concept of the world soul, which Bruno had already drawn in his Italian dialogues written in London, was that everything in the infinite universe is to be considered as imbued with soul, to a greater or less degree—an idea that led him to refute the Aristotelian distinction between different kinds of soul such as the vegetative, the animal, and the rational soul.[28] In Bruno's atomically structured universe, although the hard, impenetrable atoms are held together by the humid element in the ether, it is the element of soul that coordinates such agglomerations, transforming them into live, moving, and organic individuals: what Lucretius calls "concilia" and Bruno himself calls "marvellous artifices."

In his effort to visualize this action carried out by the ordering spirit of soul, Bruno envisages two levels of its operation within the individual body. First, calling on the traditional medical idea of *spiritus*, he sees soul as running throughout the body coordinating its movements and its local growth and decay.[29] Beyond the level of *spiritus*, however, there is another more central core of soul that resides principally in the heart. Here it is probable that Bruno is remembering the so-called fourth element of Epicurus's soul, which is defined by Lucretius as "the soul of soul." Epicurus's fourth element, which complements the soul-elements of heat, wind, and air, remains nameless. It is the center of consciousness throughout the individual agglomeration: if penetrated by severe pain, the results can be fatal. Nevertheless, Epicurus' fourth element is always material, even if he sees it as the subtlest, smoothest, and most mobile element in existence. At the moment of death, it too participates in the breaking of the vessel, flowing back into the infinite flux together with the other elements in the makeup of the individual body.[30] By transforming Epicurus's fourth element into a purely spiritual substance, Bruno attempts to make it into a center of consciousness capable of surviving the breakdown of the individuality and of achieving eternal life within the infinite universe.

Can he really sustain such an argument? Probably not, in strictly logical terms, although he makes a bold effort to do so. His argument is by analogy. The circular atoms, he claims, do not come together in linear terms, but always gather around a center in formations of multiple atoms around a central atom—along the periphery of this primary agglomeration, further atoms gather, expanding into a composite body.[31] Death can thus be seen as a compression of the body inward until all that remains is its original center—or to use another analogy developed in this chapter, the pattern of the web "converges back to its point of departure, exiting as it entered along the same path and by the same door."[32] It is this central point of "the soul of the soul" that Bruno would have survive, looking for further occasions in which to express itself as a new individuality—a term that, as Nicola Badaloni has demonstrated in a recent study, assumes an uncertain semantic status in Bruno's philosophical vocabulary.[33]

The question is undoubtedly a delicate one within Bruno's scheme of things, as for immortality to be achieved he must allow this "soul of the soul" to transit in search of a new body, even if for a mere instant, without any accompanying material atoms. However, such a possibility contradicts his Aristotelian conviction that form and matter cannot exist apart, except perhaps for very limited space or time—a concession that seems to lie behind his reference to the black-magicians (or necromancers) who dedicate their attention to the bones of the dead in the conviction of being able to communicate through them with their lost soul.[34] In the second part of his earlier Latin work the *Sigillus sigillorum*, Bruno had referred to a "space," or a limited distance, within which the "humors" of certain natural bodies conserve the structural characteristics of the bodies from which they derive, allowing the magicians to exert damaging influences on them.[35] Bruno had added, however, that he personally knew little of this phenomenon. It is thus difficult to avoid the conclusion that in order to achieve the kind of immortality of the individual consciousness that Bruno—or perhaps more ardently his reader—wish to assure themselves, there must be some kind of recourse to a transcendental principle. Bruno was well aware of the concept of immortality developed by the Platonic and Neoplatonic tradition: he quotes from the seventh book of the *Theologia platonica* of Marsilio Ficino, where the soul is the center that coordinates the five senses; the book "de animi immortalitate" of the fourth *Enneade* of Plotinus: and from the source that lay behind them both—Plato's own metaphor in the *Phaedrus* of the flight of the soul that, freed of its body, returns to its origins in "that place beyond the heavens," where assuming the perfect form of the circle, it finally contemplates "true being."[36]

Whether Bruno himself, in his post-monastic years, continued to believe in a transcendental God is one of the most warmly debated subjects in the critical tradition. I am personally of the opinion that he was agnostic on the subject, preferring to explain all known phenomena, including death and immortality, in terms of his infinite cosmology. It is precisely in terms of his infinite cosmology, however, that the difficulties concerning the survival of an individual soul arise. The problem concerns, as we have seen, Bruno's acceptance of the Aristotelian doctrine of the necessary link between matter and form. That Bruno himself was well aware of the difficulty presented by this aspect of his atomic theory is evident, in my opinion, from the fact that on the subject of immortality, he often contradicts himself, giving rise to a sense of uncertainty about the whole question, even in its form of metempsychosis, which accurately reflects his later cautious remark to his judges: "Although not certain, the opinion of Pythagoras at least seems likely."[37] In parts of the third chapter of book I of *De triplici minimo*, Bruno can even be seen referring back, at moments of crisis, to the Epicurean and Lucretian idea that death implies necessarily a complete dissolution of composite, individual entities: an idea that becomes the principal theme of the later chapter VI of book 2. In this case, only the original atomically structured substance remains, although, for Bruno, that infinite substance will at no point be without a spiritual component, or a vital power to which he continues to give the traditional name of soul. Perhaps his final attitude to the whole question is based not only on Pythagorean but also on Stoic sources when he claims that on death we progress toward an unknown light—that is to say, the individual soul finishes its cycle, not to return, as in Epicurean philosophy, into the infinite material flux, but like a transitory spark to return within the primal source of divine light that is logically prior to multiplicity. This is Bruno's monad, defined metaphysically as being, the good: that which precedes the many.

There are, as we have seen, a number of hesitations and uncertainties in Bruno's idea of an individual soul, particularly as far as its destiny after death is concerned. Nevertheless it is the introduction into his primal atoms of an element of power, in some way associated with light, and that he continues to call soul, that Bruno obtains some of his most interesting and original results.[38] His sources here are undoubtedly manifold. The clear tendency in Bruno to render matter itself divine is related by him, already in the early Italian dialogues, to two major sources: David de Dinant, who considered matter as "cosa eccellentissima e divina" (something excellent and divine), and the eleventh-century Jewish philosopher Avicebron, author of *Fons vitae* (*The Fountain of Life*).[39] Both the Old Testament account of creation and Hermes Trismegistus are also referred to, particularly where this universal soul is considered as a special form of divine light, to be distinguished from normal material light. The source of divine light is unseen and unknown, but its presence can be felt everywhere in the universe. It is what gives us the impression that things are illuminated, as it were, from within.[40]

Insofar as the point of divine light lies within the atom, it is to be seen as a contraction into the minimum of the total energy and illumination of the divine monad. The influence of Cusanus is clear here, for as the Kantian historian of philosophy, Buhle (a keen reader of Bruno) pointed out: "The divinity was for Cusanus, as for Ficino, the logical concept of the maximum being to be thought of by way of the mathematical concept of an absolute quantity, not a relative quantity, which coincides with the absolutely small, the absolutely simple and, in as far as it contains the essence of the maximum being, with the concept of the absolute good: even if, for Cusanus, this coincidence was nothing more than a concept of pure logic."[41] Bruno accepts this logical concept of the identity of the maximum and the minimum, incorporating it into his atomistic theory of matter with results that transfer the idea from a theological–logical plane to an ontological–epistemological one. The absolute light within the atom becomes the principle of intelligence within the material world that makes the world intelligible by the mind.[42]

The question that now arises is in what ways and modes Bruno foresaw the development of a new science based on this atomistic theory of matter. The *De triplici minimo* itself offers at least one clear answer to that question insofar as it finishes with a book on the subject of measure. That Bruno is not concerned here only with the theoretical concept of measure, or extension, but also with the possibility of practical measurement of the objects in the physical universe is clear from his renewed reference, in the *De triplici minimo* (III, 7), to the compass invented by Fabrizio Mordente, to whom he had dedicated two brief dialogues in Latin in 1586.[43] It is well known that pure mathematics was looked on with suspicion by Bruno, precisely because of his conviction that the unlimited energy and force contained within the minimum atom, giving rise to a world of vertiginous vicissitude, precluded the possibility that things could ever correspond to the abstract logic of pure mathematical concepts. On the other hand, the existence of the minimum atom guaranteed the possibility of practical measurement of things in their relative positions, one to another: the new science, aided by ever more perfect instruments, was going to map out the programmed order of things in time and space. Above all, it was going to develop within a new cosmology—that infinite, homogeneous, newly unified universe that Bruno had first proposed in London in *The Ash Wednesday Supper*, and would define for the last time in the final volume of the Frankfurt trilogy, *De immenso et innumerabilibus, seu de universo et mundis*.

For this mapping to be possible, however, extension must develop in some kind of ordered sequence: measure, and indeed life itself, would not be possible if the atoms were to fuse one into the other forming an indeterminate mass. Bruno was well aware that Aristotle himself, in books V and VI of *Physics*,[44] had proposed just such an objection to the atomic theory of matter, and he devotes considerable space to his reply. Developing an argument already hinted at by both Epicurus and Lucretius, Bruno defines the concept of terminal points or limits: the virtual points of contact between atom and atom.[45] These are not parts of the atoms, which by definition have no parts, but only their limits that permit them to remain distinct. Once this concept is allowed, measure becomes possible, and once again, in the opening sentence of book IV of *De triplici minimo*, titled "On the Principles of Measure ... and Figures,"[46] it is to Pythagoras that Bruno returns—with a note of pride in his own Mediterranean origins—as the philosopher who laid the conceptual foundations of the idea of number:

> The Samian Pythagoras, who lived in Latin lands, demonstrated the migration of the monad into the dyad, of the dyad into the triad, of the triad into the tetrad. He discovered the monad within the tetrad and the tetrad within the monad, defining the monad as the limit and number of things, and thus decreeing that they could be determined.[47]

Another scientific development foreseen by Bruno was that of what today we call the "life sciences": biology and botany in particular. Because the world around us contains within it an intelligible principle that tends toward the maximum life force and reproductive energy, the new scientist will observe, with the eyes of the lynx, every minute pebble and stone in an attempt to penetrate the secrets of its formation and structure. This idea is expressed by Bruno already in *The Ash Wednesday Supper*, before he proposes in detail a theory of the atom, but at the moment when he is arguing in favor of an infinite, eternal universe.[48] In the "Argument" of his second dialogue, describing the activity of his new philosopher, Bruno writes that with the eyes of Lynceus, looking here and there at one thing after another, without stopping too often on his way, as well as contemplating the great system of the universe, he will dedicate his attention to every minute stone and pebble in his path. The terms of the development of his atomic theory in the later *De triplici minimo* could only enhance his desire of a careful observation of all the minute details of natural phenomena. The secrets of our nature, Bruno already says in his earlier work, lie within the deepest recesses of material bodies: there, if anywhere, is to be found "the monad of monads," the code that contains the shape of things to come. It is for this aspect of his thought that Bruno was considered by many nineteenth-century historians of science as one of the earliest precursors of a theory of evolution.

Last, there is the attention paid by Bruno to the secret inner core of the atom itself. Although he may not have considered "the monad of monads" itself subject to investigation in any systematic way, nevertheless, as Michel pointed out, even if in a very embryonic form, Bruno intuited the idea of an atomic nucleus imbued with extraordinary power. What he usually calls with the traditional theological name of soul becomes in the course of his development of an atomistic theory of matter a kind of force. In chapter IV of book 1 of *De triplici minimo*, he writes that the natural minimum contains within it the power of the sensible world, in its aspect of extension, explicating it in wonderful ways. He adds that the minimum exceeds in energy whatever corporeal mass it gives rise to in the course of its aggregations. For Bruno, all aggregations are accidents; only the atomically structured primal matter is substance. Moreover, the power of the divine unity, as well as spreading itself out over an infinite substance as a universal principle of divine light, contracts itself into the inner core of soul or energy that inhabits each discrete, minimum atom. Indeed, in those pages of the *De triplici minimum* in which he accentuates the essential identity of maximum and minimum, Bruno appears to conceive of the atom as a kind of encapsulated monad containing within it the total power of the divinity—a concept that he will further develop in the second work of the Frankfurt trilogy, *De monade, numero e figura*.[49] Here Bruno appears clearly to anticipate Leibniz's passage from mechanistic to biological explanations of phenomena: an anticipation that there is every justification for proposing now that Leibniz's copy of Bruno's *De monade* has come to light.[50]

This may be an aspect of Bruno's thought that was evolving in rather different directions when his arrest put an end to his philosophical activity. For in a late fragment, unpublished during his lifetime, titled *De rerum principiis*, Bruno developed his thought on the composition of the universal substance in significantly modified terms.[51] Although often considered in the context of Bruno's works on magic, this fragment is really a brief treatise concerning a special kind of atomistic physics. Its object, as Bruno himself specifies, is "to contemplate nature so as to be able to act according to nature." To achieve this end, Bruno proposes to eliminate from his discourse here any reference to metaphysics or to the universal intellect. His analysis of the primary composition of the universe, and the origins of all things in it, will be limited to physical phenomena only. This should not be taken to mean that Bruno is eliminating spiritual entities from his universe. Rather, such entities are now contained within an attempt to define the origins of all things in universal nature without reference to metaphysical entities such as the divine monad. This elimination of the maximum from the universal picture modifies to some extent the composition of the picture itself, particularly as far as the atomic minimum is concerned. One of the poles of the maximum–minimum dialectic is now lacking, and the minimum quantity of matter tends consequently to be deprived of its divine fount of internal energy. Furthermore, Bruno's new treatment of the primal matter requires a new vocabulary—a problem that he solves by recovering an Aristotelian terminology and redefining his universal substance in the elemental terms of fire, air, water, and earth. This move, however, does not lead him to abandon his atomically structured universal substance. Rather, Aristotle is considered to have been mistaken in taking as the starting point of his elemental physics a stage that, in Bruno's opinion, already represents composite forms of the original entities rather than their pure or primal state. In this primal state, the four elements are not found as they are in nature but as kinds of primordial principles. The four traditional elements, in their primordial states, can thus merge with the four entities that already made up the universal substance as it had been defined in *De triplici minimo*. The universal light is the primordial fire; the universal soul is the primordial air; the universal humidity is the primordial water; the infinite number of atoms are the primordial earth.

In his effort to define the origins of things in such a way as to favor a new *praxis*,[52] Bruno concentrates his attention with particular intensity here on the primordial humid substance. For it is the primordial water that brings together and unites the fragments of now relatively inert dry stuff into an infinite number of ever-changing combinations, according to ever-varying grades of agglutination. To understand the intimate laws that govern the behavior of the primordial humid element, or universal womb of time, would therefore be to reveal the very principles that govern natural growth and change. Bruno attempts to stress the importance of this idea by multiple references to Biblical texts such as the Genesis account of creation, the waters that lie above and below the firmament, and the rites of baptism—references that assume the character of rhetorical underlining and illustration of what intends to remain a treatise of an atomisic physics. In fact, the humid element, although not itself envisaged as properly atomistic, is now considered as composed of primal "seeds" containing within them the powers of attraction and repulsion that "incline" bodies toward each other as male to female.

The fragment finishes, like the *De triplici minimo*, with a definition of the idea of measure, and the necessity of measuring all those entities that make up the sensible world. The problem raised by Bruno is that it is such a complicated world, where the order of things is revealed in infinite time as well as in infinite space, in the movements or gestures of an infinite number of bodies as well as in the meanings of an infinite number of names or words. The consequent difficulties involved in the idea of accurate measurement are underlined very explicitly by Bruno in this fragment, where he follows sections dedicated to each of the four primordial elements with sections dedicated respectively to *Time*, to *Light and Shadow*, to *The Virtues of Place*, to *The Virtues of Names*, and to *The Virtues of Gestures*. Only when all these factors are taken into consideration does he finish off briefly with a section on *Number and Measure*. Ultimately, however, Bruno, still harping on Pythagoras, thinks of measure as nothing more nor less than the discovery of numbers lying deep within a universal substance in which the ultimate limits are still marked by the monad, point, or atom.

Bruno's reference to magic, in its most "antique and noble state," in the last line of this treatise should not be seen as a final abandon to non-physical discourse. The reference should rather be interpreted in terms of the treatise of atomic physics that it concludes. Bruno appears to be thinking here in terms of an implicate or enfolded order that measurement, based conceptually on the idea of an atomic minimum, attempts to unfold, in the full awareness that the unfolding process will never tell the whole truth about that order. That is to say, the divinity never reveals itself fully to the finite mind. The idea is remarkably similar to the implicate order of the modern quantum theorist David Bohm, who explains himself in these terms:

> Well, the simplest example is that if you fold a piece of paper and make a pattern on it, and then unfold it you get all sorts of new patterns. While the paper was folded the pattern was implicit—in fact the word implicit means enfolded in Latin—and therefore we could say the pattern was enfolded. Now quantum mechanics suggest that this is the way that phenomenal reality comes about from a deeper order in which it is enfolded. Reality unfolds to produce the visible order and folds back in.

It is remarkable that these words can be so easily inserted into an account of this final expression of Bruno's atomism. The surprise, however, lessens when Bohm goes on to quote one of Bruno's major sources, Nicholas of Cusa: "He had three words: *implicatio* (enfolded), *explicatio* (unfolded) and *complicatio* (all folded together). And he was saying that reality has this enfolded structure: that eternity both enfolds and unfolds time." Thus, Bruno's final reference to ancient forms of magic in the context of an early modern discussion of the atom, already has something in common with this recent awareness of "the mysteries of quantum physics."[53]

In conclusion, Bruno's pioneering return to ancient forms of atomism in the light of a reading of the natural philosophy of Lucretius and the pre-Socratic philosophers of antiquity, mediated through Plato and the Neoplatonic philosophers of the Renaissance, anticipates, in an embryonic form, developments that will become an important part of seventeenth-century philosophy and science with figures such as Bacon, Gassendi, Descartes, Liebniz, Newton, and Boyle. Yet there are also some aspects of his sixteenth-century formulation of the problems connected with an atomistic theory of matter that appear to contradict the mechanical philosophy of the coming centuries. Some of these aspects present interesting analogies with the philosophical speculation proposed in our own era of quantum-mechanical science.

NOTES

1. For the medieval references to these texts, see John E. Murdoch, "The Medieval and Renaissance Tradition of *Minima naturalia*," in *Late Medieval and Early Modern Corpuscular Theories*, eds. Christophe Lüthy et al., Leiden, Brill, 2001.

2. See Eugenio Garin, "Ricerche sull'epicureismo del Quattrocento," in *La cultura filosofica del rinascimento italiano*, Florence, Sansoni, 1961, 72–92.

3. For Fracastoro's corpuscularianism, see G. B. Stones, "The Atomic View of Matter in the XVth, XVIth, and XVIIth Centuries," in *Isis* 10 (1928): 445–65.

4. The 1563 publication of Telesius's work was in two volumes. The definitive expansion into nine books appeared in Naples in 1586. See in particular, *De rerum natura*, book III, chapter XVI. For a general appreciation of Telesius's natural philosophy see B. Copenhaver and C. Schmitt, *Renaissance Philosophy*, Oxford, UK, Oxford University Press, 1992, 310–14.

5. The Italian works cited here are from Bruno (2002). The Latin works are in Bruno (1879–91) and Bruno (2000c).

6. See the relevant pages in Tocco (1889).

7. See Michel (1957), 251–67.

8. See the fourth dialogue of *De la causa, principio et uno* in Bruno (2002), vol. I, 713–14.

9. See Bruno (1980), 15–16, and Atanasijevic (1972).

10. See P. Duhem, *Le système du monde*, Paris, Hermann, 1956, vol. VII, p. 16: "L'infiniment petit e l'infiniment grand." The mathematical aspect of Bruno's minimum, seen as the first principle of multiplicity in the light of the problems of the passage from unity to multiplicity raised by the *Parmenides* of Plato, has recently been discussed by Angelika Bonker-Vallon, *Metaphysik und Mathematik bei Giordano Bruno*, Berlin, Akademie Verlag, 1995, chapter 4.

11. See Bruno (2000c), 18.

12. See Michel (1957), 251.

13. See, on this subject, Spruit (2003) and Mendoza (2003).

14. Bruno (2000c), 18.

15. "Based on Anaximander's *apeiron*, the infinite and eternal, perfectly homogeneous but utterly structureless and amorphous primary substance constituting everything that exists in the universe, Bruno transformed the ancient notion of the ether into a qualified void inside which atoms and bodies move." See Mendoza (1995), 243, note 3.

16. See the *De umbris idearum* with Italian translation and comment in Bruno (2004), 3–585.

17. For the importance of the element of light in Bruno's atomism, and of his atomism generally within his natural philosophy, see Gatti (1999), part II, chapter 8: "The minimum is the substance of all things."

18. For the importance of Bruno's many references to the philosophy of Averroes in his definitions of a universal intellect, see Sturlese (1992).

19. See T. Gregory, "Studi sull'atomismo del seicento," in *Giornale critico della filosofia italiana* XLIII (1964): 38–65; XLV (1966): 44–63; and XLVI (1967):

528–41. Also U. Baldini, "Il corpuscolarismo italiano del seicento: problemi di metodo e prospettive di ricerca," in *Ricerche sull'atomismo del seicento*, Florence, La Nuova Italia, 1967, 3–76. For the English aspect of the same problem, see J. Henry, "Occult Qualities and the Experimental Philosophy: Active Principles in Pre-Newtonian Matter Theory," in *History of Science* XXIV (1986): 335–81.

20. For this debate, see Martin L. Pine, *Pietro Pomponazzi: Radical Philosopher of the Renaissance*, Padua, Antinore, 1986.

21. J. Barnes, *The Presocratic Philosophers*, London, Routledge, 1982, 364. See also H. Post, "The Problem of Atomism" in *The British Journal of the Philosophy of Science* 26 (1975): 19–26: "The atomistic programme is to explain everything, but everything, in terms of a denumerable number of identical invariant units, or at least units of limited variety, i.e. of a small number of species. Ideally, we would want one species only."

22. For the logical consequences of the differences in shape and size of Democritus's atoms, see Andrew Pyle, *Atomism and Its Critics*, Bristol, UK, Thoemmes Press, 1995, 19–25. For the shape, weight, and fall of Epicurus's atoms, as defined by Plutarch, see Cyril Bailey, *The Greek Atomists and Epicurus*, Oxford, UK, Oxford University Press, 1929, 275–99.

23. For the *Phaedrus*, see Plato, *The Collected Dialogues*, op. cit., 475–525. See in particular 493. For Cicero, see *De Fato*, XI. For St. Augustine, see *De genesis ad litteram*, VII, 28. For Lull's ideas on activity *ad intra*, see the paper by Charles Lohr in *Late Medieval and Early Modern Corpuscular Matter Theories*, op. cit., 75–89. For Bacon, *On Principles and Origins according to the Fables of Cupid and Coelum* in *Philosophical Studies c. 1611–c. 1619*, ed. Graham Rees, Oxford, UK, Oxford University Press, 1996, 197. For Bacon's atomism, and the seventeenth century generally, see B. Gemelli, *Aspetti dell'atomismo classico nella filosofia di Francis Bacon e nel seicento*, Florence, Olschki, 1996.

24. The *Cabala del cavallo pegaseo* is the fourth of Bruno's Italian dialogues written and published in London. It is in Bruno (2002), vol. II, 407–84. For Bruno's statement about Pythagoras at his trial, see Firpo (1993), 285.

25. For Pythagorean unit-point atomism, see in particular G. S. Kirk and E. Raven Kirk, *The Presocratic Philosophers*, Cambridge, UK, Cambridge University Press, 1957, 245–50.

26. Walter Burkert, *Lore and Science in Ancient Pythagoreanism*, Cambridge, MA, Harvard University Press, [1962] 1972.

27. See C. A. Huffman, *Philolaus of Croton: Pythagorean and Presocratic*, Cambridge, UK, Cambridge University Press, 1993, in particular section 6, "Soul and Psychic Faculties," 307–32.

28. "*Theophilus*: 'Thus the earth and the other stars move according to the peculiar local differences of their intrinsic principle, which is their own soul.' 'Do you think,' asked Nundinio, 'that this soul is sensitive?' 'Not only sensitive,' answered the Nolan, 'but also intellective, and not only intellective as our souls, but perhaps even more so.' At this point Nundinio kept quiet and did not laugh." See Bruno (1977), 156.

29. On *spiritus*, see D. P. Walker, "Medical Spirits and God and the Soul," in *Lessico Intellettuale Europeo: IV Colloquio Internazionale*, eds. Marta Fattori

and Massimo Bianchi, Rome, 1983, 223–44. The concept of *spiritus*, in Bruno as in many other philosophers of the early modern period who developed an atomic theory of matter, may have been influenced by the *Pneumatica* of Hero of Alexandria. As Mary Boas pointed out some years ago, the *Pneumatica* was extraordinarily popular after it was first printed in 1575, and served to disseminate a non-Lucretian atomistic theory in authors such as Gassendi, Bacon, and Galileo. See M. Boas, "The Establishment of the Mechanical Philosophy," in *Osiris* X (1952): 423–42.

30. The *Letter to Herodotus* mentions only heat and wind as elements of the soul, so that the "nameless element" that is "many degrees more advanced than these in fineness of composition" appears as a third rather than a fourth element. That Epicurus nevertheless thought of four elements as composing soul (that is, heat, air, and wind as well as the nameless element) is claimed by both Plutarch and Aetius, as well as by Lucretius in book III of *De rerum natura*. On Epicurus's fourth element, see Bailey, *The Greek Atomists*, op. cit., 384–87.

31. Bruno illustrates the way in which atoms gather round a center with a figure that he calls the Area of Democritus. For the pioneering importance of Bruno's attempts to illustrate his atomistic theory, see Lüthy (1998).

32. See Bruno (1879–1891), vol. I, part 3, 143.

33. See Badaloni (1994), 31–34.

34. See *De la causa, principio e uno*, dialogue 2, in Bruno (2002), vol. I, 662. In the English translation of this work, Bruno (1998), 44.

35. The concept is attributed to Heraclitus, Epicurus, Sinesius, and Proclus. The idea expressed here is connected to the traditional doctrine of the *species intelligibilis*. Leen Spruit has pointed out that "Bruno's mnemotechnical and magical works contain elements drawn from the doctrine of species multiplication. For example, the images and species that are present in the soul are seen as effects of emanations from the surface of sensible objects. In the case of visible species, these are transported by the light." See Spruit (1995), vol. II, 209.

36. The Platonic and Neoplatonic sources of Bruno's concept *de immortalitate animae*—one of the subjects he claimed to have spoken about at Oxford—have been studied by Sturlese (1994), 89–133.

37. "Il che se non è vero, pare almeno verisimile l'opinione di Pittagora"; see note 14.

38. The importance of a "primordial force," a creating germ, or a divine spark at the core of Bruno's atomism was pointed out in the early study by Stones, "The Atomic View of Matter," op. cit., 445–65, and has more recently been repeated by Michel (1957), 63. Frances Yates, although expressing no particular interest in Bruno's atomism, nevertheless claims in a footnote that he carried out "the introduction of magical animism into the Lucretian cosmology"; see Yates (1964), 452, note 1.

39. For the references to David de Dinant and Avicebron, see Bruno (2002), vol. I, 601, 678, 687, 708, and 723. Bruno also referred to Avicebron ("author libri Fontis vitae") in his Latin work *Summum terminorum metaphysicorum*, composed at the same time as the *De triplici minimo*. See Bruno (1989), 178.

40. A not dissimilar idea is to be found in the poem of Lucretius (2.144), where Dawn, or Sun, is considered to glow with borrowed light. See D. Clay, *Lu-*

cretius and Epicurus, Ithaca, NY, Cornell University Press, 1983, 111–68: "The Philosophical Armature." The importance of Bruno's constant reference to Epicurus and Lucretius has been underlined by Monti (1994).

41. See J. G. Buhle, *Geschichte der neuern Philosophie seit der Epoche der Wiederherstellung der Wissenshaften,* Gottingen, 1800, II, 347. Quoted by M. Longo in "Presagio di modernità: August Henrich Ritter, interprete di Cusano," in *Concordia discord: studi su Niccolò Cusano e l'umanesimo europeo,* ed. Gregorio Piaia, Padua, Antenore, 1993. The importance of Cusanus as a source of Bruno's thought was already established by nineteenth-century commentators. See in particular Tocco (1892), 60 ff.

42. K. Flasch in *Die metaphysik des Einen bei Nicolaus von Kues,* Leiden, Brill, 1973, 155–77, claims that Cusanus's One, or God, is not an object, but contains also all thought and the totality of all potentiality, both his own and that of all other things and all other thoughts. The *coincidentia oppositorum* is the coincidence not only of the contraries but also of all contradictions, thus lying outside the Aristotelian principle of noncontradiction that the neoscholastics used to refute Cusanus. See E. Berti, "*Coincidentia oppositorum* e contraddizione nel *De docta ignorantia I, 1–6,*" in *Concordia discord,* op. cit., 107–27. For Bruno's doctrine of knowledge, see Spruit (1988).

43. For the importance to Bruno of Mordente's compass within the terms of his mathematical doctrine, see "Bruno e la matematica a lui contemporanea," in Aquilecchia (1993a), 311–17.

44. Book V, 3, 226b–27a and book VI, 1, 231a–31b. See Aristotle, *The Complete Works,* 2 vols., ed. and trans. J. Barnes, Princeton, NJ, Princeton University Press, 1984, 383 and 390–91.

45. Epicurus in the *Letter to Herodotus* talks of "boundary-marks," whereas Lucretius himself tends to refer more metaphorically to "boundary stones," without fully developing the distinction between minimum parts and "termini." See *De rerum natura,* I, 594–634.

46. See Bruno (1879–1891), vol. I, part III, 269.

47. Ibid.

48. See Bruno (2002), vol. I, 435. This program of precise biological and botanical investigation was to be carried out, with the use of the newly invented microscope, in the years immediately following Bruno's execution by the *Accademia dei Lincei* founded in Rome on August 17, 1603, by Prince Federico Cesi. No explicit references to Bruno survive in the early archives of the academy, which included among its friends and correspondents some of Bruno's most implacable enemies. Nevertheless, his influence can be traced in the new academy's project for scientific research, and has recently been studied by S. Ricci, "Scienza e vita civile da Giordano Bruno ai Lincei," in *"Una filosofia milizia": tre studi sull'accademia dei lincei,* Udine, Campanotto, 1994, 61–105.

49. I have argued elsewhere for an influence of this aspect of Bruno's thought on Thomas Harriot, particularly in his papers *De infinitis;* see Gatti (1989), 49–73. For recent appreciations of this soul-powered aspect of Bruno's atoms, see Mendoza (1995), 113, and Lüthy (1997), 5.

50. For details of Leibniz's copy of the *De monade,* see Sturlese (1987), 122. On Leibniz's atomism, see Benedino Gemelli, "Leibniz e l'atomismo classico:

dal meccanismo al biologismo," *Nouvelles de la république des lettres* I (1997): 49–76.

51. The *De rerum principiis*, a brief fragment of text written at Helmstedt in 1590, was first published in 1891, together with other manuscripts left unfinished by Bruno at his arrest, in volume III of Bruno (1879–1891). An Italian translation of this text by Nicoletta Tirinnanzi, ed. Michele Ciliberto, was published in Naples in 1995 by the Istituto Italiano per gli Studi Filosofici.

52. It is to Bruno's credit to have realized the necessity to proceed from the more purely speculative atomism of *De triplici minimo* toward some kind of experimental verification, or practical use, of an atomistic conception of matter. He was by no means the only early atomist to be puzzled about how this could be done, if even Boyle writing in the 1660s had to concede that: "the intelligible [i.e., corpuscular] philosophy ... seems hitherto not to have so much employed, much less produced, any store of experiments." See C. Meinel, "Early Seventeenth-Century Atomism: Theory, Epistemology, and the Insufficiency of Experiment," in *Isis* 79 (1988): 68–103.

53. That some aspects of Bruno's theory of atomism are closer to the philosophical bases of modern quantum physics than to the mechanical philosophy of the seventeenth and eighteenth centuries has been already suggested by Mendoza (1995). For some of the philosophical speculation stimulated by modern quantum physics, see Davies and Brown, *The Ghost in the Atom*, Cambridge, UK, Cambridge University Press, 1988. The interview with David Bohm is at 118–34.

4

THE MULTIPLE LANGUAGES OF THE NEW SCIENCE

An artistic methodology for expressing an abstract and
scientific idea? Why not?
—*Imre Toth*

THE NEW SCIENCE THAT BEGINS TO emerge at the end of the
sixteenth century can be seen as a search for the order that underlies the vicissitudes of the natural world. This immediately raises
the problem of the language, or languages, most appropriate for grasp-
ing and following the logic of that order. The great scientific names of
the end of the sixteenth century, Galileo, Kepler, Tycho Brahe, had no
doubts about the answer to that question: God wrote the universe in the
language of mathematics, and the new science must learn that language
in order to discover the order that defines the cosmos and the laws that
underlie its multiple movements. Galileo's formulation of this concept in
the *Saggiatore* is well known:

> Philosophy is written in this great book which lies continually open before our
> eyes (I mean, the universe), but it is impossible to understand its meaning be-
> fore learning the language and understanding the signs in which it is written.
> Its language is that of mathematics, and its characters are triangles, circles, and
> other geometrical figures, without which it is impossible humanly to under-
> stand a word of it. Without these, it is like wandering about in an obscure
> labyrinth.

These words echo equally well-known passages by Leonardo di Vinci, as
for example: "Whoever criticizes the absolute certainty of mathematics
gives himself over to confusion, and never will he be able to silence the
contradictions inherent in the sciences of the sophists, which only teach
us how to shout."[1]

Bruno repudiated this solution to the problem, taking up a position
"adversus huius tempestatis matematicos atque philosophos" (against
the mathematicians and philosophers of our time) as he writes in the
title itself of a work of 1588 published in Prague and dedicated to the
Emperor Rudolph II.[2] Furthermore, already in dialogue I of the preceding *La cena de le ceneri* (*The Ash Wednesday Supper*) of 1584, the first
of the six Italian dialogues written and published by Bruno in London,
he had stated of Copernicus that he was more a student of mathematics

than of nature, and so was unable to penetrate to the depths of things as profoundly as he might have done.[3] To Bruno, the mathematics of the modern age, beginning with Copernicus himself, appears as founded on logical abstractions, to be considered in the same light as the logical universals of Aristotle. Both of them compress a universe that Bruno thinks of as infinite, and subject to infinite vicissitudes, into linguistic formulae that are at the same time too narrow and too sophisticated.

This refusal of the ever more sophisticated developments of classical mathematics, however, gives rise to a profound crisis within Bruno's scientific thought. Clearly the problem is not restricted to the question of the communication of theoretical data and the results of observation, but involves an alternative search for the linguistic means with which to grasp such data: to elaborate an ordered picture of events. It is a search that will become gradually more challenging and complex as Bruno's picture of the natural world becomes more complicated, assuming infinite dimensions in both spatial and temporal terms: the expression of an infinite substance composed of aggregates of minimum quantities, at the same time arithmetical, geometrical, and physical. Numerically, Bruno thinks in terms of an infinite number of monads that seem to assume an original status founded on the Pythagorean concept of natural numbers, insofar as, for Bruno, the monad must remain simple and unique in order to save the possibility of quantification itself. Transposed into two dimensions, the discourse is repeated in the primal figures of a plane geometry, such as the point, the line, the circle, the triangle. In three dimensions, the spherical atoms combine to create the bodies of the physical world of phenomena.[4]

Bruno's primal numbers are like the primal letters, or alphabet, of an ordinary language. They combine in an infinitely varied number of regulated groups, just as letters do to create the languages we speak and write. What we see developing within Bruno's work is thus a search not so much for a single language as for a plurality of languages that are necessary, as well as possible, if the mind is to "look for, find, judge, order and apply" its knowledge of the world, as Bruno writes in the subtitle of his first published work to have survived, the *De umbris idearum*.[5] This search, which was long and complex, would accompany Bruno throughout his philosophical life, leading him to develop a challenging and extremely complex meditation on the principles that regulate in the mind the relationships between language, thought, and things, or, to use a more modern terminology, between signifier and signified. In this chapter, I will attempt to offer a contribution that can perhaps be defined as epistemological, taking into consideration three moments of Bruno's reflection on the multiple languages of a new science of nature: in the first place, the possibility of a non-Euclidean geometry postulated as a tool with which to describe his newly infinite universe; in the second place, the role of poetry, intended as a metrical exercise or mnemonical structure, which reduces to order the infinite vicissitudes of chaotic sensations and vain thoughts that invade the mind; and in the third place, very briefly, his extremely personal and much discussed use of the traditional art of memory. Today our rigid disciplinary barriers make it difficult for us to consider together the possible efficacy of a non-Euclidean geometry with the logical–structural function of a Petrarchan sonnet; for Bruno, on the contrary, they were part of a multifaceted yet ultimately unique linguistic experience. They both appeared to him as mnemonic tools with which to discern the subtle pattern that underlies events.

Starting from Bruno's geometry, it may be noticed that notwithstanding his lack of faith in classical mathematics, we find in his work a daring and original consideration of Euclid's *Elements*. This suggests that Euclid should increasingly be recognized as one of the major sources of Bruno's thought. There is an interesting divergence in this sense between the Italian dialogues, written and published in London between 1583 and 1585, where we find just two references to Euclid—only one of which is of any scientific relevance—and the mathemical works of the later years, where there are at least thirty explicit references to Euclid as well as a silent use of many Euclidean principles and theorems.[6] Furthermore, the explicit references are strikingly precise: both in the *Articuli adversus mathematicos* and in the *De triplici minimo*, we find references to a large number of axioms and theorems taken from Euclid's *Elements*, often with indications as to which ones are being considered. Sometimes groups of axioms or theorems are associated with a personal name that seems to lie between the historical and the mythical dimensions. At the end of the nineteenth century and beginning of the twentieth, this mixture of myth and mathematics was judged extremely negatively by positivist or neo-positivist critics such as Felice Tocco or Xénia Atanasijevitc. It may, on the contrary, seem of unusual interest today, when scientific theories are ever more often considered as kinds of particularly elaborate myths, whose logical connections happen to have been spectacularly successful in their practical applications.[7] It is impossible not be struck by Bruno's intriguing use of a chorus of geometrical voices (Orestes, Pylades, Amyntas, Hermes, Polites, Pericles, Emiclas, Arcas, Horus are only some of them), almost as if they were reciting in that remarkable "theater of space" that is the book of Euclid's *Elements*. Although he may simply have been using this technique as a didactical aid, to help the memory by reciting the basic axioms of Euclid's art, the reader is nonetheless surprised into interest by this breaking up of the unique Euclidean truth into a plurality of voices, variously denominated.[8] It is a technique that immediately raises the question: may there therefore be various geometrical truths, or even various geometries, as many as there are human voices, rather than that one unique Euclidean system sanctioned by tradition as canonical?

Until a short time ago, the question of alternative, non-Euclidean geometries was considered to originate with the work of the Jesuit Gerolamo Sacchieri in 1733. It is well known that his studies began as an effort to defend Euclid from the accusation of having introduced as a simple postulate of his geometry the proposition of the equality between the angles of any triangle to two right angles, without having offered any formal demonstration. Sacchieri's attempt to save Euclid's system from this criticism failed. On the contrary, in the course of his studies he succeeded, in spite of himself, in demonstrating that it is possible to concieve of non-Euclidean geometries in which there are triangles with angles that add up to less than 180 degrees (the so-called hypothesis of acute angles), just as it is possible to conceive of triangles whose angles add up to more than 180 degrees (the so-called hypothesis of obtuse angles). Sacchieri thus becomes a curious case of someone who constructs a hypothesis and then proceeds, albeit unwillingly, to demolish it. Nevertheless he is frequently considered the pioneer of those non-Euclidean geometries that will begin to develop in the nineteenth century and that play such an important role in science today. More recently, however, particularly in the work of Imre Toth, we find the claim that alternative geometrical speculations to those found in Euclid's *Elements* already appeared many centuries earlier, in important pages of both Plato and Aristotle. For them too. such speculations were perceived as an art of "false" rules, or "false" circles—the work of a malevolent geometrical demon, as Plato defines it in the *Philebus*.[9] Thanks to the work of Luigi Maierù, it is now known that these classical precedents were picked up and considered more positively by a number of sixteeenth-century thinkers more or less contemporary with Bruno. These sixteenth-century speculations were not immediately productive, for the development of an acceptably classical mathematics came to characterize the science of the seventeenth and eighteenth centuries, closing the door to an alternative mathematical speculation that was considered by many, then as in classical times, to be too dangerous and too daring. Bruno's name has not so far been included in this sixteenth-century discussion of alternative geometrical hypotheses.[10] In my opinion, however, he offered an extremely original contribution to it, which I shall attempt to clarify in the following paragraphs.

A question arises during Bruno's lifetime concerning what became known as the "meraviglioso problema" (or remarkable problem), which goes back to Proclus's comment on the first book of the *Elements*. The problem concerns Euclid's fifth postulate and consists in a consideration of the possibility of its negation. What is being debated here is whether there can exist lines that are not parallel but that never meet. Such a formulation of the problem obviously includes the possibility of whether there can exist lines that are parallel but that nevertheless meet at a point. The claim being made is that such propositions do not assume absolute value within the Euclidean system. On the contrary, it is possible to demonstrate that some nonparallel lines will never meet. Maierù has shown how, deriving from Proclus, the problem reaches the sixteenth century also following a Jewish path that includes the work of Maimonides, from whom it is picked up by Mosé de Narbonne in a text published in Italian in 1550. Within a few years, at least three different demonstrations were published concerning the "remarkable" fact that it is possible to postulate nonparallel lines that will never meet. In 1551, the Frenchman Oronce Finé developed a demonstration that Maierù considers superficial and obscure, but that seems to have encountered some success in the public eye. In just these same years, however—that is, between 1550 and 1551—in Italy Girolamo Cardano developed a different and more refined demonstration. Like Finé, Cardano developed his demonstration within the field of conic geometry, publishing it in book XVI of his *De subtilitate*. In 1552, Cardano met Jacques Peletier in Lyons, where the two men discussed the "remarkable problem." Some years later, Peletier developed two much simpler and better demonstrations of the same problem in the context of the contact between lines and circles, which he published in his comment on Euclid of 1557 and then in a work of 1563 titled *De contactu linearum*. Peletier, however, did not speak of the "remarkable problem" only with Cardano. He spoke of it also in the house of Montaigne, in a discussion with the already famous French essayist. Furthermore, if the mathematicians of the time seem to have limited themselves to an attitude of simple "wonder" in front of the serious paradox they had revealed in the Euclidean geometrical system, Montaigne, in his *Apologie de Raymond Sebond*, expressed a more far-reaching conclusion: "E qui sait s'il n'est plus vraysemblable que ce grand corps, que nous appellons le monde, est chose bien aultre que nous ne jugeons?" [And who knows whether this great body that we call the world may not be something quite other than that which we consider it to be?][11]

There can be little doubt that Bruno knew of Peletier's demonstration of the "remarkable problem" as we find the same diagram reproduced, if only summarily, in the *Articuli adversus mathematicos* (see figure 4.1). It will appear again in the *De triplici minimo* of 1591, together with a discussion of the Euclidean postulate that parallel lines will never meet. Once again, Bruno is reproducing the illustration used by Peletier, even if with some minor variations. Furthermore, Bruno remains, like Peletier, within the context of the contact between lines and circles, developing his argument in chapter XV of book II of *De triplici minimo*, which is titled *Conclusio, ut ex virtute consuetudinis credendi falsis, sensus etiam ipse*

Figure 4.1 Diagram from Bruno's *Articuli adversus huius tempestatis mathematicos atque philosophos*, Prague, 1588, based on the proof of the "marvellous problem" by Jacques Peletier in *De contactu linearum*, 1563. Courtesy of the Biblioteca Comunale, Como, Italy.

perturbatur (A conclusion disturbing to common sense, given our habit of believing what is false).[12] What emerges from this page is that Bruno rejects the conventional Euclidean conclusion that however large the circle becomes, the intersection can never coincide with the line AB, because there must be only one point of contact with the circle, even if it assumes infinite dimensions. This appears to be because Bruno considers the contact to be anyway linear rather than limited to minimums or points, allowing the maximum circle to merge at infinity with a straight line that coincides with AB. In this way, Bruno posits the possibility of two parallel lines meeting at infinity, conceived of as an extreme geometrical situation. Also Oronce Finé appears, with the name Orontes, as one of Bruno's geometrical voices in this same work, although not specifically in relation to the "marvellous problem."[13] As for the comment by Montaigne, it is difficult to believe that Bruno had not read his *Essays*, above all in view of his close friendship in London with John Florio, who some years later would be the first to translate them into English.[14]

It is of interest in this context to notice the one penetrating comment on Euclidean geometry that can be found in Bruno's London dialogues. This is to be found in the fifth dialogue of the *De la causa, principio et uno* (Cause, Principle and Unity), where Bruno's mouthpiece, Theophilus, claims that "the man who could reduce to a single proposition all the propositions disseminated in Euclid's principles would be the most consummate and perfect geometrician."[15] Bruno does not specify here what that one geometrical principle might be, but in the later *De triplici minimo*, where the same idea is picked up in terms of an ontology that has become based on the monad-point-atom, his principle is defined with great clarity: "moving only from the circle and the radius," Bruno writes at the end of book I of that work, "we will easily achieve the object of our research."[16]

This new geometry of Bruno's, based only on the circle and the radius, maintains intact the first three Euclidean postulates but not the last two. Above all, it abandons the famous fifth postulate, or the postulate known as "Euclid's," which is the one about parallel lines.[17] And it is precisely this development that opens the way toward a geometry that is no longer Euclidean, even if Bruno attempts to pass off this simplified geometry based on circles and radii as nothing more than a "purified" form of the Euclidean one. No doubt, he is using a strategy that plays on a note of calculated ambiguity. It is not difficult to understand his motives for this, if we remember that the "false" geometrical demon of Plato and Aristotle appeared to Christian culture for centuries as one of the many voices of the devil. Toth reminds his readers that even at the end of the nineteenth century, the tempestuous development of non-Euclidean geometries was considered by some commentators as the expression of a newly satanic theology of evil.[18] Or, as Toth writes in another work: "The non-Euclidean prehistory followed a path indicated by the unfortunate sign of *negativity*."[19]

Bruno's own awareness of this "diabolical" aspect of the question of non-Euclidean geometries appears with great clarity in some illustrations, published in the *Articuli adversus mathematicos*, which play with the difference between the orthogonal planes that represent Euclidean space and the internal space of a concave curved surface that, on the contrary, requires a non-Euclidean geometry to describe it. Bruno develops this contrast with reference to the figure of a serpent that we see in one of his drawings sadly flattened on the orthogonal Euclidean plane, where it appears not only to be dead but also to have the tip of its tail hanging out-

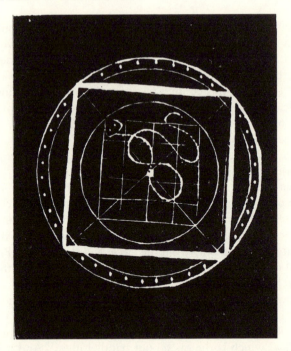

Figure 4.2 Diagram from Bruno's *Articuli adv. Math.*, Prague, 1588, showing a serpent flattened on the orthogonal Euclidean plane. Courtesy of the Biblioteca Comunale, Como, Italy.

side the pattern of squares (see figure 4.2). So Bruno seems to be saying (and this was precisely the meaning of Peletier's "marvellous problem") that not all geometrical propositions can be explained in terms of Euclid's postulates. This is presumably why, in another drawing in the same *Articuli adversus mathematicos*, we see the same serpent now "liberated" and entire within the space of a universe that is formed of a concave curvature. The double frame of this universe is decorated with points representing the atomistically fragmented space that is the speculative context in which Bruno develops his meditation on a possible alternative geometry (see figure 4.3).

It is not by chance that we find the most complex development of this aspect of Bruno's geometrical speculation at the end of the first book of *De triplici minimo*, where his atomism is given its most refined explanation. The space that interests him here is one that he calls the Area of Democritus—that is, a plane in which a minimum sphere at the center is surrounded by six minimum spheres each touching the circumference of two others as well as that of the central sphere (see figure 4.4).[20]

Figure 4.3 Diagram from Bruno's *Articuli adv. Math.*, Prague, 1588, showing a serpent liberated from the Euclidean plane. He titles this diagram "PROMETHEUS." Courtesy of the Biblioteca Comunale, Como, Italy.

Bruno's reasoning here follows a somewhat tortuous path. He begins by noting that if we surround the whole space known as the Area of Democritus with an external circle, we can see that the outer empty spaces between the circles become ever greater with respect to the inner ones. This would not be true, Bruno claims, if we were to imagine ourselves inside a spherical body filled with spherical atoms, so that only the ignorant would expect to find in the plane those properties that can be found in spherical triangles or more generally in the dimension of the sphere. Analagously, only the ignorant would expect to find in spherical triangles between convex spheres the same properties that we find in a Euclidean plane. Nevertheless, Bruno knows that it is possible to object to this formulation of the problem, insofar as a Euclidean plane folded over an ample convex globe remains locally Euclidean. This means by analogy, in Bruno's opinion, that the Area of Democritus represents lo-

Figure 4.4 Illustration of the Area of Democritus in Bruno's *De triplici minimo*, Frankfurt, 1591. Courtesy of the Biblioteca Nazionale, Rome. (Unauthorized reproduction of this image is prohibited.)

cally, or for limited dimensions, the geometry that can be applied inside a concave curved space. This line of reasoning appears to justify the claim made by Bruno a few lines earlier that the universe may be conceived of as an enormous sphere filled with spherical atoms where, as in the Area of Democritus, "the void is mixed with bodies and extends beyond the surface of the spheres, just as it extends beyond the body of the earth." In a later work, the *De rerum principiis*, Bruno will develop this idea further in order to describe a cosmos seen as a sphere that has become infinite, but within which, locally, dense masses of matter (that today we would call galaxies) are surrounded by ever vaster empty spaces, or by space filled with pure ether.[21] Perhaps the most interesting aspect of these pages, however, is the claim implied by Bruno that, no longer considered locally, these empty curved spaces correspond to non-Euclidean triangles—that is, to triangles whose angles are not equal to two right angles. Such spherical triangles can be defined using the so-called hypothesis of the obtuse angle—that is, they are triangles whose angles add up to more than two right angles. This discovery will later be made again, and developed much further, by Johann Heinrich Lambert in his *Theory of Parallel Lines* of 1766.[22]

We can thus see that what seemed at the end of the first book of *De triplici minimo* the point of departure for a non-Euclidean geometrical speculation seems in a first moment to be interrupted by a declaration, actually mistaken, of the absolute universality of Euclid's principles. Then this conclusion itself is challenged in the very last words of the book, where Bruno contradicts himself once again by postulating the possible existence of a geometrical method that has not yet been revealed to the world, but that is certainly superior, and to which, if the Greek geometer himself had known about it, he would without doubt have adhered.[23] So it is surely no coincidence if, at the very beginning of the second book of *De triplici minimo*, we find a particularly interesting development of a geometrical intuition that is now clearly non-Euclidean.[24] Here Bruno makes a telling comparison between our perception of the setting sun as it is perceived by the eyes—that is, in terms of rays of light that behave like Euclidean straight lines—and a perception proper to the intellect alone in which a point of light becomes diffused following another law that gives rise to a perpetual illumination. Although Bruno fails to elaborate on this idea, it can only be understood by postulating the so-called hypothesis of the obtuse angle, or a non-Euclidean curved world in which the angles of a triangle are greater than two right angles. In this case, from a point placed at a finite distance, which we will call P, and which Bruno identifies with the sun of the intellect, rays are emitted in every direction that illuminate perpetually the entire sphere of a non-Euclidean world. In Bruno's own words: "the mind is thus orientated toward the longed-for vision of the divine monad." The hypothesis of the obtuse angle (see figure 4.5), as Toth has pointed out, already appears with remarkable precision in Aristotle's *Eudemian Ethics* (see Aristotle 1222 b 35), which is probably where Bruno discovered such a concept. In conclusion, unorthodox geometrical principles, feared and considered disturbing both by the ancients and by most Renaissance mathematicians, are of interest to Bruno as possible descriptive tools for his new universe based on the concept of an identity between the maximum and the minimum spheres—that is, a universe whose nonorthogonal spaces are characterized by the concept of an infinitely repeated curvature.

Bruno appears to be aware of the fact that the adoption of non-Euclidean geometrical principles does nothing to invalidate the principles of Euclidean geometry itself. Euclidean geometry always remains one of the possible geometries (see figure 4.6), and in many ways can be considered our most natural geometrical language. What happens when non-Euclidean geometries are postulated is that the mind becomes freed from the necessity of seeing Euclid's geometry as the only possible one, and what follows from this is the development of a new awareness of mathematical liberty of thought. It becomes possible to choose within a field of alternative geometrical values, and this aspect of the question of non-Euclidean geometries is eloquently celebrated by Toth:

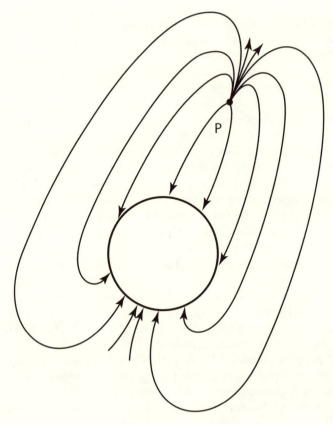

Figure 4.5 Diagram of the non-Euclidean hypothesis of an obtuse angle.

The conscience is invested with the possibility of *negating a whole World*, the given world of Euclidean geometry, and of creating—thanks to a simple negation—a new world. Furthermore, and more meaningfully, it is possible to accept both of these worlds as opposing ontological domains, perfectly autonomous one with respect to the other, and to assign truth value at the same time to the axioms of the Euclidean world and to those of the non-Euclidean world. In a word: the two geometries possess equal rights to exist and to be true.[25]

Here, perhaps, is to be found the key with which to interpret the decision made by Bruno to introduce precisely his *Articuli adversus huius tempestatis matematicos atque philosophos* with a plea for liberty of philosophical thought, addressed unfortunately to an unheeding Holy Roman Emperor, Rudolph II.[26]

Figure 4.6 Diagram in Bruno's *Articuli adv. Math.*, Prague, 1588, showing as if in a mirror image an absolute Euclidean plane (lower right) and an absolute non-Euclidean plane (upper left). Courtesy of the Biblioteca Comunale, Como, Italy.

It is within this context of libertarian thought that it can be interesting to pass from Bruno's geometry in order to discuss his position with respect to the art of poetry. As a poet, Bruno composes with a remarkably unorthodox autonomy that was most unusual in Renaissance culture. In this dimension too, the fundamental concept expressed by Bruno is that of liberty of choice on the part of the poet, who must be able to adopt whatever language he considers most appropriate to his theme.[27] This liberty of choice was claimed by Bruno also with respect to the various metrical forms and literary genres consecrated by the ancient classical cultures. These were to be used, in Bruno's opinion, without excessive respect for the classical rules, even if they had become canonical. For in Bruno's view, a language, metrical form, or literary genre should be adopted by an author only because the necessities of the work being developed require it. Bruno makes his views on this subject quite clear in a celebrated and much quoted page of the *Heroici furori* (*Heroic Frenzies*), the last of his Italian philosophical dialogues written and published in London in 1585, where he repudiates a muse that "only imitates other muses." Bruno arrives at the conclusion that "poetry is not born from rules, except in a very indirect fashion, but rather rules are derived from poetry."[28] "And I add," continues Tansillo, who is Bruno's mouthpiece in these pages, "that there are and should be as many kinds of poet as there are types of sentiment and human invention." This declaration is in line with the chorality of geometrical voices that we find in the *De triplici minimo*. In both cases, what we have is a plea for the autonomy of the new philosopher, by now not only capable of choosing the most appropriate language for the concepts being developed, but also capable of adapting it to his own particular ends.

Almost as proof of this view, the Petrarchan sonnet is chosen and used by Bruno in the *Furori* as a particularly disciplined and structured metrical instrument that he adapts to a purpose clearly opposed to the courtly tradition in which it had originally developed. Such a tradition had already been sharply and eloquently criticized in the pages of the Argument of the dialogue, dedicated to Sir Philip Sidney. For Bruno's work is not an expression of "a curious thought about or concerning the beauty of a female body," but rather:

> it demonstrates in an ordered fashion ... the causes and principles and intrinsic motives, which appear behind the names and figures of mountains, rivers, and of muses, who declare their presence not because they have been invoked, called on or searched for, but rather because they have repeatedly and insistently offered themselves: and this means that the divine light is always present, always offers itself, always calls and knocks at the doors of our senses and our other powers of apprehension and knowledge.[29]

Proceeding along this path, the fifth and final dialogue of the *Furori* leads us toward what Bruno calls "a natural contemplation" through which, as the final *Canzone de gl'illuminati* (Song of the Enlightened) recites, the new philosophers begin to inquire into "those eternal laws" that regulate "the flaming sky, where lies that blazing zone in which the eminent Chorus of your planets can be seen."[30] Nor is this cosmological conclusion to the *Furori* of any less interest because it has been arrived at through a long sequence of sonnets, commented on in the dialogues in ample passages in prose. It is these prose passages that define the intense and dangerous journey of the frenzied poet, which is both spiritual and intellectual, and whose final conclusion finds its expression in a more extended poetic form, that of the *canzone*. For the impetuous fury of the poet has now quietened, and the nine, newly enlightened philosophers "dispose themselves in the form of a ring." In the final pages of the *Furori*, the nine blind philosophers, who have now regained their sight (nine, as the commentators have often pointed out, being the number of the muses), unite the poetry of their final song to the harmonies of music, every one of them playing an instrument to accompany their words. Music and poetry are joined by geometry, in the image of the ring or wheel, to which corresponds the circularity of the poem itself. For the first verse ("Oh rocks, oh gulleys, oh thorns, oh serpents, oh stones") is repeated in the final verse ("Of rocks, of gulleys, of thorns, of serpents, of stones").[31]

In the last pages of the *Furori*, we thus find ourselves involved in a discourse that moves smoothly between poetry, music, and geometry, all of them used not according to rules consecrated by the tradition, but by following the necessities of the inquiry that is being pursued. Here the most pressing necessity, as Bruno had already announced in *The Ash Wednesday Supper*, the first of his Italian dialogues written and published in London in 1584, is that of celebrating once again the new infinite cosmology, composed of an infinity of ordered spheres in a homogeneous space, revolving around their suns according to a principle that is post-Copernican and heliocentric. It is precisely this infinity of the number of the spheres that is suggested by the internal form of the song, where the final verse of each stanza recited by the various "enlightened" philosophers is taken up and repeated by the next one. "Oh, what a fortunate journey," ends the first philosopher, singing while playing his guitar, while the second one begins, singing while playing his mandolin: "oh, what a fortunate journey, oh goddess Circe, oh what wonderful efforts we have made." Once this second philosopher has ended with the words, "Exhausted after so much effort," immediately the third philosopher begins, "Exhausted after so much effort, the tempests have prescribed this harbor for us." The circular forms of each stanza thus multiply themselves within the circular form of the final song of the *Furori* itself, the metrical structure operating an opening out toward the infinite curvature of the newly infinite universe itself. For the philosophers are nine, and the number nine will be defined by Bruno in these terms in the later *De monade*, where he is giving new life to a long tradition of Pythagoean numerology:

> Altogether, there are nine Muses corresponding to the melodic harmonies of the guitar-playing Apollo, and they dance to the sound of sublime notes. Of the same number are the powers of the soul, and the doors which lead into the mind. There is one according to which this animal sees; one by which he examines what he has perceived with his ears; one which permits him to unite what he has seen to what he has heard; one by which he reproduces images in

his imagination; one which, from these forms, produces a hidden sense; one by which he thinks; one by which he remembers; one by which he thinks discursively, one by which he proceeds towards the clearly intelligible species.

This generation takes place in nine months, after which the birth is accomplished.[32]

In the first work of the Frankfurt trilogy, the *De triplici minimo,* Bruno had already indicated a path for arriving at the clearly intelligible species, not only through the inspiration of the Muses and their music but also through a new concept of an infinite, atomistically fragmented universe. It is in this work that he conceives of a space represented as the Area of Democritus (see figure 4,4), which, as we have seen, can be described by using the axioms of a non-Euclidean geometry. For by calling upon a geometry based on the hypothesis of the obtuse angle, or of a triangle whose angles add up to more than 180 degrees, it is possible to postulate a source of light of a finite order, identified by Bruno with the sun of the intellect ("a hidden sense," according to which the mind thinks) orientated toward illuminating perpetually and simultaneously the entire surface of the innumerable spheres that make up the one and infinite world.

It is clear that the whole idea of alternative worlds, of specular realities, of universes that are possible but unseen, and indescribable in the terms of a unique Euclidean geometry, are of intense interest to Bruno. His own universe was conceived of as infinite not only in extent but also intensively—that is, as populated with infinite forms of life capable of infinite vicissitudes and change. Just as space has become infinite, so has time become eternal. As the character with the interesting name of Minutolo says in the second part of the *Heroici furori*:

> Motion is change. That which moves always becomes something else; the things which are always work new changes and turn into other things, because their concept and character follow the reasons and conditions of the subject. And that which aims to become something else again, always other than what it was, must necessarily be blind with respect to that beauty which is always one and unique, which is the unity of being, identity.[33]

This speech by Minutolo, imprisoned in time, defines one of the forms of blindness that has struck the nine philosophers who have put their faith in Circe. It expresses the desire for a geometry that defines the unity and identity of being, which may include the Euclidean geometry based on the infinite line, but also postulates an infinite number of alternative geometries capable of describing that "wheel of infinite change" to which space-time is ineluctibly subject.

It is in this context of the wheel of time that mention can be made at this point of the last, or perhaps it would be more correct to say the first, form of a universal language proposed by Bruno: his art of memory. This is the subject of the first of his published works, the *De umbris idearum* with its annexed *Ars reminiscendi* published in Paris in 1582, and again of his last work published with his consent, the *De imaginum, signorum, et idearum compositione*, which appeared in Frankfurt for the autumn book fair of 1591. It has become commonly accepted by the long tradition of prestigious studies of Bruno's art of memory that one of his most important achievements in this field was to put in motion the traditional images associated with the memory "places" in the mind. This led to a system of intersecting wheels that permitted a series of combinations of interconnecting images that could be considered as virtually infinite in number. Using as a source the picture logic of Raymond Lull, Bruno thus made his own specific contribution to an art that had originated in the context of classical rhetoric. Nor is it necessary to underline once again the importance for Bruno of the new flourishing of this art in its Thomistic version that he would have learnt at the Dominican monastery in Naples as part of his preparation as a preaching monk. Although it is true that in the *De umbris idearum*, there is still no mention of that extension of the cosmos to infinite dimensions that will characterize Bruno's natural philosophy as well as his metaphysical speculation starting from the first of the Italian dialogues written and published in London between 1584 and 1585, it remains clear that an art of memory developed around the idea of a number of intersecting wheels would inevitably find its most significant application in the context of an infinite universe filled withinfinite worlds. It is precisely the idea of intersecting wheels that turns Bruno's art of memory into a linguistic instrument that, even if only a finite number of mnemonic images or icons can be accommodated on each of the single wheels, becomes capable of potentially infinite combinations.[34]

Undoubtedly it is important to inquire into Bruno's sources for this development of his art of memory, and it is a task to which Bruno scholars have been devoting themselves for at least the last half century. It is also important, however, to inquire into what scientific use might be made of it within the context of his natural philosophy. With regard to this problem, there have been diverse reactions from the beginning. The sixteenth-century followers of Peter Ramus, for example, considered an art of memory that worked through images to be useless and obscure, and Bruno was aware of this objection that he expressed through the words of the somewhat pedantic Logifero in the *De umbris idearum*.[35] Later, in England, the Cambridge follower of Ramus, William Perkins, sharply criticized the art of memory of Alexander Dickson, who was in close personal contact with Bruno and whose art of memory was also based on imagery.[36] For Bruno, however, it was precisely the image that connected this art to our perception of the phenomena, saving it from that abstraction from the natural world that rendered contemporary logic and mathematics so odious to him. Nor is it necessary to explain this insistence on the image only with a reference to magic, due to the frequent use on Bruno's part of astrological or talismanic images on his wheels of memory. For the "groups" of memory images used by Bruno on his memory wheels are at times nothing more than simple letters of various alphabets, or groups of numbers that clearly form "codes" whose meaning is founded on the internal logic of their combinations rather than on some magical power infused into the image itself.[37] It is precisely this programming of logical languages based on alphabets, numbers, and various other systems of images that was destined to remain useless as a scientific instrument—as today we can easily understand—for as long as it lacked the necessary technological backing in the form of combinatory machinery, to be developed only in a far later age thanks to the genius of Turing.

It would clearly be absurd to propose Bruno ingenuously as a "precursor" of the scientific realities of today, which he would have been unable even to conceive of, let alone understand. On the other hand, even in a scientific sphere, a historical exercise that is able only to "photograph" the thought of previous ages within the framework of their "sources," or in static photograms relating to the "period," seems singularly inadequate for an understanding of a thinker such as Bruno, who was always attempting to project his mind forward toward still unrealized events. The major historians of the nineteenth century, who considered Bruno a "precursor" of much that was to be realized in a more modern world, were far from being as ingenuous as they are often considered. They attempted to understand the ways in which ideas suddenly appear on the scene, often to fold backward on themselves for lack of a context in which they can be developed, only to reappear unexpectedly much later in a new context of thought. They knew that a new theory never completely cancels out the theory that it replaces, and also that the development of knowledge rarely follows a straight and easily identifiable line. It remains of some significance that a philosopher of the status of a Leibniz, and following him the post-Kantian philosophers in Germany, rediscovered Bruno's combinatory art of memory with enthusiasm, and that some of the major British historians of science of the nineteenth century, both of astronomy and of the theory of evolution, dedicated substantial sections of their works to his natural philosophy.[38]

Today we can see how Bruno's refusal of the classical mathematics of his time excluded him to a considerable degree from the scientific revolution that would develop in the seventeenth century, of which he saw only a gray and cruel dawn. At the same time, we can understand how his particular way of formulating the idea of an infinite universe full of infinite and ordered life stimulated him to look for alternative linguistic solutions that, in the course of the centuries to come, would be reproposed in the context of a series of developments of extraordinary scientific importance. Without doubt, Bruno's thought was to a large extent unknown to those who were responsible for those developments: his intuitions with respect to possible non-Euclidean geometries and to possible machinery based on the logical combinations of icons remained in his work at the level of a search for alternative linguistic tools, with respect to the classical concepts that dominated his times, whose full realization he was barely able to glimpse. They are intuitions that tend inevitably to exhaust themselves in artistic and poetic flights of the imagination, at times of remarkable beauty. Nevertheless, they appear in the context of a new cosmology based on the Copernican revolution, and were completed, above all in the works of his final years, by an atomism that anticipates later scientific developments. They are intuitions that need to be considered in the context of properly scientific languages, even if of a kind that would remain almost completely misunderstood by his contemporaries.

Today, on the contrary, it is precisely Bruno's insistence on the primary importance of language that strikes us as interesting and requires our attention. For it is linked to his proposal of a newly infinite cosmology, which represented a dramatic refusal of the traditional picture of the world that had dominated the European mind from the classical era to his own day. Bruno understood the importance of forging new linguistic tools for understanding this new and dramatic picture of the world. At the same time, he was aware of an epistemological crisis that necessarily accompanied the finite mind, obliged now to come to terms with both an extensive and an intensive infinity—a newly homogeneous space that Bruno conceived of as atomistically fragmented. Only by uniting the powers of the reason with those of the imagination would it be possible, in Bruno's view, to widen the network of possible combinations of alphabets, numbers, geometrical figures, poetical structures, and images in order to catch, in ever more sophisticated mental grids susceptible of extension to virtually infinite variations, at least some fragments of a new science.

NOTES

1. See G. Galileo, *Opere,* ed. F. Flora, Milan–Naples, Riccardo Ricciardi, 1953, 121 and note 3. The translations of these passages are mine.

2. The complete title of this work is *Articuli centum et sexaginta adversus huius tempestatis mathematicos atque philosophos.* See Bruno (1879–1891), vol. 1, part III, 1–118.

3. References to Bruno's Italian works in this chapter are to the photostatic reproductions of the first editions edited by Canone (1999). For the comment

on Copernicus in *La cena de le ceneri (The Ash Wednesday Supper)*, see vol. 2, 343–44.

4. On Bruno's monads as the "'primal bricks'" of being, see Bönker Vallon (1996).

5. See Bruno (1991).

6. For the references in the Italian works, see Ciliberto (1979). For the references in the Latin works, see the entry "Euclides" in Lefons (1998).

7. For negative comment on Bruno's mixture of myth and mathematics, see Tocco (1889), and Atanasijevitč [1923] (1972).

8. See Bruno (2000c), 156–57.

9. See Imre Toth, *Aristotle e i fondamenti assiomatici della geometria*, Milan, Vita e Pensiero, 1997.

10. See on this subject the essential article by L. Maierù, quoted by Toth: "Il meraviglioso problema di Oronce Finé, Girolamo Cardano e Jacques Peletier," in *Bollettino di storia delle scienze matematiche*, 1984, 141–70. For much of the material mentioned in the following paragraph of this chapter, I am indebted to Maierù's article.

11. All the references given in the preceding paragraph can be found in the article by Luigi Maierù referred to in note 10.

12. The diagram in figure 4.1 is taken from the *Articuli*. For the same diagram and the comment in *De triplici minimo*, see Bruno (2000c), 102–4.

13. Ibid., 164–65.

14. For the relationship between Bruno and Florio, see the still valid study by Frances Yates, "Italian Teachers in Elizabethan England," *Journal of the Warburg Institute* I (1937–1938), now in *Renaissance and Reform: The Italian Contribution, Collected Essays*, vol. II, London, Routledge and Kegan Paul, 1983, 164–80. A more recent contribution to this subject, which concentrates on the linguistic dimension of their work, is in Wyatt (2002).

15. See Bruno (1998), 95.

16. Bruno (2002), 61.

17. For a discussion of Euclid's postulates, see *Euclid, the Thirteen Books of the Elements*, 2 vols., ed. T. Heath, New York, Dover Publications, 1956, in particular the note to postulate 5, or Euclid's postulate, at 202–20.

18. See I. Toth, *Aristotele e i fondamenti assiomatici della geometria*, op. cit., 559. The English translations from Toth's work on the following pages are mine.

19. See I. Toth, *No! Libertà e verità: creazione e negazione*, Milan, Rusconi, 1998, 26.

20. The argument referred to here is developed by Bruno in chapter XIV of the first book of *De triplici minimo* titled: "Rursum minimum in magnis & maximis esse perspicuum" (The minimum can be perceived both in what is great and in the maximum). See Bruno (2000c), 57–58.

21. The *De rerum principiis* was written at Helmsted in 1590, but published only in 1891, together with other incomplete manuscripts relating mostly to magic, in the third volume of Bruno (1879–1891). This work can now be consulted, complete with Italian translation and comment, in Bruno (2000b), 585–759.

22. On Lambert, see T. Heath, *Euclid, the Thirteen Books*, op. cit., 212–13.

23. Bruno (2000c), 61.

24. Ibid., 63.

25. I. Toth, *No! Libertà e verità*, op. cit., 32.

26. The by now classical comment on these pages is in Calogero (1963).

27. For a study of Bruno's language that links it to geometry, although in rather different terms from those proposed in this chapter, see Saiber (2005).

28. See Bruno (1999a), vol. 4, 1278–1280.

29. Ibid., 1245 and 1256.

30. Ibid., 1520.

31. The songs of the nine "enlightened" philosophers are at ibid., 1516–1519.

32. See Bruno (2000c), 377.

33. See Bruno (1999a), vol. 4, 1504.

34. For early studies of Bruno's art of memory, see Rossi [1960] (2000) and Yates (1966). The entire series of his memory works, with facing Italian translations and detailed comment are now available in Bruno (2004) and Bruno (2008).

35. The *De umbris* is Bruno's first published work to have survived. It appeared in Paris in 1582 and remains one of his major works. It is now available with an exhaustive comment in Bruno (2004), 2–585. The cultural background to Bruno's works on the art of memory is discussed in Ricci (2000), 150–64. In English, it has been the subject of a number of papers by Stephen Clucas; see in particular, for the subject of memory and imagery, Clucas (2002).

36. The importance of this polemical discussion was first underlined by Frances Yates in the chapter "Conflict between Brunian and Ramist Memory," in Yates (1966), 260–78.

37. For this much discussed subject in Bruno studies, see for differing points of view, Vasoli (1958) and Sturlese (1990) and (1993).

38. On the many and varied readings of Bruno in nineteenth-century culture, see Canone (1998a). Considerations on the thought of Leibniz in relation to Bruno's art of memory are to be found in the already cited books by Rossi [1960] (2000) and Yates (1966). For the post-Kantian discovery of Bruno's art of memory, see the introduction to Bruno (1993).

PART 2

BRUNO IN BRITAIN

5

PETRARCH, SIDNEY, BRUNO

E per mia fede, se io voglio adattarmi a defendere per nobile
l'ingegno di quel tosco poeta che si mostrò tanto spasimare
alle rive di Sorga per una di Valclusa, e non voglio dire che sia
stato un pazzo da catene, donarommi a credere, e forzarommi
di persuader ad altri, che lui per non aver ingegno atto a
cose megliori, volse studiosamente nodrir quella melancolia,
per celebrar non meno il proprio ingegno su quella matassa,
con esplicar gli affetti d'un ostinato amor volgare, animale
e bestiale, ch'abbiano fatto gli altri ch'han parlato delle lodi
della mosca, del scarafone, de l'asino, de Sileno, de Priapo,
scimmie de quali son coloro ch'han poetato a' nostri tempi
delle lodi de gli orinali, de la piva, della fava, del letto, delle
bugie, del disonore, del forno, del martello, della carestia,
de la peste; le quali non meno forse sen denno gir altere e
superbe per la celebre bocca de canzonieri suoi, che debbano
e possano le prefate et altre dame per gli suoi.

[And, in truth, if I wish to assume the defense of the noble
spirit of that Tuscan poet who displayed so much anguish
on the banks of the Sorgue in adoration of a woman from
Valcluse, and if I want to refrain from saying that he was
as mad as a hatter, then you must allow me to believe, and
oblige me to persuade others, that because he had no ability
to cultivate better things, he wished studiously to nourish
such melancholy in order nonetheless to celebrate his own
wit with respect to such a quandry. So what he did was to
explain the effects caused by an obstinate and vulgar love
of an animal and bestial kind, just as others have done by
praising flies, beetles, asses, Silenus, or Priapus. Slavishly
imitating such things, some poets of our own times have sung
the praises of urinals, peas, beans, beds, lies, dishonor, the
oven, the hammer, famine, and plagues. And indeed, perhaps
those things have as much right to move proudly
and disdainfully through the verses of their celebrated poets
as the aforementioned and other ladies do in his.]
—GIORDANO BRUNO, *Author's translation*

THESE WORDS ARE FROM ONE of the final pages of Giordano Bruno's dedicatory letter of his *Heroici furori* to Sir Philip Sidney.[1] The *Furori*, composed of a Petrarchan sonnet sequence interspersed with long passages of philosophical comment in prose, was the last of the six dialogues in Italian written by Bruno in London and published by the printer John Charlewood between 1584 and 1585. The words quoted come toward the end of the long and complex dedicatory letter to Sir Philip Sidney.[2] This remarkable document is noteworthy for many reasons. Here I am above all concerned with its definition of a critical stance toward not only Petrarch but also Sidney himself. It has earned Bruno words of harsh criticism, such as those of Thomas P. Roche Jr., in his chapter on "Annotators, Spritualisers, and Giordano Bruno" in his otherwise useful volume on *Petrarch and the English Sonnet Sequences*.[3] Roche considers Bruno's letter "an act of presumption," finding nothing in the *Furori* to distinguish him from the Petrarchan discussion that had already developed during the sixteenth century, and even Gordon Braden, in a more recent and far more sympathetic comment on the passage quoted earlier, considers it "rude."[4] Both these judgments ignore the scintillating linguistic construction of the verse and the prose of the *Furori*, unequaled in virtuosity and brilliance by both the Petrarchan and the anti-Petrarchan poets of his age. As for the conceptual content of Bruno's work, I shall be arguing, on the contrary, that his contribution to the Petrarchan discussion is original for two reasons: first because he brings a long Italian experience of Petrarchan and anti-Petrarchan debate to the banks of the river Thames, developing it in terms of a direct confrontation with the principal English Petrarchan poet of his time, Sir Philip Sidney; and second, because Bruno proposes to maintain the Petrarchan sonnet as a valid form of expression in the early modern world by developing it as a linguistic instrument in philosophical debate. This is coherent with the position already defined by Bruno in one of his earlier Latin works, which claimed that the quests of the artist, the poet, and the philosopher are intimately linked insofar as they are all involved in a unique pursuit of truth: "philosophers are in some ways painters and poets; poets are painters and philosophers; painters are philosophers and poets. So true poets, true painters, and true philosophers recognize and admire one another."[5] This claim has far-reaching implications: it proposes an important collaboration, rather than a conflictual opposition, between imagination and reason, between intuition and logic, between magic and science. In the light of these preliminary considerations, I shall now return to the quotation with which I started this chapter, in an attempt to understand what Bruno was trying to say about Petrarch.

In my opinion, Bruno is saying that there is no essential difference between Petrarch, the Petrarchans, and the sixteenth-century anti-

Petrarchans—all of whom are in error with respect to the objects praised in their sonnets. All of them are rejected. Bruno's own strategy is not to deride Petrarch by reducing to its minimal dimensions the physical object of adoration, but rather to adapt the Petrarchan linguistic and metrical code to a far larger subject, which Bruno himself calls "the contemplation of divinity." The dialogue will reveal that by this Bruno means something quite different from a quest for a Christian vision of God, such as that famously proposed to Petrarch's ghost by the Franciscan friar Hieronimo Malipiero in his *Il Petrarca spirituale* of 1536.[6] It is probable, nevertheless, that Bruno knew Malipiero's work, of which at least six editions were published before the end of the sixteenth century. In it, Malipiero makes a pilgrimage to the tomb of Petrarch in Arquà and in a neighboring forest meets Petrarch's ghost, which has remained in the purgatory of a spiritual body because of his "youthful error" in loving Laura. They converse together. Malipiero assures Petrarch that his poetry is fundamentally chaste. Petrarch himself, however, sees his poetry as a confession of the anguished passion of a sinful lover. Malipiero proposes to resolve Petrarch's dilemma by rewriting his sonnet sequence as virtuous praise of God and the Virgin Mary, so that it will not incite the young to follow in Petrarch's footsteps. As far as Petrarch himself is concerned, this reformulation of his poetry, according to Malipiero, will act as a liberation from his earthly sins and a support in the journey of his soul toward Paradise and celestial love. During the conversation, in which Petrarch's ghost approves of Malipiero's intention of rewriting his poems, he is informed by Malipiero of the sensual satire of the sixteenth-century Tuscan anti-Petrarchans: Malipiero names no names, but is probably thinking above all of Berni. This reduction of the Petrarchan tradition of Italian love poetry to what Malipiero calls "vain and dishonest praise," or "impudent" poems of carnal lust, is rejected by Malipiero with pious horror. His own solution of purification of Petrarch from his carnal love is situated at exactly the opposite end of the sixteenth-century anti-Petrarchan spectrum.

Although using some of Malipiero's vocabulary, Bruno himself is putting forward something quite different from Malipiero, and indeed from Petrarch himself . He is proposing a philosophical quest for truth within a natural world that, in the final *canzone* of the *Furori,* will culminate in an ecstatic vision of an infinite universe conceived of in its essential infinity and ordered unity. The Petrarchan sonnet is conserved by Bruno, but only insofar as it is adapted to the definition of a natural philosophy, becoming, in the process, less of an artificial mode of expression in which the poet "celebrates his own wit," and more of what Bruno himself calls a "true and natural form of discourse." A major problem raised by this line of approach, which Bruno himself comments on in the final part of his letter to Sidney, is whether there is any room left for the presence in his Neopetrarchan work of the figure of the woman. Bruno claims at once that the feminine figure will not be eliminated, although she can appear in his poetical scheme only at two precise levels of expression. Both of them are rigorously distinguished from what Bruno thinks of as Petrarch's exaggerated attitude of adoration of Laura. One of these levels is what Bruno calls "love of an ordinary kind." Bruno insists that this has nothing to do with a vulgar, venereal exercise in sexual intercourse, which he condemns as a form of "disorder." What he is thinking of is rather an ordinary falling in love, such as he himself had experienced in his youth in his native Nola with his own Laura, who appears briefly at the end of the *Furori* with the name of Giulia. She seems to have been a cousin, and the dialogue tells us that she repudiated him.[7] This, however, as she and her companion, Laodomia, agree in the final lines of the *Furori,* is not to be thought of as food for tragedy. In spite of the precedent represented by Petrarch and his followers, this unrequited love turns out to have been beneficent. It is what initially set Giulia's rejected Nolan suitor off on his philosophical quest, which throughout the work has been at the center of attention in what Bruno evidently thinks of as a new phase in the fortunes of the Petrarchan tradition: the sonnet used as a vehicle of philosophical inquiry and debate.

The second level at which the feminine figure is admitted into Bruno's Neopetrarchan discourse is that of myth. Bruno is drawing here on the already well established Renaissance use of classical myth as part of a discourse seen, with clearly Platonic echoes, as a quest for philosophical truth. The mythical figure who dominates Bruno's *Heroici furori* is that of the moon-goddess Diana. In the earlier part of the work, she is seen in relation to Acteon, who stands here for the solitary hunter, or intensely mystical Neoplatonic philosopher. His impetutous intellectual quest for truth permits him to glimpse the goddess in her nakedness, bathing in a pool in the midst of a thickly wooded forest in central Italy, where he is immediately devoured by the hounds of his own thoughts.[8] At the end of the work, however, Bruno centers his reader's attention on a group of nine more tried and experienced philosophers, who, in the course of a long journey through sixteenth-century Europe, have (at the other end of the philosophical spectrum) been blinded by the natural magic of Circe— that is, by their adherence to a crass materialism. They finally arrive on the banks of the river Thames, where they are liberated from their blindness by the chief nymph of the gently flowing river, who is explicitly praised as an English Diana. It is she who pours healing waters on their eyes, initiating them into a vision of the infinite universe that is no longer ennervating, no longer an endless wandering through a blind labyrinth (a *"lungo error in cieco labirinto,"* as Petrarch famously expressed his own plight in sonnet 224). Rather, in Bruno, the new experience ushered in by

the English Diana is seen as energizing. It opens up for the nine philosophers (who are also poets, nine being the number of the muses) an entirely new world composed both of a lower sphere symbolized by Father
Ocean and of a higher, celestial sphere, which is that of Jove.[9] This composite, infinite, and infinitely vital universe is open both to poetical and
musical celebration of its intimate harmonies, magically conceived, and
to a more rational or scientific definition through a methodical inquiry
into the laws that regulate its infinite vicissitudes. This complex image of
his infinite universe is the subject of the final *canzone* of Bruno's *Furori*,
which he calls the Song of the Enlightened (*Il canzone degli illuminati*). It
is clearly an act of homage to England's Virgin Queen, written in contrast
to Petrarch's final *canzone* that resolves his long years of despair due to
his all too human love for Laura by celebrating a now purely spiritual
love for the Virgin Mary. On the contrary, in Bruno's *Heroici furori*, history, both natural and political, is never abandoned for a celestial realm
beyond this world. Rather, an infinite universe, thought of as the habitat
of an immanent divinity, becomes the proper object of Bruno's philosophical quest for truth. The earthly realm reflects the celestial, absorbing
within itself both its infinity and its spiritual potencies.

It is the English Diana who grants the nine philosophers this vision of
the *sommo bene in terra*, or the greatest good on earth. So we may say
that Bruno, at a very early stage, understoood and appreciated the British
empirical mode.[10] Sidney too elaborates his Petrarchan sonnet sequence
in a strictly terrestrial dimension, refusing, in a clearly Protestant stand,
to follow Petrarch in his final metamorphosis of his earthly Madonna
into a heavenly one. Nevertheless, Bruno's final joyous *Song of the Enlightened* can be equally well contrasted to the English nobleman's final
sonnet of *Astrophel and Stella*, with its leaden sorrow, its "most rude
dispaire," its lament for the physical "annoy" that no prayers suffice to
eliminate, thus forbidding the poet to enjoy undisturbed the illuminating
vision of Stella's perfect beauty.[11] Bruno's strategy in the final pages of the
dedicatory letter of the *Heroici furori* thus becomes clear. Sidney, no less
than Petrarch, is evidently being chided for having adored in his sonnet
sequence a mere woman such as Stella, or Penelope Rich, rather than the
true Astraea or mythical English Diana, to whom Bruno thus pays his political as well as philosophical homage.[12] Yet precisely because this letter
culminates in such fulsome praise of Sidney's own queen, of which he was
one of the principal and most celebrated courtiers, he could hardly have
refused to associate himself with Bruno's Neopetrarchan work. It should
not be forgotten that Bruno covered in London a diplomatic position as a
gentleman attendant to the French ambassador, Mauvissière, which took
him frequently to the English court. He was by no means the obscure
upstart that English and American commentators so often depict him as.
Furthermore, his printer, John Charlewood, is known to have had close

ties with Sidney and his circle.[13] It is unthinkable that Sidney should not have known about, and allowed, Bruno's dedication, although whether he actually read it is perhaps another matter.

French influences are clearly at play in Bruno's considerations concerning both Petrarch and Sidney, as well as Italian ones. Bruno had arrived directly in London from Paris, where he had moved in the courtly circles surrounding Henri III, and would surely have known and read the poets of the Pléiade. Ronsard, it may be remembered, died in 1585, the year of publication of the *Heroici furori*. His beautiful *Sonnets pour Hélène*, first published in 1578, are more faithful than Bruno is to the Petrarchan ending by dissolving the physical object of adoration into a vision of a heavenly fountain of Christian truth.[14] Nevertheless, they may well have influenced Bruno in his final multiplication of the lovers from one into many, as well as in the introduction of a courtly element by appealing to a princely sponsor of his ultimate spiritual apeotheosis—in Ronsard's case, the militantly Catholic King Charles IX of France. Charles IX had died in 1574 after licensing the anti-Protestant massacres of Saint Bartholomew's night together with his mother, Caterina dei Medici. Bruno's portrayal of the moderately Protestant Elizabeth as his spiritual sponsor must surely have been made as a conscious choice against the orthodox Catholic resolutions of their sonnet sequences by both Petrarch and Ronsard. The dedication of the *Furori* to Sir Philip Sidney would be consistent with such a choice.

It is not, however, the historical-religious implications of Elizabeth I's presence as the presiding spirit over the ending of Bruno's *Furori* that I wish to inquire into here. Nor do I intend to look further into the French influence on Bruno's reference to the Petrarchan tradition, although it is probable that he knew, besides Ronsard's sonnets, also Du Bellay's *Olive*, which has some interesting cosmological imagery in it. Nor shall I comment any further on Bruno's lively spirit of anti-Petrarchism, which was such a prevailing theme in his time. Rather, what I wish to underline in the concluding remarks of my contribution is Bruno's ultimate faithfulness to the Petrarchan poetical code. For, when all is said and done, the *Heroici furori* remains, formally, a Petrarchan sonnet sequence that, in its way, does "assume the defense" of the Tuscan poet by reproposing his metrical mode of expression in Italian as a valid linguistic tool for philosophical inquiry within the early modern world. It is true that in order to do this (in order to direct his poetical discourse to this end), Bruno feels the need to call on numerous external forms of support, such as the more metaphysical sonnets of Luigi Tansillo, who was a fellow Nolan of his father's generation, and who figures as one of the speakers in his dialogue. Then there are the poetical jokes of the modern *strambottisti* such as Serafino, the Renaissance emblem books that are used as sources for highly wrought imagery in certain parts of the *Furori*, the Biblical *Song of Songs* that was such a favorite source for the Renaissance theorists of love.[15] All these, as well as other more or less important figures, whom Bruno calls to his aid, have been the subject of recent study; so that it can now be claimed that the extraordinary complexity of Bruno's text, at the poetical and linguistic as well as the conceptual level, is satisfactorily established.[16] What remains surprising (in spite of a now classic essay by Frances Yates first published in 1943) is how little attention has been paid to his use of Petrarch as his ultimate source, and to his sonnets as deriving directly from the *Canzoniere*.[17] In my remaining remarks, I shall attempt to remedy this situation by comparing Petrarch's sonnet XIX with the penultimate sonnet of dialogue I, part I of the *Heroici furori*, extending the comment briefly to include a reference to Sidney's final sonnet in *Astrophil and Stella*.

This particular sonnet of Bruno's has been chosen because in the *Metrical Table* included in the recent edition of Bruno's *Opere italiane*, edited by Giovanni Aquilecchia and Nuccio Ordine, it appears as the only strictly regular Petrarchan sonnet in the *Furori*, assuming as regular Petrarch's favorite rhyme scheme ABBA ABBA CDE CDE.[18] The few pages dedicated by Pasquale Sabbatino, and later by Aquilecchia himself, to Bruno's metrical schemes have demonstrated the remarkable poetical self-consciousness with which he experiments in anomalous sonnet forms; so that when he produces just one entirely regular sonnet it can surely be assumed that he does so deliberately.[19] That this sonnet was an important one for Bruno is further demonstrated by the fact that he had already used it, albeit with some minor differences, as the introductory poem to the most metaphysical of his Italian dialogues in prose, the *De la causa, principio et uno*.[20]

> Amor per cui tant'alto il ver discerno,
> ch'apre le porte di diamante nere,
> per gli occhi entra il mio nume, e per vedere
> nasce, vive, si nutre, ha regno eterno;
>
> fa scorger quant'ha 'l ciel, terr', et inferno;
> fa presenti d'absenti effigie vere,
> repiglia forze, e col trar diritto, fere;
> e impiaga sempr'il cor, scuopre l'interno.
>
> O dumque volgo vile, al ver attendi,
> porgi l'orecchio al mio dir non fallace,
> apri, apri, se puoi, gli occhi, insano e bieco:

fanciullo il credi perché poco intendi,
perché ratto ti cangi ei par fugace,
per esser orbo tu lo chiami cieco.[21]

(a) Love, which bids me see the truth on high
(b) Opens doors of black diamond, making them bright,
(b) Through my eyes it enters my mind, and by sight,
(a) Is born, lives, eats: its kingdom is ever nigh.

(a) It shows me the earth and hell, and the sky;
(b) True images gives of things absent from sight,
(b) Gathers strength, and gains ever in might,
(a) Wounding the heart, where the inmost thoughts do lie.

(c) Lend your ears to these truths, you ignorant crowd,
(d) And mind the not unworthy things I say,
(e) Open your eyes, obtuse and foolish, with all your kind.

(c) You think love a boy because you're ignorant and proud,
(d) You find him inconstant because you change each day,
(e) Because you yourselves cannot see, you call him blind.[22]

Not surprisingly, Bruno's regular sonnet takes the form of an ortho-dox Petrarchan exhortation to the classical figure of Love as Cupid. Its regularity appears strictly related to the fact that Bruno's Cupid has nothing to do with love of any Laura, but only with a philosophical love of truth. It is the philosopher's attempt to raise his mind to a higher level of truth than a purely animal one that opens for him, with Cupid's aid, the doors of black diamond that had previously impeded his vision, allowing his intellect to expand into new regions of both external and internal experience. In the final sestet, the poet turns from his celebration of philosophical truth to expostulate with the reader, who traditionally blames Cupid for all possible ills. On the contrary, according to Bruno's poem, the reader's blindness is not Cupid's fault, for Cupid would be-come an ally if only the mind of the reader were directed toward love of the highest kind.

If we compare Bruno's sonnet to sonnet XIX in Petrarch's *Canzoniere*, we find that it too is concerned with an attempt to move beyond a purely animal vision of truth, which the poem proposes to achieve in its adora-tion of the lady:

Son animali al mondo de sì altera
vista che 'ncontra 'l sol pur si difende;
altri, però che 'l gran lume gli offende,
non escon fuor se non verso la sera;

et altri, col desio folle che spera
gioir forse nel foco, perché splende,
provan l'altra vertù, quella che 'ncende:
lasso, e 'l mio loco è 'n questa ultima schera.

Ch'i' non son forte ad aspectar la luce
di questa donna, et non so fare schermi
di luoghi tenebrosi, o d'ore tarde:

però con gli occhi lagrimosi e 'nfermi
mio destino a vederla mi conduce;
et so ben ch'i' vo dietro a quel che m'arde.[23]

(a) There are beasts in this world with powerful sight,
(b) Whose gaze can meet the sun.
(b) Others go out when the day is done,
(a) For their eyes are wounded by the light;

(a) Others again, with a fool's delight,
(b) Wish to bask in the fire and burn;
(b) And I, alas, of these am one,
(a) Drawn by the power which sets alight.

(c) I am not strong enough to gaze
(d) On this woman's light, nor know how to use
(e) The cooling shade, or the hours of night.

(d) So now my eyes with tears do ooze.
(c) My desire to see her will never wane,
(e) But draws me into the flames so bright.

The image of the truth that the lady represents here is expressed in terms that will later be used for the image of the highest truth by Bruno: a liberating and splendid light. The problem Petrarch's sonnet poses is that the poet's physicality makes him too weak to allow him to approach unwounded such a pure and ethereal fire, and so he finally envisages himself as tearfully following a lady whose unsullied splendor can only destroy him. Petrarch's sonnet is slightly irregular in the rhyme scheme of the final sestet: CDE DCE rather than CDE CDE. The effect produced is that of a sob, which is actually the final image conjured up by the poem. It is precisely that sob that Bruno's sonnet aims at eliminating, both metrically and conceptually—Bruno's sonnet, placed near the end of the first dialogue of the *Furori*, thus announces the theme of the work, and the sense of its joyous ending in the illuminating vision of his infinite universe, conceived of as the ultimate good on earth.

Sidney, on the other hand, tends to exasperate Petrarch's tragic vision in sonnet XIX, even in his ending. He uses almost the same image as Bruno for the impediment to his vision, although his are iron doors rather than black diamond ones—the image is anyway of Petrarchan origin.

> When sorrow (using mine owne fiers might)
> Melts down his lead into my boyling breast,
> Through that darke fornace to my hart opprest,
> There shines a joy from thee my onely light;
>
> But soone as thought of thee breeds my delight,
> And my young soule flutters to thee his nest,
> Most rude dispaire, my daily unbidden guest,
> Clips streight my wings, streight wraps me in his night,
>
> And makes me then bow downe my head, and say,
> Ah what doth Phoebus gold that wretch availe,
> Whom iron doores do keepe from use of day?
> So strangely (alas) thy works in me prevaile,
>
> That in my woes for thee thou art my joy,
> And in my joyes for thee my only annoy.

As in both Petrarch and Bruno, the closed doors must be opened if Phoebus's golden light is to penetrate his heart. But in the Protestant Sidney, that remains an impossibility up to the end, and the whole poem—indeed the whole collection of Astrophel's sonnets to Stella—bears down on that final word of his volume, the physical weakness or "annoy" that impedes his final joy. The emphatic rhythmical irregularity of Sidney's last line ("And in my joyes for thee, my only annoy") indicates how self-consciously he too was referring to Petrarch as his model. So although there are anti-Petrarchan elements in both Sidney and Bruno, it would be an oversimplification to enclose the Petrarch–Sidney–Bruno connection I have been following in this chapter entirely within the schemes of sixteenth-century anti-Petrarchism.

What both Sidney and Bruno are doing, but perhaps Bruno in particular, is to propose the Petrarchan sonnet as a still valid form of expression in the early modern world. For Sidney, it can narrate a more dramatically realistic and naturalistic love story. For Bruno, it can do more than that: it can transmute the idea of love into a search for order and truth within a post-Copernican, infinite universe, making the rigorous linguistic discipline of metrical order and form into a form of philosophical discourse. Bruno's proposal of a natural philosophy written in songs and sonnets is not one that many modern philosophers have taken up. But then few philosophers have been poets of the calibre of Giordano Bruno.

NOTES

1. Bruno (2002), vol. 2, 498. For the argument (Argomento del Nolano sopra *Gli eroici furori* scritto al Molto Illustre Signor Filippo Sidneo) discussed on the following pages, see ibid. 487–521.

2. Bruno's relationship to Sidney and the circle of his uncle, the earl of Leicester, was an important part of his experience in London between the spring of 1583 and the autumn of 1585. The relationship is discussed in Aquilecchia (1991), Ciliberto (1991), and the pages on Bruno's stay in England in Ricci (2000). For a discussion of this letter in English, see Farley-Hills (1992).

3. Roche (1989).

4. Braden (1999).

5. This claim is put forward in one of Bruno's earlier Latin works on the art of memory, the *Explicatio triginta sigillorum*. It appears in Bruno's explication of his twelfth seal referring to the images of painters. See Bruno (2009), 120–21, and the comment by Marco Matteoli at 388–89.

6. Frate Hieronymo Marepetro [Malipiero] Venetiano del Sacro Ordine de' Minori, *Il Petrarca Spirituale*, Venice, Marcolini, 1536.

7. The possible family relationship was first discussed by Spampanato (1921).

8. A great deal of attention has been dedicated in recent years to Bruno's use of the Acteon myth in the *Furori*. See, for example, the relevant pages in Ciliberto (2002b).

9. On Bruno's use of apocalyptic imagery in the final sequence of the *Furori*, see chapter 6, "'The Sense of an Ending in Bruno's *Heroici furori*,'" in this volume.

10. On this subject, see Gatti (2002).

11. For the sonnets from Astrophel to Stella, see Sir Philip Sidney, *Poems,* ed. by William A. Ringler Jr., Oxford, UK, Clarendon Press, 1962.

12. For the mythical images that surrounded the figure of Queen Elizabeth I, and especially that of Astraea, see Frances Yates, "Queen Elizabeth as Astraea," *Journal of the Warburg and Courtauld Institutes* 10 (1947): 27–82, now in *Astraea. The Imperial Theme in the Sixteenth Century,* London, Routledge and Kegan Paul, 1975.

13. On this subject, see Provvidera (2002).

14. On the sixteenth-century French Petrarchans, see Lionello Sozzi, "Presenza del Petrarca nella Letteratura Francese," in *Petrarca e la cultura europea,* eds. Luisa Rotondi Secchi Tarugi, Milan, Nuovi Orizzonti, 1997, 243–62.

15. On these subjects, see Rowland (2003a) and (2003b), Maggi (2003), and, for the *Song of Songs,* the "Introduzione" by Nicoletta Tirinnanzi in Bruno (1999b).

16. A major study of the theoretical sources of the *Furori* remains Nelson (1958). More recently, see Canone and Rowland eds. (2007).

17. See Yates (1943).

18. The "Tavola metrica," prepared by Zaira Sorrenti, is in Bruno (2002), vol. 2, 777–81.

19. See Sabbatino (1993) and Aquilecchia (1996).

20. For a recent English translation of this work, see Bruno (1998).

21. For this sonnet, see Bruno (2002), vol. 2, 539.

22. The translations of this sonnet and, later, of Petrarch's sonnet XIX are mine, made in an attempt to keep to the original rhyme schemes.

23. Petrarch's sonnet is in Francesco Petrarca, *Canzoniere*, ed. Marco Santa-gata, Milan, Mondadori, 2004, 79.

6

THE SENSE OF AN ENDING

IN BRUNO'S *HEROICI FURORI*

AN UNDENIABLE CHARACTERISTIC OF books is that they come to an end. This was the aspect of books investigated by Frank Kermode in his *The Sense of an Ending*, which has given me the title for this chapter. Kermode—nowadays Sir Frank, and Britain's most prestigious living literary critic—is not concerned at all with Bruno. His *Sense of an Ending*, however, was considered by a distinguished colleague, on publication in 1966, to be "a very beautiful book"—a judgment with which I can only agree. As well as my title, it has given me many of the ideas about endings that I shall be developing in this chapter.[1]

The fact that books end is indeed one of their defining characteristics. It is what assimilates them to music. On the other hand, it is what differentiates them from works of art. For although, as Kermode points out, Sir Ernst Gombrich in *The Art of Illusion* has insisted that works of art are also looked at in time, it is clear that pictures can be looked at in many directions, from bottom to top or from right to left, while sculpture can even be looked around. Only narrative cycles of frescoes can really be looked at from beginning to end, and even then one's eye may well be caught at length by a picture somewhere in the middle. Books, on the other hand, are unidirectional, and as the pages to be read dwindle in the reader's hand, expectations are raised about "how it will end."

The author can play with these expectations in many ways. Aristotle sanctioned the idea, at least for tragic texts, of building up the plot to terminate it in a great moment of catharsis, or spiritual purification, but not all tragedians have followed his advice.[2] Think, for example, of the ending of Shakespeare's *King Lear*, when an exhausted Edgar wearily accepts the crown of an utterly ruined and desolate kingdom with a few laconic and subdued verses telling us how difficult and unheroic the reconstruction is going to be, and that the only important thing is to say what you really feel. Verses of enormous historical and ideological importance, in my opinion, but hardly cathartic—certainly not, and surely deliberately not, a grand *finale*.[3]

The weight of this sad time we must obey;
Speak what we feel, not what we ought to say.
The oldest hath borne most: we that are young
Shall never see so much, nor live so long.

Bruno himself was against endings on principle, at least the endings of those longer time-processes of history, and even of time itself. Like Epicurus and Lucretius before him, he attempted to dissolve fears of personal death by seeing it as a mere moment of transition toward further atomistic agglomerations, within a process of infinite and eternal vicissitude. For Bruno, there are no real endings, only transformations into something new.[4] Even so, the transient objects, the accidental formations that make up the differing species and their exemplars at any one moment, do clearly come to an end, and Bruno understood the dramatic impact of these individual end-moments within the eternal processes of time. Think of the way in which he organized his own last terrible moments as a scenario of more than passing significance.[5] In my opinion, he also organized some of his individual works, and in particular the *Heroici furori*, by building up the text deliberately toward a dramatically orchestrated ending, of which the musical component is certainly not coincidental. On the contrary, the instruments on which the newly illuminated philosophers play while singing their final choric song unite music to poetry in what I see as a deliberately cathartic ending in an Aristotelian sense: a moment of spiritual cleansing and reawakening after blindness and tragic tribulation. For their intellectual journey has led the philosophers, with what are clearly also Dantean echoes, through various types of error and of hell. To read this text as if it finds its climax in the middle, where Bruno places the myth of Diana and Actaeon as a moment of transition between the first and second part of his work, is in my opinion to misunderstand the structural principle as well as the intellectual journey around which the *Furori* is organized.[6]

In the Christian tradition, the idea of endings is indissolubly linked with the idea of Apocalypse: that extraordinary vision of a sudden ending of the world in a fearful blaze of symbolic splendor that closes the central and sacred book of the tradition, the Holy Bible. The visionary ending of the *Heroici furori* clearly raises the question of whether Bruno was influenced by the apocalyptic vision—a question that, to my knowledge, has been seriously considered only by Michele Ciliberto.[7] His essay on this subject, however, is explicitly limited to the alternative Apocalypse that can be found in the *Hermetica*, and particularly in the *Asclepius*, which delineates the disaster represented by the end of the natural religion of the Egyptians and the flight of the gods to an unknowable region in the sky, beyond the limits of human vision. Ciliberto's contribution on this subject is essential reading, concerned as it is with a text that, as Frances Yates had already eloquently pointed out, Bruno not only knew but quoted from extensively, in the *Heroici furori* as elsewhere. Yet surely this represents only one strand, and a particularly erudite one at that, of a subject that the wider public of Bruno's time, as indeed of ours, would associate more readily and directly with the Biblical Apocalypse, and with the prophetic and millenarian spirit associated with it from the Middle Ages to our own.

It is the Biblical Apocalypse that Bruno quotes from in a significant discussion of the millenarian spirit that he develops in the introductory letter of the *Furori* addressed to Sir Philip Sidney. Not surprisingly this discussion occurs in just that part of the introduction in which Bruno is describing to Sidney the ending of his work. The passage is of great importance and needs to be quoted in full:

> ... é detto in revelatione [cioè nel libro biblico dell'Apocalisse] che il drago stará avvinto nelle cathene per mille anni, e passati quelli, sará disciolto. A' cotal significatione voglion che mirino molti altri luoghi dove il millenario hora é espresso, hora é significato per uno anno, hora per una etade, hora per un cubito, hora per una et un'altra maniera. Oltre che certo il millenario istesso non si prende secondo le revolutioni definite da gl'anni del sole, ma secondo le diverse raggioni delle diverse misure et ordini con li quali son dispensate diverse cose: perche cossì son differenti gl'anni de gl'astri, come le specie de particolari non son medesime.

> [... it is said in Revelations that the dragon will remain in chains for a thousand years, and when they have passed, it will be freed. This is the meaning of many other passages too where the millennium is mentioned, sometimes referring to a year, sometimes to an age, sometimes to a cubit, sometimes to one or other of these things. What is certain is that the millennium itself should not be considered according to the revolutions defined by the years of the sun, but according to the different ways in which different degrees and orders are ordained in different things: because the years of the stars are as different as are the particular kinds.] [8]

In recent years, Bruno has often been celebrated as the philosopher of a vision of unity, of the divine One as the principle that, with Cusanian echoes, lies behind his conception of infinity. This is only true, however, of the metaphysical principle, the divine paradigm of which the infinite universe is the seal or shadow. Bruno's infinite universe itself is based on the idea of diversity, of the differences between the species. It is this idea of the diversity of the objects of knowledge that is celebrated at the end of the *Furori*, where the *sommo bene in terra* (the ultimate good on earth) is announced as a new understanding of these differences between the particular species of things. Such ideas of multiplicity and diversity lie behind Bruno's contribution to the new science, which was not of secondary importance. It is what distinguishes him from the Neoplatonic tradition of Ficino, who had encouraged the philosophical mind, in the introductory pages of his Latin translation of the *Hermetica*, to look away from the world of multiplicity in order to contemplate the divine One. So it is of particular interest to see Bruno insisting here on the idea of the diversity of species within the infinite whole, and to see him doing it in relation to that most central of apocalyptic texts, Revelations 20, 7–8, where it is written that "when the thousand years are expired, Satan shall be loosed out of his prison, and shall go out to deceive the nations in the four quarters of the earth."

These verses announce the millennium in terms of the freeing of the beast of evil, which will bring darkness and tribulation before the final light of salvation destroys the created world in a blaze that announces the Last Judgment. Bruno's reading of them, as so often, is quite unorthodox. For what he is doing is to claim that there is no one Apocalypse within a world of differing species, but as many Apocalypses as the number of species themselves. Each species follows its own time scheme and reaches its own form of salvation. In the *Furori*, with respect to his own time, Bruno places the beast in the region of Rome in the form of the blinding magic of Circe, surely interpreted correctly by Giovanni Gentile as the corrupt element within the Catholic Church.[9] The moment of illumination is situated on the banks of the river Thames and achieved at the hands of a chief nymph who is clearly Queen Elizabeth I. Bruno was writing in a Protestant context and addressing his work to one of the most enthusiastic supporters of the Anglican settlement, Sir Philip Sidney. The ending of the *Furori* is subordinated to his present situation, for Bruno was a political realist, a gentleman attendant to an ambassador, and not a solitary visionary or a recluse.[10]

If we ask ourselves how original to Bruno such a vision of the ending of his heroic story is, we need, in my opinion, to divide our answer into two parts. The idea of interpreting the apocalyptic vision in pluralistic terms, as dependent on the individual position of each species within the eternal and infinite whole, is dependent on Bruno's newly infinite cosmology, and undoubtedly anomalous with respect to the dominant ideas of his times. The delineation of his own specific apocalyptic vision that closes the *Furori*, on the other hand, is clearly in line with ideas that were circulating throughout Europe at that time, and particularly with the English version of them as we find it developed in numerous sixteenth-century texts.

In sixteenth-century Europe, the idea of Apocalypse was closely associated with the name of Joachim of Fiore. Even if we do not find Joachim explicitly mentioned in his works, it was not a name that Bruno could possibly have ignored, if only because severe objections to Joachim's apocalyptic prophecies had been voiced by St. Thomas Aquinas in his *Summa theologica*—the work that lay at the center of Bruno's studies in theology during his years in Aquinas's own previous stronghold, the Dominican Monastery in Naples.[11] In spite of St. Thomas's objections, however, Joachim, a Cistercian monk of high rank closely associated with the southern Italian region of Calabria, was widely read and his ideas respected. His idea of the monastic life, in a purified contemplative form, as fulfilling the apocalyptic vision of the Bible, had exercised a considerable influence on all the monastic orders in Italy, particularly in the south.

Joachim himself was very much a figure of the Middle Ages (he died in 1202) and is known to have influenced Dante, who praises him as "il calavrese abate Giovacchino / di spirito profetico dotato" (the Abbot Joachim from Calabria, endowed with a prophetic spirit) in the 12th canto of the *Paradiso*. Furthermore, since the final decades of the last century, scholars such as Bernard McGinn, Marjorie Reeves, and Riccardo Rusconi have been underlining the importance of the revival of Joachim's thought during the sixteenth century, for most of his major works appeared in print in Venice between 1504 and 1527, including his comment on the Apocalypse. Rusconi, in his recent book on prophecy and prophets in the Middle Ages, underlines how the influence of Joachim's prophetical vision was by no means limited to Catholic Italy, but penetrated deeply into the mentality of the Protestant Reformation, as well as many of the heretical and mystical currents of sixteenth-century thought.[12]

If we are going to consider Joachim of Fiore as a major source for Bruno's idea of history, particularly as it is articulated in the final episodes of the *Heroici furori*, we can expect that there will be only a highly selective use of his thought, at times accompanied by harsh criticism. That was Bruno's way of using his sources, for he was always as concerned with specifying what he repudiated in them as he was with making clear what he intended to take from them. So it is no surprise to find the character called Armesso, in the first dialogue of *De la causa, principio et uno*, declaring: "Io non parlerò come santo profeta, come astratto divino, come assumpto apocaliptico" [I shall not be speaking like a saintly prophet, like an abstract theologian, like an assumed apocalyptic], preferring to the abstractions of apocalyptic prophecy a use of language that is more colloquial and realistic.[13] On the other hand, in the first dialogue of the second part of the *Furori* itself, the character called Cesarino states quite clearly that prophetic writings can be useful, particularly for civilizations that find themselves in the shadows of a period of obscure evils, for they can predict the certain return of a phase of prosperity and light.[14] This is precisely Joachim of Fiore's attitude in his comment on the Biblical Apocalypse. For in spite of discussion concerning certain aspects of Joachim's apocalyptic vision, all commentators agree that the distinguishing mark of Joachite prophecy is his vision of a final period, called a "sabbath" of divine light, to be consummated within history in the form of a "sommo bene in terra," or an ultimate good on earth.

Joachim's own version of the coming sabbath is complex and at times rather obscure. Although his historical scheme varied in his different writings, he tended to divide Christian history into three phases, corresponding to the seven seals of the Apocalypse. The first phase, and the first seal symbolized by the white horse, corresponds to the innocence of the primitive Christian era, from Christ's resurrection through the centuries of the early Christian Church. After the millennium, when the beast of the Apocalypse is freed to bring evil back into the world, we get various forms of sinners and heretics worshipping Antichrist. This is the period of the second to the sixth seal, and it is an era of blindness and of corruption within the Roman Church. Joachim's own time is that of the sixth seal, when the empire and the church are at war with one another and confusion and violence reign. But soon the Angel of Apocalypse will open the seventh seal, bringing in a final era of peace and light, which will correspond to a Reformation within the church and the victory of contemplation of the divine over religious dogma and dominion.[15] For Joachim, this third age of light, which is the New Jerusalem of the Apocalypse, is to be lived *in* the world and *in* time, *within* the historical process.[16] This interpretation of the Biblical message defies the readings sanctioned both by St. Augustine and, later, by St. Thomas Aquinas, for whom the heavenly state of the saints, accompanied by the music of the planetary spheres, is situated, after the Last Judgment, in a trascendental heaven.[17]

The revival of Joachite prophecy in Catholic Italy, in the early sixteenth century, which gave rise to the printed editions of Joachim's works published in Venice already mentioned earlier, involved figures such as Egidio di Viterbo, whose strong and stimulating presence in the Neapolitan culture of Bruno's youth has already been made the subject of a significant study by Ingrid Rowland.[18] Egidio's eschatological expectations of a new role within a purified and more spiritual world for the order of Augustinian hermits over which he presided for many years were explicitly characterized as of Joachite origin by the Augustinian Silvestro Meucci, whose meeting with Egidio in Venice, during the years in which Joachim's works were being published there, is narrated by Silvestro himself in 1527 in his dedication of his edition of Joachim's *Expositio in Apocalypsim*. The Abbott Joachim, in his *Expositio*, had prophesied in exalted terms the rise of an order of hermits living the life of angels who, in the last age of the world, would revive the collapsed church and restore glory to all things. Silvestro sees his own Augustinian order as dominating this future age of spiritual men, and he invokes the authoritative figure of Egidio to support his vision.[19]

Since the second half of the last century, the influence of Joachim's prophetical vision on Protestant, and particularly English, culture during the reign of Elizabeth I has also been much studied and is now considered to have been of considerable importance. English interest in Joachim's ideas went back a long way, to his famous meeting in 1190 with Richard the Lionheart in Messina, where the king had stopped off on his way to the crusades. In this carefully recorded conversation, the English king is known to have asked Joachim the standard question: "When and where will the Antichrist be born?" and to have received the astonishing answer that he had already been born and lived in Rome. This was obviously just what, later, the sixteenth-century Protestants wanted to hear, and it is no surprise to find that a prominent English Protestant, John Bale, back from exile in Europe after the reign of the Catholic Mary, published a report of this conversation in the middle of the century. Bale was a friend of John Foxe, who together with many other British Protestants of the period, published works of Joachite prophecy in which they searched through the history of the church during the Catholic period for divine signals pointing to a future Reformation, which they obviously identified with Protestantism itself. Marjorie Reeves, in a paper that attempts an overview of the English apocalyptic thinkers active between 1540 and 1620, finds a significant increase of such publications in the late 1570s and early 1580s, when the Protestant Jacopo Brocardo visited England. Important comments on the Revelation of St. John were published in English in the years 1581–1582. Marjorie Reeves notes that in all these works, Joachim of Fiore "stands out as a prime witness."[20]

It is worth attending briefly to the figure of Brocardo, whose Latin comment on the Apocalypse was translated into English by James Sanford in 1582 and was dedicated to the earl of Leicester. Brocardo's dedicatee was one of Elizabeth I's most powerful courtiers and uncle to Sir Philip Sidney, to whom Bruno would dedicate the *Heroici furori*. Early in his work, Brocardo interprets the apocalyptic writings of his fellow Italian, "the Abbot Ioachim," in the light of the Lutheran Reformation, which he sees as terminating Joachim's age of the sixth seal and bringing in the final age of wisdom and true faith.[21] According to a frequent topos of Protestant propaganda, the Pope is declared to be Antichrist. Brocardo later attempts to strengthen his violent criticism of the Papacy by calling in as his witnesses many of the most celebrated Italian writers and preachers, who had often criticized their own church from within. These "witnesses" include Petrarch, Boccaccio, and San Bernardino of Siena. Coming to his own times, Brocardo predicts that the Roman Catholic wars against the Gospellers will eventually be turned against the Papists themselves, who will ultimately be defeated. According to Brocardo, who again cites Joachim to support him, apocalyptic prophecy is a historical interpretation of the Scriptures, not a theological one, which can be left to the divines.[22] The English Reformation does not play a large part in Brocardo's historical scheme, but it is mentioned with enough approval to justify the English translation of his work. Henry VIII is praised as the king of England who "withstoode" the Papacy, while on an earlier page Brocardo had noticed that the Gospellers finally dominated the religious scene in England, when the Anglican settlement was confirmed by Elizabeth I, during the years that in continental Europe saw the Counter-Reformation being elaborated by the Council of Trent.[23] Brocardo is known to have visited England, where he may have met the Queen, although his visit seems to have preceeded that of Bruno.[24] He certainly met (or was seen by) Bruno personally somewhere, according to a marginal note in the manuscript of the *De rerum principiis*, where Brocardo is referred to above all for his use of cabbalistic number symbolism, considered by Buno to be a major element of the prophetical art. Bruno, who refers to him as "Brocardus Noribergae," writes that his prophetical calculations were not always turning out to be exact, but that Brocardo himself put that down to the fact that he was still deficient in the art, whose validity he was clearly not disposed to put in doubt.[25]

Writing the *Heroici furori* in London in 1585, Bruno appears to have been aware of the recent prophetic writings in English such as Brocardo's, helped perhaps by his bilingual friend, the Protestant John Florio. Obviously he made use of them in his own terms, adapting them rigorously to his own ends, and his own end. For the final pages of the *Furori* amount to a reversal of all that has gone before, in terms that do not seem to me to have been sufficiently appreciated in the critical discussion. Both the sonnet sequences and the emblematic images that in the earlier and more Neoplatonic parts of the text had constituted the literary structures around which Bruno articulates his philosophical discourse, culminating in the mystical death of Actaeon devoured by the hounds of his own thoughts, are now replaced by the account of what is clearly a renewed spiritual as well as a physical pilgrimage. A major literary source for the accounts of the journeys of the nine philosophers that terminate the *Furori* was already identified in the nineteenth century by Francesco Fiorentino as a drama titled *Cecaria* by the Neapolitan poet Marc Antonio Epicuro, first represented in Venice in 1525, and has been frequently discussed in more recent years.[26] Its importance for a reading of Bruno's work, and particularly for the last two dialogues of the *Furori*, is undeniable. Yet it fails to account for the English dimension of Bruno's final point of arrival, or for the significance of the opening of the illuminating vial by the chief English nymph on the banks of the river Thames.

For this aspect of the ending of the *Furori*, a reference to Joachite prophecy becomes essential. This is because Bruno, in the final canto of his work, narrates a tripartite journey of the nine philosophers from the slopes of Vesuvius, which stands for Nola itself, to the banks of the river Thames, passing through a traumatic meeting with Circe in the precincts of Rome. This journey culminates in an ultimate era of illumination *in this life*, which is precisely how Joachim of Fiore used the great Biblical figures of pilgimage. It has been pointed out by commentators how often Joachim quotes in his comment on the Apocalypse from the Biblical passage in 2 Corinthians, 3:18 ("euntes de claritate in claritatem"), his theme being "the urgent necessity of pressing on from the dark into the light of the moon, that at last we may reach the full sunlight."[27] In Bruno's sixteenth-century version of the journey, this culmination is reached only when the philosophers are illuminated by the English Diana. Using typically Joachite imagery, Bruno puts into her hands the vase or vial, whose opening corresponds to the breaking of the seventh seal of apocalyptic vision. According to Joachim, the first six vials are poured out by the vial angels onto the heads of various types of sinners and heretics, but the seventh vial is poured into the air and signifies the cleansing of the spiritual church. Bruno keeps the images of the opening of the vial, the cleansing of the polluted air, and the music of the psalter that accompanies it in Joachim's vision. But he turns his English angel into a river nymph, and his philosophers into her nine muses, transferring the whole moment of illumination not only into a historical but also into a specifically naturalistic dimension. The waters above the firmament are now irrevocably joined to the waters beneath the firmament, and the eternal vicissitudes of an infinite world, which formerly were hidden from the understanding of the human mind, are now open to a new inquiry into natural things:

> In questa mole immensa
> Quant'occulto si rende e aperto stassi.

> In this immense space
> The hidden things appear before us, and stand in the open.[28]

It needs to be emphasized that the last songs of the illuminated philosophers represent the end of the phase of heroic fury that Bruno's journey—partly a sophisticated fiction and partly autobiography—has led up to. The nine philosophers have now reached a stage of their inquiry that will need to be carried out in more ordered and tranquil meditation ("tranquillato essendo alquanto l'impeto del furore"). It is surely a mistake to read this work as representing a single state of mind, to be identified in purely Neoplatonic terms with the raptus of Acteon's self-destructive vision of the iconic Diana. Rather, it is a fiction that represents an intellectual journey during which that vision is left behind—not without regrets, gratitude, soul-searching, and a remarkable exercise in creative poetic composition. At the end of the book, however, that phase of Bruno's life, and of Europe's intellectual history, seems to be considered exhausted, and a new era to be celebrated of a consciously less heroic kind. The individual is replaced by the group, the fury by a newly disciplined intellectual discourse concerning natural things that are no longer occult, but open to rational inquiry and research.

A reading of the *Furori* that takes into account Bruno's point of arrival necessitates asking the question: what significance is he giving to the English Diana under whose protection the nine illuminated philosophers initiate their new inquiry? The question is clearly of no small importance if we bear in mind that the work has brought the reader in a circle back to the starting point of the Italian dialogues—that is, to *La cena de le ceneri (The Ash Wednesday Supper)*, which is precisely where Bruno develops for the first time the definition of his newly infinite, post-Copernican universe. In what terms did the Nolan philosopher who argued for that universe, during the supper held in the house of Sir Fulke Greville, think of his relationship with the chief English nymph who breaks the seal of the illuminating vase at the end of the *Furori*? Certainly in positive terms, or it would have been senseless to select Sir Philip Sidney as the dedicatee and privileged reader of his work. But then both the English Queen and her famous courtier Sidney had already been celebrated by Bruno in *The Ash Wednesday Supper* itself, in terms that would seem to explain his decision to travel to England in the first place.[29] Sidney is praised as a man of clear, rare, and singular intelligence, while the Queen's judgment and wisdom in government are considered second to no other princes' on earth. These are political rather than religious virtues, and Bruno was probably sincere when he told his judges during his trial that he had not praised Elizabeth I for any religious attribute, but rather for her way of being a prince as the ancients understood that word.[30] This, of course, included the control of the sphere of religion by the Prince in the name of public security and morals, without which the new natural philosophy was destined to fail before it had even begun.

What Bruno was looking for at the end of the *Furori* was a way of institutionalizing the new research into natural things that his philosophy, from then on, intended to sanction and support. The English sequel to the *Furori* would be written thirty years later by Francis Bacon, whose *New Atlantis* is still in the form of a Utopia, but which already foresees the flourishing of the natural sciences within the security of a publicly protected and recognized institution. This would later find concrete form in the Royal Society founded in London at the end of the seventeenth century.[31] Salomon's House, Bacon called it, and the Biblical echo was by no means incidental. The sciences too can sing to the glory of God—perhaps Bruno would have said of the gods, who reside in the shadows of the natural world as well as in the human mind and heart.

NOTES

1. Frank Kermode, *The Sense of an Ending: Studies in the Theory of Fiction,* Oxford, UK, Oxford University Press, 1966.

2. For Aristotle's well-known idea of "catharsis," see *The Poetics,* in *The Complete Works of Aristotle,* ed. Jonathan Barnes, Princeton, NJ, Princeton University Press, Bollingen Series, LXXI, vol. 2, 2316–2340.

3. See William Shakespeare, *King Lear,* ed. Kenneth Muir, London–New York, Methuen, 1972, 206.

4. For a recent treatment of this subject, see Mendoza (2002).

5. See Firpo (1993), 333–58. For a comparative reading of Bruno's and Galileo's trials, see Finocchiaro (2002). See also the final essay in this book.

6. This chapter is centered on dialogue 5 of the second part of the *Heroici furori.* All quotations from Bruno's Italian dialogues use the anastatic reprint of the first edition edited by Canone (1999). For the *Heroici furori,* see vol. 4.

7. See "Bruno e l'Apocalisse," in Ciliberto (2002b), 63–94.

8. Bruno (1999a), vol. 4, 1266. My translation.

9. See Gentile's note to his edition of the *Furori* in Bruno (1958), 1168–69, note 4.

10. For Bruno's years in London, where he served in the French Embassy under Michel de Castelnau, Lord of Mauvissière, see the relevant chapters in Ricci (2000) and Rowland (2008).

11. See Thomas Aquinas, *Summa teologica,* op. cit., I–II, question 106, article 4, where Joachim is not mentioned by name. However, commentators assume that it is Joachim's idea of the ultimate good on earth that is being repudiated when St. Thomas, citing St. Augustine, claims that the third state, that of the Holy Ghost, is not to be expected in this life but in the reign of the saints—that is, in the heavenly life prophesied by Christ.

12. Roberto Rusconi, *Profezia e profeti alla fine del Medioevo,* Viella, San Giovanni in Fiore, 1999, 234.

13. See Bruno (1999a), vol. 2, 496.

14. Ibid., vol. 4, 1406: "Et *quanto* á gli stati del mondo quando ne ritroviamo nelle tenebre, et male, possiamo sicuramente profetizzare la luce et prosperitade" [And as for the state of the world, when we find ourselves in the shadows, and suffer for it, we can prophecy with confidence a return to light and prosperity.] Although Cesarino has in mind here the Hermetic Apocalypse discussed by Ciliberto (see note 7 earlier), it is clear that his words can be applied to the Biblical Apocalypse as well, particularly if read in Joachim of Fiore's terms.

15. Joachim seems to have started work on his comment on the Apocalypse in 1183. His apocalyptic visions were written down by his scribe, Rogerius. His *Expositio in Apocalypsim,* in its final form, consists of a prologue, an introductory book, and six parts. For the history of the difficulties Joachim had to face

during its composition, partly due to the complexity of his apocalyptic vision and partly due to the suspicions of unorthodoxy already voiced in his own time, see the *Prefazione* by Kurt-Victor Selge in Gioacchino da Fiore, *Introduzione all'Apocalisse*, Rome, Viella, 1995, 7–25.

16. Joachim's insistence on this point can be found in all his works. For example, in his *Concordia novi ac veteris testamenti*, he writes that he is certain of only one thing: that one day the seventh angel will sound his trumpet, and all the mysteries will have been revealed and an age of peace will reign on earth. See *Concordia*, book V, 118.

17. "Joachim broke decisively with the Augustinian tradition of interpreting the Apocalypse allegorically and instead interpreted it historically." See E. Randolph Daniel, "Joachim of Fiore," in *The Apocalypse in the Middle Ages*, eds. Richard K. Emmerson and Bernard McGinn, Ithaca, NY, Cornell University Press, 1992, 72–88: 87. See also "Radical Views of the *renovatio mundi*" in Marjorie Reeves, *The Influence of Prophecy in the Later Middle Ages*, Oxford, UK, Oxford University Press, 1969, 473–504.

18. See Rowland (2002). The known documents referring to Bruno's Neapolitan years have recently been published and commented on by Miele (2003).

19. For a study that explicitly links Egidio with Joachite prophecy, see Marjorie Reeves, "Cardinal Egidio of Viterbo and the Abbot Joachim," in *Il profetismo gioachimita tra Quattrocento e Cinquecento*, ed. Gian Luca Potestà, Genoa, Marietti, 1991, 139–55.

20. See Marjorie Reeves, "English Apocalyptic Thinkers (c. 1540–1620)," in *Congresso internazionale di studi gioachimiti* 4, San Giovanni in Fiore (1994): 259–73.

21. *The Revelation of S. Ihon revealed, written in Latine by Iames Brocard and Englished by Iames Sanford*, London, Thomas Parishe, 1582, fol. 7.

22. Ibid., fol. 18.

23. Ibid., fols. 139 and 50. Brocardo wrote more than once in this text that he was personally present at the early sessions of the Council of Trent but then left as a sign of protest against the dominating influence of the Roman curia. He suggests that his conversion to Protestantism was a direct result of his experience at Trent.

24. An interesting testimony of the bitter resentment felt in Catholic Italy for Brocardo's conversion to Protestantism can be found in the *Syllabus scriptorum pedimontii* of Andreae Rossotti (Monteregali, 1667), where he is defined as a "vir ineptus" whose works "non sunt confutatione digna reperta, sed ferro, ligna, & flammnis purganda" [a stupid man whose works are not worth searching for to confute, but rather should be purged in iron, wood, and flames]. I am grateful to Margherita Palumbo of the Biblioteca Casanatense in Rome for this reference. As was to be expected, Brocardo's works finished immediately in the Index of forbidden books. For biographical details, see the useful entry by Antonio Rotondò in the *Dizionario biografico degli italiani*.

25. See Bruno (2000b), 702–3. The marginal note indicating that the reference is to "Brocardus Noribergae" is in the hand of Bruno's scribe, Hieronomous Besler, not that of Bruno himself. This mention of Brocardo, which does not appear in any of the relevant indexes, seems to have gone unnoticed by both Bruno

and Brocardo scholars. I am grateful to Eugenio Canone for pointing it out to me. According to Antonio Rotondò, *Dizionario biografico italiano* (DBI), Brocardo arrived in Nuremberg only in 1591, so this episode must have taken place in Bruno's last months before his return to Italy. It is not clear what prophecies exactly Bruno is referring to here, although many of them can be gleaned from Brocardo's work. The British Library has a manuscript list of 16 prophecies of Brocardo's (COTTON VESP. F V. folio 130, r & v) that predict terrific calamities for all the Catholic parts of Europe, and particularly France, where the atrocities of St. Bartholomew's night were long remembered.

26. This drama was well known throughout the sixteenth century. It can be consulted in M. A. Epicuro, *I drammi e le poesie italiane,* ed. A. Parenti, Bari, Laterza, 1942, 1–50.

27. See the introduction to the important modern edition of Joachim's *Figurae,* eds. Marjorie Reeves and Beatrice Hirsch-Reich, Oxford, UK, Oxford University Press, 1972.

28. For the way in which Bruno, particularly in the *Furori,* constantly supports his cosmological inquiry with Biblical references, see chapter 13, "Bruno's Use of the Bible in His Italian Philosophical Dialogues," in this volume.

29. *La cena de le ceneri (The Ash Wednesday Supper)* is in Bruno (1999a), vol. 2.

30. Firpo (1993), 188–89. The conclusion in this chapter agrees in many respects with the analysis of the essentially political-religious relationship between Bruno and Queen Elizabeth I put forward in Sacerdoti (2002).

31. For Bruno as a precursor of Bacon in this sense, see Gatti (2002), 145–66.

7

BRUNO AND SHAKESPEARE: *HAMLET*

THE HISTORICAL CONTEXT

*H*AMLET'S CENTRAL POSITION AS A moment of transition between the early period of Shakespeare's more brilliant and happy mood toward the years of his mature tragic art can be considered as an acquired fact in almost any modern reading of his best known and most celebrated play. Those who wish to underline Shakespeare's position in the course of British history between the end of the sixteenth century and the beginning of the seventeenth, when *Hamlet* was written and acted for the first time (1600–1601), often explain this dramatic change of mood by pointing to the final years of the long and fertile reign of Queen Elizabeth I, which would soon, in 1603, give way to the beginning of the more unpopular and conflictual story of the Stuart dynasty. For it seems difficult to deny that Shakespeare was possessed of an almost prophetic vision when, in the first decade of the seventeenth century, he elaborated his remarkable sequence of tragic stories of failed princes, who drag with them in their personal disaster the whole kingdom over which they should have reigned (King Lear, Macbeth, Antony and Cleopatra). Such tragedies surely have something to do with the coronation of James VI of Scotland as James I of England, once Elizabeth had indicated him, *in extremis*, as heir to her throne. For it would be James's son, Charles I, who would eventually plunge England into a civil war that would radically change the course of British history.

On the other hand, justice is not done to Shakespeare's work if it is considered as enclosed within a historical context limited to British affairs. For, from the outset, his plays show signs of a broader stance, which takes into account both in historical-political and in more general cultural terms the late Renaissance world of the Europe in which he lived. Moreover in the early years of the new century, European history as a whole appears far from serene, tensed as it is within the radical polarization of the conflict between the Reformation and the Counter-Reformation, accompanied by the progressive weakening of the Holy Roman Empire. The gradual exasperation of a crisis that was together religious, political, social, and economic would lead eventually to the Thirty Years War (1618–1648) that, if its principle theater was to be Germany, nevertheless involved most of the countries of continental Europe, not without repercussions in Britain. A growing hostility toward a British monarchy that was becoming ever more absolute would lead to a rapid strengthening of the Protestant and Parliamentarian areas of the bourgeoisie and common people who, in 1640, challenged the power of Charles I and, proposing the Parliament as the most valid center of political power, lay the bases of the modern British state.

At the root of this growing situation of unease, both in a European and in a British context, lay an increasingly dramatic conflict between the principles of authority and of liberty, evident already in the early years of the century. In the light of this situation, the tragedy of the young and cultured prince of Denmark (the most "philosophical" of Shakespeare's dramatic heroes, as Coleridge would note), who is deprived of his throne and with it of his liberty by his crafty uncle, murderer of Hamlet's father to become all too soon the lover of his mother, appears as a remarkable anticipation of the crises of violence, corruption, and alienation that will soon plunge the entire continent of Europe into one of the darkest periods of its long history.

In the light of this wider historical picture, some few but significant voices in the critical discussion of *Hamlet,* which begin to be raised at the end of the nineteenth century, have noted an interesting coincidence of dates between the composition of Shakespeare's first fully mature tragedy in 1600–1601 and the death at the stake of Giordano Bruno on February 17, 1600, in Campo dei Fiori in Rome.[1]

PERSONAL HISTORIES

Giordano Bruno's dramatic life story included, as is well known, a stay in England from the spring of 1583 to the autumn of 1585 that was neither peaceful nor untroubled. We know from his own testimony that after the publication in London in 1584 of his first philosophical dialogue in Italian, *La cena de le ceneri (The Ash Wednesday Supper),* which proposes the new post-Copernican and infinite cosmology that Bruno had already tried unsuccessfully to talk about in Oxford, he had to take refuge in the French Embassy in London because he was considered a revolutionary who was attempting to subvert "a whole city, a whole province, a whole kingdom."[2] The verbal revenge that he developed in his Italian dialogues was bitter and at times violent in its castigation of English culture as a patient refusing to be treated by a foreign doctor who was attempting to apply remedies unknown to the natives. Bruno was merciless in his satire both of the "obtuse" academics and of the "uncouth" populace of England, but he remained constant in his admiration of England's principal figure of a Renaissance courtier, Sir Philip Sidney, to whom he dedicated two of those dialogues, as well as of Queen Elizabeth herself. In his own words, he considered her:

> superior to all the kings of this world, for she is second to none of the sceptered princes for her judgement, her wisdom, her advice and her government. As for her knowledge of the arts, her notions of science, her intelligence and expertise in the use of those European languages which are spoken by the erudite and the ignorant, there is no doubt that she compares favourably with all the other princes of our time.

Considered by him as a new Astraea, as she was frequently called by admirers both British and foreign, Bruno's public admiration of the English Queen was held against him during his long trial at the hands of the Roman Catholic Inquisition. On June 13, 1592, in Venice:

> he was interrogated as to whether he had ever praised foreign or heretical princes, given that he had lived for so long under their rule. And if he had praised them, what was his intention in doing so. He replied: "I have praised many heretics, and heretical princes among them. But I have never praised them because they were heretics, or in any religious way because of their religion or piety, but only for their moral virtues. In particular, in my book *Of the cause, principle, and one*, I have praised the Queen of England and called her 'diva', not as a religious attribute but as a kind of epithet which the ancients used for their princes, for in England, where I wrote that book, they are in the habit of using such an epithet."[3]

The date of Bruno's stay in England, in the first half of the 1580s, renders extremely improbable a personal meeting with Shakespeare, who seems not to have arrived in London before the end of the decade, or even possibly at the beginning of the 1590s. Nor can we be certain that Shakespeare had a firsthand knowledge of Bruno's philosophical dialogues, published in London but written in a notoriously difficult Italian and not translated until more recent centuries. He would probably have had fewer problems with the Latin works, found in the libraries of some of the cultured aristocrats of the period, protectors of the arts as well as the sciences, such as the ninth earl of Northumberland. One of the major figures of the Elizabethan court, Northumberland was very soon imprisoned for high treason in the Tower of London by James I, who suspected him of participation in the Gunpowder Plot, although most modern scholars consider the trial to have been stacked unfairly against him. It is documented that Northumberland was reading Bruno's works, of which he held a major collection in his private library.[4] Bruno's stay in London thus seems to have left a mark on some of the most cultured people in England at that time. Furthermore, if only through his friend John Florio, Bruno must have known the London community of Italian refugees from the tentacles of the Roman Catholic Inquisition, which Shakespeare seems to have frequented. It is from them that Shakespeare is supposed to have gathered his news of the places, people, and customs of Italy that form the Italian setting of so many of his plays, as well as his knowledge of the numerous Italian texts identified as sources of both his early and his mature dramas.

There can be no doubt of the importance, indeed the centrality, in this context of the figure of John Florio. Bruno's closest friend in England, and his constant companion during his stay in the French Embassy in London, Florio remained an active Anglo-Italian exponent of London society and culture throughout the period of Shakespeare's theatrical activity. Son of a Tuscan, Michelangelo Florio, who had converted to the Protestant religion, John Florio's role of master of Italian to more than one generation of English students and writers between the end of the sixteenth century and the beginning of the seventeenth has been amply documented in a still valid book by Frances Yates.[5] Her text underlines the friendship between Bruno and Florio, who figures in *La Cena de le ceneri (The Ash Wednesday Supper)* as one of the messengers who accompany Bruno to the supper hosted by Sir Fulke Greville, friend and future biographer of Sir Philip Sidney. In Bruno's work, Florio accompanies him on the evening journey in a boat down the Thames "singing, as if in rememberance of his early loves, *Dove senza me, dolce mia vita (Where without me, my sweet life),*" verses from Ariosto's *Orlando furioso*. Later Florio will return the compliment by introducing the figure of Bruno into his bilingual dialogues *The Second Fruits*, calling him "the Nolan" in memory of his origins as a citizen of Nola, near Naples. The portrait Florio paints of Bruno, who appears in his pages as a severe chider of pedants and loafers, leaves no doubt that he was a friend. Florio was not to forget him even after the long years of Bruno's trial and its tragic ending. In 1603, in the note *To the curteous reader* prefixed to his translation into English of Montaigne's *Essays*, Florio remembered his old friend the Nolan who had taught him the value of translations. Later, in 1611, in the second and enlarged edition of his English–Italian dictionary titled *The New World of Words*, which Florio dedicated to Anne of Denmark, queen of James I, he includes a list of the Italian works of Bruno in the pages of acknowledgment of the books used by him in the compilation of his work.

As far as Shakespeare is concerned, the link with Florio is highly probable but not documented. It is, however, considered certain by most commentators that he made use in his dramas, and particularly in *The Tempest*, of numerous pages of the English version of Montaigne's *Essays* in Florio's translation. There, as we have seen, he would have found Bruno's name mentioned favorably in the preface. Furthermore many commentators consider that Florio provided Shakespeare as well as Ben Jonson and other playwrights of the period with their knowledge of Italian topology and their habit of introducing a few phrases of the language into their plays. Indeed, it is possible to go deeper into this aspect of the question, remembering the primary importance in this period of Italian as the language of high culture. In a recent study of the baroque abundance and richness of Florio's Italian as it appears in his dictionary, Michael Wyatt has seen this as the linguistic trait that links Bruno's Italian dialogues to the extraordinary literary flourishing of Shakespeare's England.[6]

It is generally considered that neither Shakespeare nor Ben Jonson ever made a journey to Italy. However that may be, it is more than possible that *chez Florio* there were readings of Italian texts such as the Siennese comedy *Gli ingannati (The Deceived)*, which Shakespeare is thought to have used as a source for *Twelfth Night*. These readings could have included philosophical works by Bruno, as well as his only drama, the comedy *Candelaio*, published in Paris in 1582 shortly before his arrival in London. Echoes of this play have been found in Shakespeare's *Love's Labour's Lost* as well as in Jonson's *The Alchemist*.[7] It is in *Love's Labour's Lost* that we find the figure of the somewhat pedantic schoolmaster, Holofernes, sometimes considered a satirical portrait of Florio himself, who was undoubtedly a notable pedagogue, linguist, and man of letters, but lacking in the creative imagination of an artist.

A convincing basis for a knowledge of Bruno on the part of Shakespeare, probably mediated through John Florio, thus undoubtedly exists. This seems indeed likely in view of Bruno's tragic story and his audacious ideas, which caused him to enter (Hamlet-like) into dramatic conflict with both the cultural and the religious authorities of his time. Given the inflexible dominion of those authorities at the end of the sixteenth century, Bruno's thought often found violently polemical forms of expression in his works. His use of dialogue, which was so popular with the poets and philosophers of the Renaissance, was far from being purely rhetorical. It indicated a profound dissatisfaction with a culture that was often founded on rigidly dogmatic parameters. Bruno's theater of ideas thus becomes a drama of universal proportions: "these are dialogues," he wrote in the dedicatory letter to Sir Philip Sidney of the fourth of his Italian works written and published in London, the *Spaccio della bestia trionfante (The Expulsion of the Triumphant Beast)*, "and in them the speakers raise their voices in relation to the speeches of many others, who also have things to say, putting forward their point of view with as much energy and conviction as possible."[8] These are words that indicate how Bruno saw the speakers in his dialogues as *dramatis personae* complete with their passions and emotions, but also as profoundly engaged in a process of innovative thought that aimed at liberating the life and culture of his times from the chains of a suffocating, and at times criminal, authoritarian intransigence.[9]

The possibility that Shakespeare knew and made use of his knowledge of Bruno not only for the development of certain scenes and themes in his dramas but also, and perhaps above all, for an idea of the solitary drama involved in thinking philosophically (and it is precisely here that the connection with *Hamlet* appears most convincing) has been advanced since the second half of the nineteenth century. It was precisely at that moment that the study of Bruno was beginning to assume a new intensity, with the publication of ever more sophisticated modern editions of both his Italian and his Latin works, as well as the appearance of the first seriously documented biographies. This is also the period that sees the publication of critical works on Bruno that remain of fundamental importance even today. Major Italian historians of philosophy such as Bertrando Spaventa, Francesco Fiorentino, and Felice Tocco were intent on offering systematic comment on a thinker who, due to their work, had finally emerged at the end of the nineteenth century as one of the finest minds of the European Renaissance.[10] It is also, however, the moment of an increasingly intense debate around the name of Bruno, with some assuming him as a hero of modern thought and an ideal founder of a newly free, secular, and increasingly anti-clerical Europe, while others find in him little more than an example of licentious impiety. On the whole, in the second half of the nineteenth century, the former position prevailed, giving rise to what some critics have called a "Brunomania," which transformed his name into a flag or symbol of free thought, arousing an often passionate enthusiasm.[11]

The fact that the proposal linking the name of Shakespeare to that of Bruno was put forward at that time of rising libertarian sentiment has not always been propitious to a serious study of the phenomenon. For some years, the problem remained at the center of an intense discussion. Ultimately, however, it was condemned by a number of prestigious commentators, such as Robert Beyersdorff in Germany and Benedetto Croce in Italy, who were disturbed by the exaggerated notions to which it was beginning to give rise: the idea, for example, that Shakespeare was primarily a "disciple" of Bruno, and for that reason of a "Mediterranean" rather than a "northern" mind-set. Their combined efforts, added to those of some of the major Shakespearean editors and critics of the time, led to Bruno's name being virtually banished from the Shakespearean discussion. It reappears in more recent times only occasionally, here and there. On the whole, it can be said that a discussion that involved some of the major commentators of both Bruno and Shakespeare at the end of the nineteenth century and the beginning of the twentieth has now almost died out. No recent edition of Shakespeare's plays, even of *Hamlet*, which from the beginning was the drama on which most of the Bruno-orientated attention centered, refers to it with anything more than a passing glance. The Bruno–Shakespeare discussion has become a historical curiosity, of which many Shakespearean scholars of today are no longer even aware.[12] The following pages raise the subject of this relationship once again in the light of recent studies of both Shakespeare's texts and those of Bruno, concentrating on a theme central to the work of both Bruno and Shakespeare: their adoption of an intellectual stance toward a deeply lacerated and violent world that they both often define in terms of "madness."

ANTICYRAM NAVIGAT, OR THE SEARCH FOR A CURE TO MADNESS

The theme of madness in this period is closely associated with the name of Erasmus of Rotterdam, whose famous pamphlet *In Praise of Folly* was written in London while he was a guest in the house of Sir Thomas More and published in 1511. In Erasmus's text, Folly assumes a female form, presenting herself paradoxically as the only voice of reason in a world devoted to the unreason of an unbridled ambition for power. Developing this brilliant strategy, Erasmus succeeds in delving deep into the contradictions and absurdities of a world constructed on the basis of false appearances, where the rich are poor in the values that really matter, and the luxury of power covers serious crimes and the unlimited egoism of ambition. Erasmus's men of power, who are both the terrestrial and the ecclesiastical princes of the day, reappear in Bruno's *La cena de le ceneri (The Ash Wednesday Supper)*, the first of his Italian dilaogues written and published in London in 1584, as so many "Mercuries and Apollos" sent from heaven, who "with impostures of many kinds have filled the world with infinite forms of madness, bestiality and vice, as if they were virtues, divinities and disciplines: dimming the light which made god-like and heroic the souls of our fathers of old, while approving and promoting the sooty shadows of sophists and fools."[13] In a later passage of the same dialogue, the explicit reference to Erasmus merges with one to Ariosto, whose satirical treatment of the oppression of Reason takes the form, in the first two lines of canto XXV of the *Orlando furioso*, of a question asked to his loved one by Astolfo. This sane and reasoning knight, with whom Bruno clearly identifies himself, wants to know who is going to go up to the moon to retrieve his reason, given that his love, like that of Orlando, has sent him mad. "Chi salirò per me, Madonna, in cielo, a riportarne il mio perduto ingegno?" [Who will rise up to the sky, madonna, to bring back my lost wits?][14]

As Lina Bolzoni has pointed out in a major study of Bruno's use of Ariosto in *The Ash Wednesday Supper*, the contrast developed by Bruno between life on the earth and in the moon is designed to show how limited and relative our perspective on things is. For after flying through the heavens to the moon, Astolfo discovers (like Ariosto and Bruno himself) that heaven is on earth and that the divine reason dwells within each one of us.[15] This concept is closely related to the cosmological thesis that Bruno is putting forward in *The Ash Wednesday Supper*, where the universe becomes unique and homogeneous, the earth reflecting the heavens instead of being considered as a heavy mass ineluctably severed from a heavenly quintessence in which it can never participate. Bruno's newly infinite cosmology, however, is rudely repudiated by the guests at Sir Fulke Greville's supper, who openly accuse him of being mad. It is in response to that accusation that Bruno creates, in the passage cited earlier, a dramatic contrast between a lost past, when great minds such as those of Epicurus and Lucretius already knew that the universe was homogeneous and infinite, and an obscure present, in which falsity and deceit reign supreme. That "light which rendered divine and heroic the souls of our fathers of old" is compared in an image of bitter accusation with the "sooty shadows" of the contemporary world—a play between past and present just as we find it in the first act of *Hamlet*. For already before he has heard about the presence of the ghost of his assassinated father on the castle battlements of Elsinore, Hamlet has put on his suit of solemn black, to walk in the "unweeded garden, which grows to seed" of his first monologue. It is one of the many images of "things rank and gross in nature" that constitute his vision of the present. It is evident, however, that already in this first monologue, the sense of living in a historical period of deepening darkness depends on the strong contrast created by Shakespeare's dramatic imagery with a previous time, illuminated intensely by a godlike presence. "So excellent a king," Hamlet says of his father, "that was to this *Hyperion* to a satyr."[16]

It is now possible to see more clearly how Bruno, for his part, in the same page of *The Ash Wednesday Supper* that has already been cited, puts this dramatic picture of a reality conceived of as a tragic loss of ancient purity in relation to a mental state of madness:

> Human reason, for so long oppressed, and at times weeping in her newly humble condition, addresses this lament to the divine and merciful mind, who never fails to whisper to her in her inner ear
>> Who will rise up to the sky, madonna,
>> To bring back my lost wits?[17]

It is precisely this journey through the sky, evoked by the quotation from the celebrated verses of Ariosto, which will become for Bruno the true, indeed the only, solution to the problem of a seriously compromised world, which has itself gone mad. For Bruno will propose a new post-Copernican cosmology in which the suns reassume their ancient positions at the center of their stellar systems, and the universe extends to the vast dimensions of an infinite space filled with infinite forms of unknown life. Bruno's new cosmology aims at readjusting the severely distorted axes of the Aristotelian–Ptolemaic world—that closed universe that found its unique center in the earth, and that over the centuries had become fixed in concepts that, in Bruno's opinion, were profoundly "out of joint."

Shakespeare himself would clearly not develop his tragedy around a new cosmology in a technical-scientific sense. Even if he had wished to do so, it would not have been possible for him to expound in a public theater of his time those things that Bruno had not been able to explain at the University of Oxford. For Bruno had attempted without success, during his visit to Oxford in the summer of 1583, to speak to a scandalized public of academics about his new infinite version of the Copernican cosmology.[18] On the other hand, Shakespeare, as Gilberto Sacerdoti has claimed with respect to another tragedy, *Antony and Cleopatra*, is clearly aware of a newly immense space of human experience that opens up new vistas of both passion and reason.[19] Hamlet himself, after his meeting with his father's ghost under the stars that shine down on the castle ramparts of Elsinore, will start to understand that "there are more things in heaven and earth, Horatio, than are dreamt of in your philosophy."[20] Furthermore, it will be precisely in relation to this new intuition of the elusive nature of truth within the immense spaces of a now infinite universe that Hamlet will deliberately "put an antic disposition on" while waiting for a world that he too defines as "out of joint" to recover its sense of a clear and sane rationality.

To reach this end, Shakespeare, as the critics have often pointed out, develops Hamlet's behavior, after his meeting with his father's ghost, by following the model of the figure of the Fool, or the court jester, which he had already introduced with such success into some of his major comedies. Feste in *Twelfth Night* and Touchstone in *As You Like It* immediately come to mind, while the anonymous Fool of the tragedy of *King Lear* is already lurking in the background. Furthermore it is clear that after the murder of his father and the usurpation of his throne by his uncle, Hamlet remains without anything that he can call his own, dispossessed, just like the court Fools. Only his intelligence and his wit remain for him to use as weapons to protect himself in a world that he perceives as profoundly corrupt and false. Hamlet's role within the new court of Elsinore can usefully be compared with that of Momus in the court of Jove in Bruno's *Lo spaccio della bestia trionfante (The Expulsion of the Triumphant Beast)*, written and published in London in 1584. This is the fourth of the

Italian dialogues written by Bruno in London, and it narrates the story of a macroscopic, universal reform undertaken through the transformation of the signs of the zodiac from bestial vices into reformed virtues: the entire operation being carried out by a Jove who considers himself an absolute prince, both in a political and a religious sense. Bruno, however, reminds his readers that even Jove, like all things that are a part of the material world, remains subject to the laws of vicissitude, suggesting that he is far from infallible, as he wishes to be considered. In order to underline this point, Bruno sees him as being accompanied throughout his long and meticulously organized reform by the suggestions of an ironic and satirical Momus, who gets dangerously close to appearing as the real hero of the story. Momus, in the classical world, was known as the god of satire, and was expelled from Olympus by the gods because of his witty and caustic tongue. His figure had been used to good effect by classical writers such as Lucian in his satires and had already been revived by humanists such as Leon Battista Alberti and Erasmus himself. In the *Spaccio*, Bruno claims that Momus's role in the celestial court of Jove is very similar to that of the Fools in the courts of earthly princes: "where each one offers to the ear of his Prince more truths about his estate than the rest of the court together; inducing many of those who fear to say things openly to speak as if in a game, and in that way to change the course of events."[21] It is a definition that corresponds closely to the way in which Hamlet refers to his own madness while speaking to Rosencrantz and Guildenstern: "I am but mad north-north-west. When the wind is southerly, I know a hawk from a handsaw."[22]

The Erasmian precedent for this concept of a lucid madness is underlined by Bruno himself in the fourth dialogue of *The Ash Wednesday Supper*. There we find an explicit reference to one of Erasmus's *Adages*, published many times during the sixteenth century and titled *Anticyram navigat*. Anticyram was the name given in ancient times to a group of cities known for their production of hellebore, a plant that is a strong purgative and was thought to be a cure against madness. Those who "traveled to Anticyram" were said to be mad and were thought of as going there to seek treatment for their distressed minds. It is precisely this metaphor that is used by the Neoaristotelian academics in their attack on the philosopher Theophilus, who represents Bruno himself in *The Ash Wednesday Supper*.

In this work, Theophilus is trying to explain to the guests of Sir Fulke Greville, the friend and future biographer of Sir Philip Sidney, his version of the infinite universe and the plurality of worlds that he was developing in the wake of the new Copernican cosmology. As the proposer of this new cosmology, Bruno was widely considered mad by the English intellectuals of the time. The future archbishop of Canterbury, George

Abbott, would write of Bruno in 1604 that he must have been raving mad, given that he had dared to discuss the new Copernican cosmology during the lectures he gave at Oxford. Abbott's heavy-handed criticism claims that Bruno's "head did go round, while his brains did stand still." In the *Supper*, however, Bruno, in the person of Theophilus, had already turned such an accusation on its head. In his opinion, the "Nolan philosopher" (Bruno himself) is traveling to Anticyram to gather hellebore in order to cure the madness of some foolish barbarians—that is, the English academics.[23] It has already been noted how Bruno, in the fifth and last dialogue of the *Supper*, defines his task as that of a foreign doctor who brings to England the medicines of which the local doctors are still ignorant—an image that will find expression again in the last words of the work, where a reference is made to "the venerable beard of Asclepius," the mythical healer of ancient Greece.[24] Bruno was convinced that he had to carry out a mission, a historical and philosophical task, in this sense. Since the first dialogue of this work, he had written of his own philosophical activity, with what are clearly Epicurean echoes: "one man, although alone, can and will win the race, and in the end he will be victorious and triumph over the general ignorance."[25] Hamlet too, faced in his solitude by a world that he increasingly perceives, in its madness, as "out of joint," will exclaim: "Oh, cursed spite, that ever I was born to set it right."[26]

MADNESS SUPREME

In Bruno's philosophy, the proposal of an infinite universe filled with an infinite number of solar systems like our own is not only seen as a technical-scientific phenomenon. It also has value as an essential element in his "treatment" for a diseased society. Infinite space, as it was conceived of by Bruno, is the place in which all dimensions and all values become relative, in which no place is an absolute center, while every point acquires its value in relation to other points. This new cosmological thesis was used by Bruno as a powerful instrument of thought with which to oppose a Christian-theological absolutism, supported as it had been for centuries by the Neoaristotelian, Ptolemaic picture of an earth-centered universe, which was still the dominant cultural paradigm of his day. For Bruno, on the contrary, no celestial body is paramount—none of them can be said to be at the center except in the purely relative sense of being the center of their particular solar system. No body is perfect, but rather all are made of a homogenous substance that unites the universe in every single part and degree. So it is in the human cosmos as well. Bruno thus confers a new value on the individual, no longer subject to external powers or gods who must only be adored and obeyed, but now fully mature in virtue of an interior divinity. This new responsibility as a person is seen as consciousness of oneself and a new dignity as an individual, as well as the power of the individual's mind and thought:

> And our doctrine says that we must not look for God outside of ourselves, given that we hold Him within, more closely, as a part of us, than we are close to ourselves. In the same way, the worshippers of other worlds should not look for Him among us, insofar as they too have Him within themselves. For the moon is no more the sky for us than we are sky to the moon.[27]

Bruno was intensely aware of the profoundly heretical implications of this new doctrine from the point of view of the theology of his time. In one of Theophilus/Bruno's speeches in *The Ash Wednesday Supper*, he conjures up a vision of the "fifty or a hundred torches" that he foresees "will not be lacking if he happens to die in a Roman Catholic country" (and the prophecy will reveal itself to be true, when on February 17, 1600, he is led in a torchlight procession before dawn, to die at the stake in Rome).[28]

A philosophical vision remarkably similar to this, and difficult to find in other thinkers of the time—at least formulated in such radical and uncompromising terms—is expressed by Hamlet in a verbal exchange with Rosencrantz and Guildenstern. The result of this exchange will be the unmasking by Hamlet of their deceptive behavior, now that they have become spies in the service of Hamlet's enemy, the new king. "O God," exclaims Hamlet, by now well aware of the dangerous situation in which he has been placed, "I could be bounded in a nutshell and count myself king of infinite space—were it not that I have bad dreams." Guildenstern immediately brands this speech as a sign of unlimited ambition, and by doing so, he reacts instinctively against the new note of autonomous free thought implied by this vision of an infinite universe—as had the English enemies of Bruno at Oxford and elsewhere. "Which dreams indeed are ambition," Guildenstern comments, and "the substance of the ambitious is merely the shadow of a dream."[29]

The theme of shadows itself was one dear to Bruno, who considered the infinite universe a seal or shadow of an infinite divinity, and the shadows of thought in the human mind as susceptible of infinite combinations within the schemes of an art of memory that would open up new spaces of potentially unlimited knowledge. This aspect of his philosophy was well known to the culture of Shakespeare's time, for Bruno's first work was a treatise on the then popular art of memory, the *De umbris idearum (The Shadows of Ideas)*, published in Paris in 1582. This work had stimulated a lively discussion in England. Supporters of Bruno's iconographical art of memory, such as Alexander Dickson, who appears as one of the speakers in Bruno's Italian dialogues, found themselves involved in a violent polemical exchange of published works with William Perkins of Cambridge, who supported the more abstract art of memory proposed by the French logician Ramus.[30] It is therefore probably no coincidence that this exchange between Hamlet and Rosencrantz and Guildenstern in Shakespeare's text had been immediately preceded by another, where two arguments were raised that are of central importance as themes in Bruno's philosophy: the image of the contemporary world as a prison, and the conviction of the relativity of every truth that can be perceived by the human mind:

> HAMLET. Denmark's a prison.
> ROSENCRANTZ. Then is the world one.
> HAMLET. A goodly one, in which there are many confines, wards, and dungeons, Denmark being one o'th'worst.
> ROSENCRANTZ. We think not so, my lord.
> HAMLET. Why, then 'tis none to you, for there is nothing either good or bad but thinking makes it so. To me it is a prison.[31]

So, if it is the specific context that defines the space in which every true thinker acts, creating his own truths on the basis of his personal situation in time and space, then universal truths—the truths of metaphysics, and even more those of the theological tradition—lose their absolute value. In this respect, Bruno and the other *novatores* of the end of the sixteenth century had already opened the door leading into the modern age. Hamlet is on their side, and perhaps most particularly on Bruno's side. Against him, the murderous king, who bases his unique and absolute authority within the microcosm of his court on the traditional philosophy that was being questioned by the *novatores*, immediately realizes the necessity of protecting himself against Hamlet's "madness." "And can you by no drift of conference / Get from him why he puts on this confusion, / Grating so harshly all his days of quiet / With turbulent and dangerous lunacy?" he asks Rosencrantz and Guildenstern.[32] In the end, it will not be they who unmask Hamlet but rather Hamlet who unmasks the secret and criminal doings of the King. He does it by manipulating words into a form of truth that he himself, in the monologue that closes the second act of the tragedy, recognizes as the most appropriate vehicle for creative human thought— that is to say, the drama itself. As he later tells the actors, exhorting them not to waver from the discipline of a meditated and controlled rhetoric, a good play shows "virtue her feature, scorn her own image, and the very age and body of the time his form and pressure." That is why "the play's the thing" wherewith to "catch the conscience of the King."[33]

The theater thus becomes during the course of Shakespeare's *Hamlet* a theater of conscience. That is precisely the definition that can be found too in Bruno, who writes in the opening pages of his dedication of the *Heroici furori* to Sir Philip Sidney of "this theater of the world, this scene of our consciences." Even the particular context in which we find this definition can be seen as relevant to Shakespeare's tragedy. For in those pages dedicated to Sidney, the celebrated poet of courtly love, Bruno launches his famous and ferocious attack on women, seen as the worst perpetrators of falsehood and deceit: "and where can you find more pride, arrogance, adamance, wrath, indignation, falsity, lewdness, greed, ingratitude and other putrid crimes"?[34] Much as Hamlet, speaking to Ophelia after she has betrayed his faith in her by collaborating in her father's attempt to lay a trap for him, will hurl his indignation at her: "God hath given you one face and you make yourselves another. You jig and amble, and you lisp, you nick-name God's creatures, and make your wantonness your ignorance. Go to, I'll no more on't, it hath made me mad."[35]

It is no coincidence that the page of the *Heroici furori* cited earlier is part of a dedicatory letter addressed to "that most illustrious knight, Sir Philip Sidney"—that is, precisely the poet who, with his Neopetrarchan sonnets of Astrophel to Stella, was proposing in Elizabethan England a cult of spiritually pure womanhood learned from the Italian poetic tradition. Bruno himself, however, refuses to play that game. He calls it "a curious thought around or about the beauty of a female body," considering it a particularly serious form of madness:

> . . . someone who spends the better part of his time and the ripest fruits of his present life distilling elixir from his brain by elaborating conceptually, then writing and sealing in published works, those continual tortures, grave torments, rational discourses, exhausted thoughts and bitter meditations conceived of under the tyrannical influence of such unworthy, imbecile, stupid and lurid filth.[36]

Many other sources, besides this passage from Bruno, were available to Shakespeare as models for Hamlet's violent attack on what the Middle Ages had presumed to be the extraordinary purity and spirituality of women. Nevertheless the specific link with Sidney, and the English context in which Bruno composed the *Furori*, appear as elements that should not be undervalued. Take, for example, the particular bitterness with which Hamlet later in his drama accuses his mother, now that she is married to her dead husband's brother, of living "in the rank sweat of an enseamed bed, stew'd in corruption, honeying and making love over the nasty sty." Here we seem to have a precise echo of the "stupid and lurid filth" of Bruno's violent attack on the inconstant female figure quoted earlier. In precisely the moment of pronouncing these terrible words, moreover, Hamlet thinks he sees once again the ghost of his father, come to protect his wife, unfaithful as she has been to him, from his son's unbridled fury. Hamlet's mother, at that point, is quick to accuse him of being completely mad: "This is the very coinage of your brain. This bodiless creation ecstasy." Ophelia, faced by a similar attack, had found "the noble reason" of Hamlet, "like sweet bells jangled out of tune and harsh." Hamlet himself, however, thinks otherwise. He replies to his mother:

> My pulse like yours doth temperately keep time,
> And makes as healthful music. It is not madness
> That I have uttered. Bring me to the test,
> And I the matter will reword, which madness
> Would gambol from.[37]

In his constant and patient search for a thread of reason that could serve to reestablish sanity in a world gone mad, Bruno, for his part, does not direct his anger so much at women as such, as at what he considers a profoundly mistaken way in which men understand the feminine presence and role in the world. As David Farley-Hills has correctly pointed out in a study of the dedicatory letter to Sidney of the *Heroici furori*, according to Bruno, women should be honored not as goddesses but as women: "What I mean is that women, even if sometimes not satisfied even with the honours and respect attributed to gods, should not for that reason be honoured and respected as gods. I would like to see women honoured and loved as they should be honoured and loved."[38] With precisely just such a change of mood, Hamlet, once he has exhausted the fury of his reaction to the unexpected falsity of both Ophelia and his mother, shows that he does know how to play the part of a true lover. At Ophelia's funeral, after her suicide, he derides as crocodile tears the theatrical show of despair that her brother Laertes indulges in, declaring with simplicity and conviction: "I loved Ophelia. Forty thousand brothers could not with all their quantity of love make up my sum."[39]

One of the themes that runs through Bruno's works is that of a society that he sees as profoundly corrupt and compromised in its sentiments and affections, as well as in its thought and love of power. There is clearly a need for radical renewal and reform. The instrument of thought proposed by Bruno to carry out that renewal is a sound and sane form of skepticism. This is to be distinguished from a corrosive cynicism, which leads to inaction and desperation. It must take the form of a rigorous process of doubt that addresses every kind of canonical and preestablished attitude to life, lacking in originality and verve. It is not surprising that Bruno's doctrine of doubt is expressed most strongly in the opening pages of the so-called Frankfurt trilogy (*De triplici minimo, De monade, De immenso et innumerabilibus*): the three Latin works of 1591 in which he constructs the most mature and complete expression of his new cosmology. The idea he puts forward is that of an infinite universe composed of an infinite number of solar systems, united by a substance, both material and spiritual, of an atomistic nature. Even the furthest stars, traditionally thought to be composed of pure fire, or an ethereal quintessence, are really only particular agglomerations of those same atoms that compose the infinite whole. It is in this cosmological context that Bruno develops his most articulate expression of a philosophy of doubt:

> Whoever wishes to philosophise, doubting all things at first, must never assume a position in debate before having listened to the opinions of all sides, and before carefully weighing the arguments for and against. He must judge and take up a position not on the basis of what he has heard said, according to the opinion of the majority, their age or merits, or their prestige. But he must form his own opinion according to how persuasive the doctrine is, how organically related and adherent to real things, and to how well it agrees with the dictates of reason.[40]

While opening philosophically toward the modern world, such a doctrine could obviously appear socially subversive and favorable to heresy, both political and religious. Such, at least, was the opinion of the Inquisition, which condemned Bruno as "a particularly obstinate heretic ... author of a number of enormously dangerous opinions."[41] Precisely the same reaction can be found from the very beginning to Shakespeare's tragic hero within the closed circuit of the court of Elsinore. An example can be found in the reaction to the letter written by the young Hamlet to Ophelia. A self-satisfied Polonius shows it to the king as evident confirmation of the young prince's madness:

> Doubt thou the stars are fire,
> Doubt that the sun doth move,
> Doubt truth to be a liar,
> But never doubt I love.[42]

Thus Hamlet, like Bruno, finds comfort in a systematic exercise of skepticism that modulates into love. During the course of Hamlet's drama, this will modulate from love of a woman to become love of an undying principle of truth, transparent and clear. Such is Hamlet's answer to the question of whether "to be or not to be," to live or not to live—for the truth must be pursued, and nothing lie hidden in the obscure shadows of deceit. Precisely through this doctrine of doubt, Hamlet will become aware of his mother's infidelity toward the memory of his father; Ophelia's double-dealing when she helps to lay the trap thought up by her father, Polonius; of Polonius's own deceit as he spies on Hamlet during his showdown with his mother; of the double-dealing of Rosencrantz and Guildenstern once they have become spies in the pay of the king; and above all of the king, Claudius himself, caught in the mousetrap of the "play within a play." When all this is said and done, it becomes clear that what lies at the center of Shakespeare's drama is not so much the murder of a king as the murder of truth itself. Only the faithful friend Horatio remains secure in his integrity and faithfulness, in spite of the surrounding deceits and the boundless ambition of the court. Hamlet can only praise admiringly this man that "Fortune's buffets and rewards has ta'en with equal thanks; and blest are those whose blood and judgement are so well commeddled that they are not a pipe for Fortune's finger to sound what stop she please."[43] Such men who say little but say it with clarity and conviction are equally praised by Bruno, once more in the Frankfurt trilogy, in a passage that follows almost immediately after his enunciation of his doctrine of doubt:

> Truth and knowledge emerge from the simplicity of words: only laziness and cunning are pleased by redundant words, while their variety, when accompaned by self-interested avarice, gives rise to vanity.[44]

In Bruno, as in Shakespeare, there is a strong vein of criticism of the courtly adulation that expressed itself in terms of a frivolous and fatuous adoration of the prince, uncaring of the corruption that had become a characteristic of many of the courts of the period. "I shall never learn how to slip emeralds on to my rough fingers," Bruno wrote in the final pages of his Frankfurt trilogy, "how to curl my hair, paint my face with rouge, adorn my head with perfumed hyacinths, assume a foppish pose or move smoothly. I speak as a man, and cannot falsify my voice, so that it seems to come from the throat of a babe, behave as if I were still a boy, or from a man seem to become a woman."[45] It is hardly possible to read this passage without thinking of the contrast created in the words of Hamlet between the true man he finds in Horatio, whom he is talking to at the time, and the frivolous falsity of a courtier like Osric, who brings him the challenge to a duel on the part of Laertes—the duel that will prove fatal to them both. For Osric "and many more of the same bevy that I know the drossy age dotes on," in Hamlet's words to Horatio:

> only got the tune of the time and, out of an habit of encounter, a kind of yeasty collection, which carries them through and through the most fanned and winnowed opinions; and do but blow them to their trial, the bubbles are out.[46]

In the years when Shakespeare was writing his major tragedies—which followed closely on the years of Bruno's final period of liberty—one of the problems that their society attempted to address was what kind of power structure should guide a new order of events. It was a question that mattered deeply to those who wished to move toward better times compared to the "sordid age," such as that in which both Bruno and Shakespeare felt they were living. In both of them, we find a solution that seems curiously circular and self-defeating, and that they themselves appear to have considered unsatisfactory. In Bruno's case, after an initial period of exile from Italy in which he seems to search above all for a university chair from which to impart his philosophy, we find in the last years a growing attention toward the courts of the period and toward the figure of the prince. For Bruno, a new kind of prince should be capable not only of guaranteeing the material interests of his people but also of becoming their cultural leader—the model is clearly the ancient sacerdotal prince chosen from among the wise. His triple wisdom derives from his knowledge, his power, and his authority. This is what Bruno writes in his dedicatory letter of the Frankfurt trilogy addressed to Prince Henry Julius, Duke of Brunswick.[47] Nevertheless there is a clear note of anxiety in the messages Bruno addresses in these years to the princes of the period—for example, in the famous letter with which he dedicated to the Emperor Rudolph II his mathematical work *Articuli centum et sexaginta adversus huius tempestatis mathematicos, atque philosophos*, published in Prague in 1588. Here Bruno affirms the right to carry out research into natural things in complete freedom with respect to the power of the prince himself, but the emphasis with which he attempts to assert that right demonstrates how fragile and inconstant he judged the cultural interests of the contemporary princes to be.[48] Similarly Hamlet considers the young Norwegian Fortinbraccio a fragile and inconstant prince, in spite of his name, when he meets him at the head of an army intent on conquering an arid piece of Polish land: his spirit "with divine ambition puff'd, makes mouths at the invisible event, exposing what is mortal and unsure to all that fortune, death, and danger dare, even for an eggshell." In spite of this less than glowing opinion, however, Hamlet, with his dying breath, agrees to elect that same Fortinbraccio as the new prince of Denmark, saying to Horatio as he dies: "He has my dying voice."[49]

Both Bruno and Shakespeare lived before the definitive establishment of new political forces such as the English Parliament, which would assume power over the nation by subjecting the king himself to its laws in the not too distant future. We see them bowing before the monarchs of their time, although not without doubts, exasperation, and hesitations. From their words, it becomes clear that they harbored uncertain hopes of witnessing the rise of new and wiser princes, less corrupt than those who were decreeing the solitutude and desperation of men who attempted to follow new and daring paths of thought. Attacking the "madmen" who opposed them was precisely one of the instruments used by the princes of the time to control and exile new ways of thinking. For they saw the danger of those who denounced the old world and its ways, gradually corroding its structure of tattered concepts and of power. Faced by such a strategy, Bruno and Hamlet in Shakespeare's tragedy reverse the terms of play, making their own assumed folly into a weapon for counterattack. The subtlety of their thought and the sharpness of their wit constitute an assault on everything rotten and corrupt that they find blocking their path. It is clear that they will have to pay a price for this with their lives. Yet even in the supreme moment of death, there is no question of renunciation. On the contrary, they desire, with their story and their words, to make their mark on the new century appearing on the horizon. In Rome, Bruno, faced by the secure prospect of being burnt at the stake, will refuse to retract, declaring publicly to his judges on February 9, 1600, on being handed over to the secular authorities for his execution: "Perhaps your fears in pronouncing this sentence on me are greater than mine in receiving it."[50] Not many months afterward, a dying Hamlet, on the London stage, will trust his story to his friend Horatio, so that his "madness" should not be forgotten:

> If thou didst ever hold me in thy heart,
> Absent thee from felicity awhile,
> And in this harsh world draw thy breath in pain
> To tell my story.[51]

NOTES

1. For a critical bibliography of the Shakespeare–Bruno discussion from its initial stages in the nineteenth century to the present day, see Gatti (1989), appendix II, 168–88.

2. See Bruno (2002), vol. 1, 625.

3. For Queen Elizabeth as Astraea, see Frances Yates, *Astraea. The Imperial Theme in the Sixteenth Century*, London, Routledge and Kegan Paul, 1975. For Bruno's remarks about Queen Elizabeth I during his trial, see Firpo (1993), 188–89.

4. For the Northumberland collection of Bruno texts, see Gatti (1989), 35–48.

5. Yates (1934), where the Bruno–Florio references mentioned in the following paragraph are documented.

6. See Wyatt (2002) and Michael Wyatt, *The Italian Encounter with Tudor England: A Cultural Politics of Translation*, Cambridge, UK, Cambridge University Press, 2005.

7. See Buono Hodgart (1978) and chapter 8, "Bruno's *Candelaio* and Ben Jonson's *The Alchemist*," in this volume.

8. See Bruno (2002), vol. 2, 178.

9. For a discussion of these pages of Bruno in relation to Shakespeare's idea of a play, see Gatti (2008b).

10. On various aspects concerning the study of Bruno in the nineteenth century, see Canone (1998a).

11. For a modern comment on this phenomenon, see Barbera (1980).

12. See note 1 earlier.

13. See Bruno (2002), vol. 1, 453.

14. Ibid.

15. See Bolzoni (2002).

16. See William Shakespeare, *Hamlet,* ed. Harold Jenkins, New York, Methuen, 1982, Act I, scene ii.

17. See Bruno (2002), vol. 1, 453.

18. For this important aspect of Bruno's English experience, see chapter 1, "Between Magic and Magnetism: Bruno's Cosmology at Oxford," in this volume.

19. See Sacerdoti (1998).

20. See Shakespeare, *Hamlet,* op. cit., Act I, scene v.

21. See Bruno (2002), vol. 2, 212, and Hilary Gatti, " Nonsense and Liberty: The Language Games of the Fool in Shakespeare's *King Lear,*" in *Nonsense and Other Senses: Regulated Absurdity in Literature,* ed. Elisabetta Tarantino, Newcastle upon Tyne, UK, Cambridge Scholars Publishing, 2009, 147–60.

22. See Shakespeare, *Hamlet.,* op. cit., Act II, scene ii.

23. See Bruno (2002), vol. 1, 533.

24. Ibid., 571.

25. Ibid., 456.

26. Shakespeare, *Hamlet,* op. cit., end of Act I, scene v.

27. See Bruno (2002), vol. 1, 455–56.

28. See ibid., 569, and the pages on Bruno's execution in Rowland (2008), 277–78.

29. See Shakespeare, *Hamlet,* op. cit., Act II, scene ii.

30. The canonical account of this discussion is in Yates (1966), 260–78.

31. Shakespeare, *Hamlet,* op. cit., Act II, scene ii.

32. Ibid., Act III, scene i.

33. The comments in the preceding paragraph are based on words pronounced by Hamlet in his speech to the players in Act III, scene ii, and in the closing monologue of Act II, scene ii.

34. The dedicatory letter to Sidney describing the *Argomento* of the *Heroici furori* is in Bruno (2002), vol. 2, 487–500.

35. See Shakespeare, *Hamlet,* op. cit., Act III, scene i.

36. Bruno (2002), vol. 2, 488.

37. See Shakespeare, *Hamlet,* op. cit., Act III, scene iv.

38. See Farley-Hills (1992) and Bruno (2002), vol. 2, 492–93.

39. See Shakespeare, *Hamlet,* op. cit., Act V, scene i. For further possible echoes of Bruno's *Furori* in Shakespeare's *Hamlet,* see Tarantino (2007).

40. See Bruno (2000c), 16.

41. These precise words appear in a Notice of the public reading of Bruno's sentence of condemnation that appeared in a news-sheet published in Rome on February 12, 1600, only a few days before his execution. See Firpo (1993), 347.

42. Shakespeare, *Hamlet,* op. cit., Act II, scene ii.

43. Ibid., Act III, scene ii.

44. Bruno (2000c), 16.

45. Ibid., 907.

46. Shakespeare, *Hamlet,* op. cit., Act V, scene ii.

47. This letter was written by Bruno as a dedication of the entire trilogy but was not published in its first volume, the *De triplici minimo*, which seems to have been hurriedly prepared for the press to coincide with the spring book fair at Frankfurt in 1591. It appeared only the following autumn in the volume that contained the final two volumes of the trilogy, the *De monade* and the *De immenso*. See Bruno (2000c), 231–39.

48. For an Italian translation and comment on this dedicatory letter, see Calogero (1963).

49. See Shakespeare, *Hamlet,* op. cit., Act IV, scene iv, and Act V, scene ii.

50. Bruno's famous last known words were recorded in the highly critical pages of the letter describing the reading of his sentence as well as his execution, written by Kaspar Schoppe on the day of the execution itself. See Firpo (1993), 351.

51. See Shakespeare, *Hamlet,* op. cit., Act V, scene ii.

8

BRUNO'S *CANDELAIO* AND BEN JONSON'S *THE ALCHEMIST*

> The Doctrine of Generation and Corruption unfoldeth
> to our understandings the method general of all atomical
> combinations possible in homogeneous substances ... which
> part of philosophy the practice of Alchemy does much
> further, and in itself is incredibly enlarged, being a mere
> mechanical broiling trade without this philosophical project.

IN THIS PASSAGE FROM THE NINTH earl of Northumberland's *Instructions* to his son, written in the Tower of London, where he was imprisoned in the early years of the seventeenth century, we find the expression of a deeply ambiguous attitude toward alchemy.[1] In the context of the impetuous developments in the new sciences that characterize the early seventeenth century, alchemy was rapidly assuming the role of an outworn discipline, pervaded by ritualistic and linguistic practices of antique origin. Furthermore, it appeared surrounded by mystery due to its obscure and occult symbolism, partly derived from magical and Hermetical influences, and partly developed as a form of defense against ecclesiastical censure. Nevertheless, alchemy was still widely practiced, often supported by the vain hope of transforming base metals into gold. On the other hand, traditional alchemy was showing itself to be susceptible to new, more rational and scientific developments, above all when it became associated with new philosophical projects, such as the reproposal of ancient atomism that is specifically mentioned by Northumberland. It was only when it was unsupported by new and more advanced theoretical doctrine that it appeared to the late humanistic culture of the beginning of the seventeenth century as some deteriorated form of pseudo-science, or "a mere mechanical broiling trade," to use the words of Northumberland.[2]

It is precisely this inferior form of alchemy, bearing within it something not only venial but also intimately false and deceptive, that is proposed by Ben Jonson as the dramatic theme of his *The Alchemist* of 1610. For his suspect "broiling trade" is intimately connected, from the opening verses of his *Argument*, with an uninhibited sexual commerce, only occasionally camouflaged with the high-sounding name of "love." Jonson's comedy thus represents an urban scene dominated by a materialistic exuberance, where there is little time for penitence, and where the flagrantly decadent corruption becomes the sign or symbol of a deeply rooted social disease. "The sickness hot" are the opening words of Jonson's *Argument*, which immediately evokes a scene in which the "cheaters and the punks"—that is, the small-time thieves and petty criminals—succeed in taking over for a time a city infested with the plague, as the London of 1609–1610 actually was. And this thick web of deception in a plague-ridden city becomes even more dense and obscure if it is borne in mind, as Johnson's public was probably aware, that the house in which, in his comedy, so many impossible dreams were dreamt of a magical transformation from squalor and poverty into sudden wealth—the house of so many false metamorphoses—was geographically situated in precisely that part of the city occupied by the Blackfriars Theatre. This was the theater of the King's Men, the company of William Shakespeare that had incorporated Jonson's comedy into its repertoire. All of which means, as David Riggs points out in his biography of Jonson, that this comedy is not only rich in contemporary references, but that it becomes a piece of metatheater in which the "deceivers" are a company of actors who have transformed "the house of their Lord" into a theater, and the "victims" that they attract make up the public that runs to Blackfriars in search of an illusory escape from the plague, only to find themselves faced by a grim representation, however amusing, of their own disease.[3]

It is no difficult task, at this point, to observe how the same characteristics that have just been outlined in reference to Jonson's comedy of 1610 were also present in Giordano Bruno's *Candelaio,* a comedy written in Italian and set in contemporary Naples but published in Paris in 1582. The weaving together of a false alchemy and even falser forms of "love" is underlined by Bruno as his theme too, in his *Argomento ed ordine della commedia (Subject and Outline of the Comedy),* which specifies that "the principal subjects which are developed together are the doings of the elderly Bonifacio, 'the insipid lover,' and those of the alchemist Bartolomeo, 'the sordid miser.'" Furthermore, in the *Antiprologo* of his comedy, Bruno too underlines the element of metatheater in his representation of a corrupt society, presenting his play as:

> ... this discarded old boat, ruined, broken, imperfectly tarred, which seems to have been dragged from a profound abyss with boathooks, ramps and pulleys; water seeps into it from every side, for it is quite unvarnished: and you want to put out to sea? You want to leave this safe port of Mantracchio? You want to leave this silent quay?[4]

An ancient and delapidated boat is used as an image to represent a world that is chided (albeit through mirth) for its flagrant decadence, its moral and intellectual degeneration. The same image will return in the second dialogue of Bruno's *The Ash Wednesday Supper* of 1584, where it has become a boat transporting people down the Thames in London, where Bruno had arrived in 1583. The creaking boat appears to him now as "one of the relics of the flood," piloted by "an ancient helmsman from the Tartarean reign." It is symbolic that this leaking London boat fails even to set down its passengers at the required place, but leaves them stranded halfway so that they are obliged to reach their journey's end on foot.[5]

These already considerable similarities in theme and dramatic intent suggest that it would be worthwhile to attempt a more detailed comparison between these two comedies, written little more than twenty years apart, and perhaps even to hypothesize a direct relationship between them. It is true that when this attempt was first made at the beginning of the last century by C. G. Child, the comparison was judged to be untenable by figures as authoritative as the editors of Jonson's complete works. Herford's considerations on the subject in the second volume of the *Complete Works* remain, however, extremely generic and unconvincing, which makes it surprising that they were accepted unquestioningly by Mario Praz in his essay on "Ben Jonson's Italy" published in *The Flaming Heart* in 1958.[6] Praz himself, however, when writing in the same essay on the Italian elements in Jonson's better known comedy, *Volpone*, insisted on underlining the importance, for evaluating the Italian element in Jonson's work generally, of the close friendship between Jonson and John Florio, the Anglo-Italian author of bilingual language dialogues as well as of the first English–Italian Dictionary. Praz fails to mention that Florio had also been Bruno's companion and friend during the years he had passed in the French embassy in London.[7] To this common friendship with Florio, it is necessary to add the strong link betweeen Jonson and the Sidney family, which in the person of the still mythical Sir Philip (although long dead by the time Jonson wrote his comedy), appears to have helped and supported Bruno during his difficult London years. We thus find ourselves faced by an interesting series of personal relationships that make it at least possible, if not probable, that Jonson had some knowledge of Bruno and perhaps of his works, even if only through conversation with those who had known him personally in London.[8]

There is also a suggestive similarity between the intentions that underline the dedication of Jonson's comedy to Lady Mary Wroth, none other than the niece of Sir Philip Sidney, and the dedication of Bruno's comedy to Signora Morgana B., "his ever honorable lady": both dedications to revered women of plays that deal with scandalous subject matter (written, as Bruno declares of his *Candelaio*, "in these hottest times, and in the most oppressive hours, of what they call the dog-days"), almost as if they wished to evoke a higher and more noble concept of love and honor before descending into the "unbelievable chaos" of the lowest quarters of city life.[9] These dedications and notices to the reader placed before the texts of Bruno's and Jonson's comedies are also important where they underline the necessity for a cultured public to approach the obscure subject matter of these plays in the light of an intelligent wisdom and power of judgment. Not that this should be seen as diminishing the comic element in these plays, which aim without the moralizing intentions of a sermon at a lively and realistic representation of behavior and language that is not only crude but often tendentiously criminal. Nevertheless, as Bruno underlines on the title page of the text of his play: "*In tristitia hilaris, in hilaritate tristis*" (In sadness mirth, in mirth sadness). So that what he presents to the Signora Morgana is not only "this *Candelaio* which has issued from me," (a formula that does not refrain from slyly alluding to the obscene implications of such a title, given that the "candela" in Italian suggests the male sexual organ) but also, and primarily, the "candle," or the flame of intelligence and wisdom with which to judge the sense and meaning of his drama.[10]

Bruno goes on to claim that just that candle can illuminate certain "Shadows of Ideas" that at that moment were "terrifying the very beasts and, as if they were Dantean devils, leaving the asses far behind."[11] This phrase forges the well-known link that Bruno establishes between his comedy and the first of his philosophical works, *De umbris idearum*, also published in Paris in 1582—both of them texts based on the idea of a world of unceasing vicissitude whose shadowy metamorphoses constantly threaten to degenerate into folly, deceit, and crime. In the verses to Merlin, the Sober Judge, in the opening pages of the *De umbris idearum*, this vicissitude is represented through the image of the great river of life: "if you drink of it unsoberly," Merlin warns, "it will possess you to such a point that you will vomit your soul, and you will never drink of it again."[12] In terms very similar to these, Ben Jonson, in the verses of his prologue to his play, invites his public to watch with attention the flow of the river of human folly in London, without drawing back from its most obscure pools and eddies, because—Jonson claims in a notice to his readers—the light of a balanced judgment illuminates with greater clarity a deep and troubling darkness than it does a weak shadow.[13]

It is precisely with respect to the Renaissance perception of a universe by now extended to infinite dimensions and involved in a process of eternal and disturbing vicissitude, that alchemy appears in the dramas of this period as the source of innumerable metaphors of the continuing transformation of the changing and often deceitful appearances of the phenomena—metaphors that involve the theatrical "transformation" of reality itself. The theme has been well treated by Charles Nicholl in his book on *The Chemical Theatre*.[14] In his pages on *The Alchemist*, Nicholl insists on the central importance for Jonson of the book by the medieval alchemist George Ripley titled *The Compound of Alchemy*, where the "great work" of the alchemists is likened to the image of the Philosophical Wheel: "to win to thy desire thou needst not be in doubt, / For the wheele of our Philosophie thou hast turned about."[15] Jonson's alchemist also, to whom he gives the suggestive name of Subtle, precisely to ironize on the continual alchemical effort to render the base metals more "subtle," and therefore more precious, describes his work as one that turns about the "Philosophers Wheel." Furthermore, it is precisely this image of the wheel of a universe involved in a process of interminable vicissitude (a process that Jonson in a later dramatic interlude written for the court will represent as the figure of an elusive Mercury) that underlies the often quoted final lines of Bruno's dedication of the *Candelaio* to Signora Morgana B.: "if the mutation is true, I who am in the night, am waiting for the day, and those who are in the day, are waiting for the night. Everything that is, is either here or there, either near or far, either now or then, either early or late. Enjoy it, therefore."[16] And it is perhaps this new sense of finding oneself involved in an obscure natural mutation, always continuing and never concluded, and the reference to alchemy as the traditional discipline that had attempted most intensely to capture the secrets of its transformations, that explains the considerable alchemical culture of both Bruno and Jonson, in spite of their criticisms of its more shady aspects and deceptions.

Bruno's characters quote with ease from the *Tractatus aureus* of Hermes Trismegistus, the *Liber mineralium* attributed to Albertus Magnus, as well as the alchemical texts of Avicenna and the pseudo-Geber. Jonson's Subtle appears more up-to-date, citing also less scholastic authors such as Paracelsus (known obviously to Bruno too), John Dee and his assistant Edward Kelley, as well as some alchemical works published from 1598 onward by Lazar Zetzner of Strasburg, who was responsible also for the republication of those works of Bruno that were inspired by the Catalan mystic Raymond Lull.[17] In fact, Jonson seems to have searched for alchemical and magical texts for his library, given that he possessed a fifteenth-century manuscript titled *Opus de arte magica* attributed to King Solomon himself, which is mentioned in his play as a book of profound wisdom by Sir Epicure Mammon—the name given by Jonson to the character who is most easily duped by the false promises of Subtle and his companions.[18] Perhaps Nicholl is not entirely justified when he affirms that Jonson dug deep into the literature of alchemy only to furnish himself with weapons to deride it, although a similar accusation has been made with respect to Bruno by Massimo Bianchi in a paper on the presence of Paracelsus in Bruno's works.[19] It remains nevertheless undeniable that the practical aim of alchemy is considered in these two plays to reside in the absurd attempt to produce gold and silver from nothing—a theme that permits the two authors to develop a ferocious satire of the greed inherent in the emerging capitalism of their age.

Jonson would surely have had in mind Chaucer's *Canon Yeoman's Tale* as an authoritative earlier expression of this inferior alchemy, or "slyding science," as Chaucer calls it—a synonym of what Jonson himself, using an antique English vocabulary, calls "cosenage," and Bruno, remaining faithful to the Italian, or rather the Neopolitan, vulgate calls "mariuoleria." Another literary source, very probably common to both Bruno and Jonson, can be found in one of the *Colloquies* of Erasmus titled *Alchemia*.[20] Here too a so-called alchemist sells to an ingenuous buyer, who has "gone mad" with love of alchemy, the secrets of his "sacred art." For much gold is necessary, Erasmus comments sarcastically, in order to produce the—so to say—"secondary" gold of the alchemists: a comment that is repeated almost literally by Bruno's artist, Gio. Bernardo, in scene 11, act I, of the *Candelaio*. Here he says to Cencio (whose name means "ragamuffin" in Italian), the alchemist Bartolomeo's lurid assistant:

> I would like to see gold made and you better dressed than you are now. I am convinced, however, that if you knew how to make gold, you would not sell the recipe for making it, but you would make it in earnest.[21]

It is more important here, however, to underline those moments in the two comedies where Bruno and Jonson go beyond the moralistic dimension to which Erasmus's colloquium is confined in order to give expression to an almost poetical dimension of dream that underlies the undoubtedly opaque commerce taking place in the dark places of their theatrical cities. Their characters dream fantastical dreams of a new material wealth that opens up new historical horizons, conjuring up visions of a general affluence so far unknown—or at any rate known only to those whose lives placed them well outside the social dimensions of Renaissance comedy.[22]

It is in the first scene of the third act of the *Candelaio* that Bruno's Bartolomeo sees his alchemy, albeit so far of little productive value, as part of an almost poetical vision of a better future:

> Metals such as gold and silver are the source of everything: these and only these give rise to words, herbs, stones, linen, wool, silk, fruit, corn, wine, oil; everything desirable upon earth depends on them. For this reason I say they are totally necessary, for without them none of these things can be known or possessed.[23]

This celebration of the potential residing in what Bartolomeo calls "il denaio" (which means simply "money") can be compared to the remarkable lyrical outbreak of Sir Epicure Mammon in the second scene of the second act of *The Alchemist*. Here the unrepentant materialism of this merry knight, convinced by Subtle that he has already found the elixir of a richer and more pleasurable future, is expressed in an irrepressible wish for succulent things to eat out of precious dishes—a wish that indicates how, in the society of that time, hunger was still a factor to be reckoned with. Sir Epicure's enthusiasm for food is made even more lyrical by his intention to share his new wealth with his page and his cook:

> My footboy shall eat pheasants, calvered salmons,
> Knots, godwits, lampreys: I myself will have
> The beards of barbels, served instead of salads,
> Oiled mushrooms; and the swelling unctuous paps
> Of a fat pregnant sow, newly cut off,
> Dressed with an exquisite and poignant sauce;
> For which, I'll say unto my cook, there's gold,
> Go forth and be a knight.[24]

The fact is that it would be a vain undertaking to search either in Bruno or in Jonson for the almost monastic rigor of the spirituality of Erasmus. They already partake of a changing cultural atmosphere that will soon be dominated by Francis Bacon who, in his *New Atlantis*, published posthumously in 1627, will dream of a new society created by an emerging science. It was to be a society based on a desire for affluence and for a general rise in standards of living, which Bacon intentionally contrasts with the decorous poverty proposed as an ethical ideal a century earlier by the great friend of Erasmus, Thomas More, in his *Utopia*. Like Bacon, neither Bruno nor Jonson appear to be perturbed by the desire for a more wealthy future as such, which inspires their characters with a new and inventive energy.[25] Rather, they use the space of the theater to satirize the negative aspect of such desires: an ever more crowded city in which the most pressing urge is to achieve wealth at once, not through a truly productive process but through complicated deceptions practiced on those who already have something to lose. These are what Bruno calls "the stratagems of cheaters, or criminal affairs."[26]

In these "quartan fevers," or "spiritual cancers" (the terms are again those of the *Proprologo*, or the first of two prologues to the *Candelaio*), a major element is to be found, in Jonson's comedy as well as in Bruno's, at the level of false language, or in the characters' use of words to cover up the outrageous lack of, or falsity, of facts. Here the art of alchemy offers the two dramatists a vast reservoir of terms, symbols, and metaphors, at the same time colorful and obscure, with which to develop what Jonson's Subtle calls with a flagrant boast: "Alchemy ... a pretty kind of game, / Somewhat like tricks o' the cards, to cheat a man, / With charming."[27] However, other languages are also accused by the two dramatists of an intimate and hypocritical falsity. In Bruno's comedy, it is above all the absurdly antiquated, often pseudo-Latinized, language of the late humanist pedant Manfurio, with his artficial Petrarchisms—a type that had by that time become a familiar theme in Italian sixteenth-century satire.[28] Nonetheless Bruno represents this kind of pedantry with masterly linguistic verve, and it will remain as a comic element of success in his Italian philosophical dialogues written later in London. According to a critical tradition of some standing, Shakespeare may have had Bruno in mind when creating some of his own most ludicrous pedants, such as Polonius in *Hamlet* or Malvolio in *Twelfth Night*.[29] Yet in spite of these many Renaissance precedents, Jonson in *The Alchemist* turns his back on the figure of the humanist pedant, who had perhaps by that time exhausted his theatrical role, and proposes instead, as an alternative pair of London scoundrels, the two radical Protestants, indicated specifically as Anabaptists: Tribulation Wholesome and Ananias.

Ben Jonson shared with Shakespeare a profound resentment toward those Protestant Puritans who were ever more violently attempting to put an end to the theatrical life of Jacobean England. For his part, Jonson had remained a Catholic up to 1610, the year of composition of *The Alchemist*, before moving over to the moderately Protestant Anglican Church—a choice clearly dictated by motives of political convenience. So it is no surprise to find his two Puritans satirized mercilessly for their wordy and hypocritical spirituality that fails to save them from falling prey to the facile promises of quick riches that the false alchemists circulate throughout the city. They are figures that find no counterparts in the Neapolitan setting of the *Candelaio*. Moreover the Protestant cultures of the north had by then expressed an ample gallery of literary models that Jonson could look to in formulating the figures of these two religious fanatics—it is enough to think of some pages of Thomas Nash. Nevertheless, it cannot be excluded that they derive at least in part from the ferocious caricature of a Protestantism ignorant of every text except that of the Bible that Bruno develops in some passages of his Italian dialogues written and published later in London, between 1583 and 1585.[30]

The final part of the plot of the *Candelaio* is somewhat fragmented and even at times confused. Nevertheless, Bruno develops with great clarity the theme of the necessary punishment of the various dishonest characters on the part of those who possess the "candle" of moral judgment. The character who now emerges as the moralist is the artist Gio. Bernardo, who knows how to represent in his pictures, and therefore how to judge, the "unbelievable chaos" that surrounds him. On the other hand, not even Gio. Bernardo can be considered as completely extraneous with respect to the relativization of moral values that dominates the chaotic urban scene, heedless of rules and regulations. For once the punishments are over, he attempts and succeeds in persuading Carubina, the youthful wife of the aging Bonifacio, to yield to his "fervent love." Gio Bernardo succeeds in his seduction by indulging in a witty "deconstruction" of the traditional idea of public "honor" in a speech that can be compared with the later development of the same theme on the part of Shakespeare's Falstaff. As for Jonson, he goes well beyond this hint of Bruno's concerning the imperfections of all human attempts to judge others. For Jonson writes a last act that completely upsets the audience's expectations of a "poetic justice" in the final moments of the play. His closing scene unexpectedly brings back to London the owner of the by now ill-famed house in which the shady dealings of "love" and "alchemy" had been perpetrated. As soon as the owner arrives, the small-time thieves who had populated the scene during the comedy are immediately thrown out into the street where they had come from, together with their ingenuous victims. But a very different treatment is meted out to the owner's principal servant, with the suggestive name of Face, who had elaborated the entire criminal plot, as well as to the richest of the widows whom the hypocritical Face had managed to bring into his shady orbit. Instead of being punished, they are promoted, respectively, to the right-arm man and wife of the owner—a gesture that clearly signals a rising tide in the already widespread corruption rife throughout the city. From the obscure places of petty crime, corruption now enters the palaces of power and becomes a smiling part of the system—and once established there, Jonson is clearly warning his public, it is not going to be easy to uproot it.

NOTES

1. See H. Percy, Ninth Earl of Northumberland, *Advice to his Son,* ed. G. B. Harrison, London, Ernest Benn, 1930, 70. For a detailed study of the Renaissance discussion about alchemy, see *Alchimie e philosophie à la renaissance,* eds. J. C. Margolin and S. Matton, Paris, J. Vrin, 1993.

2. It should be remembered that Northumberland held in his library one of the most notable contemporary collections of Bruno texts, which undoubtedly played their part in the scientific investigations that were being carried out by some prominent members of his household. For their relationship with Bruno's thought and works, see Gatti (1989), 35–73.

3. David Riggs, *Ben Jonson: A Life,* Cambridge, MA, Harvard University Press, 1989, 170–71.

4. All English quotations from the *Candelaio* are my translations established on the basis of the text in Bruno (2002), vol. 1, 259–424.

5. See Bruno (1977), 111–12.

6. Child's article, published in *The Nation,* New York, July 28, 1904, is dismissed as of little or no value by the editors of *The Complete Works of Ben*

Jonson during a discussion of the *Candelaio* as a possible source for Jonson's play. The *Candelaio* itself is criticized negatively in these pages as a play showing "neither originality nor mastery": not an opinion that Bruno scholars would accept today. See *The Complete Works of Ben Jonson*, eds. C. H. Herford and Percy and Evelyn Simpson, Oxford, UK, Clarendon Press, 1925–1952, vol. II, 94–98. Mario Praz adds in *The Flaming Heart*, New York, Doubleday, 1958, 177, note 3, that the similarities between the two plays are very vague and the plots very different, without illustrating his point.

7. M. Praz, "Ben Jonson's Italy," in *The Flaming Heart*, op. cit., 168–85.

8. We are told by the Scottish humanist William Drummond, a contemporary and friend of Jonson, that the English playwright could not read Italian. Nevertheless, the particular esteem and respect felt by Jonson for John Florio is expressed in his dedication written in the copy presented to Florio of his best known drama *Volpone*: "To his loving Father, and worthy friend Mr. John Florio: the ayde of his Muses. Ben: Jonson seales this testimony of Friendship, & Love" (quoted by Praz in *Teatro elisabettiano*, Florence, Sansoni, 1948). It is well known that Florio refers repeatedly to Bruno in his various works. Also well known is the link between Bruno and Sir Philip Sidney, to whom he dedicated two of his philosophical dialogues written in London between 1583 and 1585. For further details of these relationships, see Yates, (1934).

9. Vincenzo Spampanato, in the *Introduzione* of his text of the *Candelaio*, Bari, Laterza, 1909, identified the Signora Morgana B. as a Nolan woman whom Bruno may have loved as a young man. She eventually married Gian Tomaso Borzello. This identification is accepted by the editors of Bruno (2002), even if Giovanni Aquilecchia has expressed some doubts about it in other places, claiming that the dedication is ambiguous and leaves it uncertain whether Signora Morgana B. is to be considered as a lady or a courtesan; see Aquilecchia (1993a), 339. Lady Mary Wroth, for her part, would become a writer herself, publishing in 1621 the volume: *The Countesse of Montgomeries Urania, Written by the Right honourable the Lady Mary Wroath, Daughter to the right Noble Robert Earle of Leicester and Neece to the ever famous and renowned Sr. Phillips Sidney.*

10. In his "Saggio di un commento letterale al testo critico del *Candelaio*" of 1991, now in Aquilecchia (1993a), 327–66, Giovanni Aquilecchia claims that the title of the comedy creates a semantic equivocation insofar as the meaning of "candeliere" in a metaphorical sense of a bearer of the light of truth alternates with a euphemistic sexual metaphor in which the word "candela" signifying the male sexual organ refers both to the scandalous subject matter of the comedy itself as well as to the pederasty of its principal character. See 339, note 1.

11. Many commentators have written on the link established by Bruno himself between the text of his play and the first of his philosophical works, a treatise on the art of memory, published in Paris in the same year. See in particular Badaloni (1988), 33–34.

12. See the poem titled *Merlinus iudici sobrio (Merlin to the sober judge)* in Bruno (2004), 10–13, and the comment by Nicoletta Tirinnanzi at p. 386.

13. "... lights are more discerned in a thick darkness than a faint shadow." All quotations from Jonson's play are taken from *Ben Jonson: Five Plays*, Oxford, UK–New York, Oxford University Press, 1988, 349–482.

14. C. Nicholl, *The Chemical Theatre*, London, Routledge and Kegan Paul, 1980.

15. Ibid., 99–102.

16. Bruno (2002), vol. 1, 264. For this aspect of Bruno's work, see in particular Ciliberto (1986).

17. The alchemical sources used by Jonson have been studied extensively by the editors of his complete works. See *Ben Jonson: The Complete Works*, op. cit., vol. X, 1950, 46–116.

18. The manuscript is held by the British Library in London, press mark BL SLOANE 313.

19. "He dug deeply into the literature of alchemy, almost exclusively for ammunition to use against it," in Nicholl, *The Chemical Theatre*, op. cit., 102. Bianchi's note is in Canone ed. (1992), 143–46. For a more nuanced view of Bruno's position, see Vedrine (1993).

20. This colloquy, together with others that comment on the falsity of alchemy, was first published in the edition of the colloquies printed between August and September of 1524 by Johannes Froben at Basle.

21. See Bruno (2002), vol. 1, 300.

22. For this aspect of Jonson's drama, see Anne Barton, "The Alchemist," in *Ben Jonson Dramatist*, Cambridge, UK, Cambridge University Press, 1984.

23. Bruno (2002), vol. 1, 323.

24. In *Ben Jonson: Five Plays*, op. cit., 386.

25. A debatable but interesting interpretation of the whole of Renaissance culture in terms of the desire for worldly goods is developed by Lisa Jardine, *Worldly Goods: A New History of the Renaissance*, London, Macmillan, 1996.

26. In the *Proprologo* to the *Candelaio*. See Bruno (2002), vol. 1, 276–81.

27. *Ben Jonson: Five Plays*, op. cit., 394.

28. Numerous comments on the pedants who populate the Italian comedies of this period can be found in Nino Borsellino, ed., *Commedie del cinquecento*, Milan, Feltrinelli, 1967, 2 vols. For a comment on Bruno's Neapolitan pedants in particular, see the section on the *Candelaio* titled "Il drama dell'ignoranza: il *Candelaio* tra realtà e apparenza," in Ordine (2003), 45–70.

29. A bibliography of the Bruno–Shakespeare relationship can be found in Gatti (1989), appendix II, 168–88.

30. Bruno's attack on English pedantry is a constant theme of the pages on Bruno's English experience in Rowland (2008), 139–87.

9

BRUNO AND THE STUART COURT MASQUES

IT HAS LONG BEEN KNOWN THAT Bruno's fourth Italian dialogue, *Lo spaccio della bestia trionfante*, written and published in
London in 1584, was used as a source by Thomas Carew for his
masque *Coelum britannicum*.[1] This was Carew's only masque but it was
by no means a minor event within the Stuart calendar of court entertainments. However, in spite of general agreement on the quality of *Coelum
britannicum* as one of the major entertainments of the Stuart Court, the
use by Carew of Bruno's dialogue has never been extensively or satisfactorily commented on. Both Bruno and Carew scholars have clearly been
ill at ease with the relationship and have tended to dismiss it with a few
brief and evasive remarks.

There are few contributions of any significance to what is still a very
fragmentary discussion. In 1949, Rhodes Dunlap, in his edition of *The
Poems of Thomas Carew*, supplied in his commentary on the masque a
useful, if partial, list of the passages in the *Spaccio* that Carew took over
and integrated into his text.[2] In 1964, Frances Yates referred to Dunlap's
work in her book on *Giordano Bruno and the Hermetic Tradition*, raising the question of the Bruno–Carew relationship briefly in a chapter
on Bruno and Tommaso Campanella. Yates thought that the two Italian
philosophers shared a common mystical cult of the French and British
monarchies that Carew incorporated into his masque.[3] In 1973, Orgel
and Strong raised the question of Carew's source in their essay "Platonic
Politics" in the edition of *Inigo Jones: The Theatre of the Stuart Court*.[4]
Orgel and Strong accepted without question Yates's Hermetic interpretation of a Bruno with pronounced mystical leanings. Carew's masque, on
the other hand, was read by them as a powerfully poetical but also keenly
intellectual Machiavellian as well as Neoplatonic celebration of absolute monarchy, far removed from Bruno's esoteric and occult mysticism.
Orgel and Strong concluded that Carew was only superficially interested
in Bruno's text; he took from it only the fable, while the meanings of the
masque remained original to Carew and Inigo Jones.

The widespread influence of this much-quoted essay by the major authorities on the Stuart masque has had the effect of quelling further discussion of Carew's use of Bruno. Subsequent scholars have rarely bothered
any longer even to name Bruno as a source in their discussions of Carew

and his masque. Among the few notable exceptions are Annabel Patterson's interesting pages on "Thomas Carew: 'A Priviledged Scoffer?'" in *Censorship and Interpretation: The Conditions of Writing and Reading in Early Modern England*, where the ambiguities of Carew's masque are underlined. A few years later, John Kerrigan's British Academy lecture on Carew recognized the political sympathies expressed in Bruno's *Spaccio* as a classical republicanism filtered through Machiavelli. Kerrigan notes that Bruno's sense of universal vicissitude renders Jove and the Olympian gods subject to fate and decay. Added to the openly heretical and anti-Christian polemic that is such a notable aspect of Bruno's dialogue, Carew's choice of source, in Kerrigan's opinion, hardly promises "a celebration of that royal asterism, the King as Defender of the Faith."[5] To these examples may be added the brief remarks by Joanne Altieri in her essay "Carew's Momus: A Caroline Response to Platonic Politics," where the "prevailingly unpanegyric eye" of Bruno's Momus is seen as the inspiration for what the author considers as Carew's brilliant but at the same time ambiguous undercutting of Mercury's Platonic idealizations. Italian scholars have almost completely ignored the relationship.[6]

This chapter will attempt a more searching and widespread inquiry into the presence of Bruno's philosophical dialogues in the fragile, refined, and essentially illusory world of the Stuart court masque. What political considerations, literary choices, or possible misreadings led to the unlikely intrusion of the Nolan philosophy into the elegant vistas of Inigo Jones's royal banqueting hall in Whitehall? For Bruno's attitude to the courts and monarchs of his time was disturbingly ambivalent. It is known that during his years in London between 1583 and 1585, Bruno moved, a little uneasily, on the outer edges of the radically Protestant, aristocratic circle of the earl of Leicester and his nephew Sir Philip Sidney.[7] Bruno's first Italian dialogue written in England, *La cena de le ceneri* (*The Ash Wednesday Supper*), which argues for a post-Copernican infinite universe, mentions Leicester's kindness and generosity and warmly praises the cultural and courtly brilliance of Sidney and his friend Fulke Greville.[8] The *Spaccio* itself, and Bruno's last Italian dialogue written in London, *Heroici furori* (*Heroic Frenzies*), both carried important dedicatory letters to Sidney.[9] Bruno further tells us that he was received more than once at court by Elizabeth I as one of the gentlemen attendants of the French Ambassador Michel de Castelnau, and his praise of the "diva Elizabetta" was so ardent that it was brought up against him, in Italy, at his trial.[10] Some years after his departure from England in 1585, in the first work of his Latin masterpiece known as the Frankfurt trilogy, the *De triplici minimo* published in 1591, Bruno makes a glowing reference to the marriage of James VI of Scotland with Anne of Denmark in 1589, and he presents the future James I of England as one of the heroes of his times.[11] On the other hand, Bruno is careful not to present himself in a courtly role in his works. Rather he tends to underline his humble origins and his clerical poverty. In *La cena*, he presents himself as unassuming and unkempt, with buttons missing on his jerkin, creating a satirical contrast with his bejeweled and begowned opponents from Oxford in the cosmological debate that the dialogue describes. In the last work of the Frankfurt trilogy, the *De immenso*, he takes leave of his reader in the guise of a virile satyr, claiming that nature made him rough and shaggy, and that he will never learn how to speak in affected tones.[12] These are words that seem to define Bruno as a critic of the decadent and declining Renaissance court. They suggest that, had he survived into the seventeenth century, he might even have had some sympathies with those unquiet spirits who in 1633, when Carew wrote his masque, were already pressing in around the magic circle of Whitehall, and would later claim the head of its defeated king.

This ambivalence on the part of Bruno would make him a strange presence indeed in the world of the masque if its Platonic politics exhausted themselves in a linear and ever-repeated celebration of royal power and its divine authority, as the interpretations of Orgel and Strong tend to suggest. Recent readings of many of the masques, however, especially of the Caroline period, have tended to notice the subversive tensions playing under the surface of the apparently smooth and unclouded fabric of the masquers' world. It is necessary to bear in mind the severe pressures to which poets and artists were subjected in those years of absolute monarchical rule, when criticism and political discussion could only be attempted in oblique and muted forms. And to bear in mind too the particular characteristics of the masque form, with its codified moves and messages, and elaborate, spectacular rituals that nevertheless could be, and sometimes were, stretched at the seams to include indications and variants strangely at odds with the necessary celebration of monarchical power. Seen in these terms, it seems to me possible to see the reference to Bruno and the Nolan philosophy as a deliberate and significant one: and not only on the part of Carew. I suggest in this chapter that Bruno's presence can be detected in other Stuart masques as well, going back to their beginnings in the reign of James I. I also try to show that Bruno's ethical concept of correctly wielded political power—his reforming zeal that lay behind the purge of the corrupt constellations of the zodiac in the *Spaccio de la bestia trionfante* or the heroic pursuit of knowledge in *Heroici furori*—provided the masquers who turned to his texts with powerfully suggestive imagery and themes.

In an early but important and influential essay on "The Emblematic Conceit in Giordano Bruno's *Eroici furori* and in the Elizabethan Sonnet Sequences," Frances Yates was one of the first to point out that parts of

Bruno's *De gli eroici furori* read like a Renaissance emblem book, with the difference that the emblems are described in words instead of in pictures.[13] It could equally well be pointed out that the final sequences of the *Furori* are cast in a form that seems to derive directly from the masque. I refer to the fifth dialogue of the second part, where the nine heroic lovers who are searching for divine truth, after being blinded by the enchantress Circe, finally reach the temperate and healing shores of the river Thames. Here the chief nymph of the region—who can certainly be identified with Queen Elizabeth I, although the identification is not made explicitly by Bruno in his text—breaks the seal of a vase given to the blind men by Circe and sprinkles their eyes with its healing liquid. The blind men's sight is restored, and they gaze into the two eyes of the nymph who reveals to them the double truth of heaven and earth, body and mind, the physical and the metaphysical perceived as a single principle of unity and truth. This is Bruno's version of the ultimate good that, as he underlines, is an ultimate good *on earth*, for, as he had already insisted in *La cena de le ceneri*, the divinity is within us, closer to us than we are to ourselves. Uplifted by this revelation of a divine truth that lies about and within them, the nine lovers sing a song that expresses their new sense of the harmony pervading all things. This is not the harmony of the spheres of Neoplatonic philosophy, but an earthly harmony that the lovers create themselves, each on his chosen instrument and each singing a song of his own composition. Finally to song they add dance, wheeling around in a circle of ecstatic praise of the unique nymph and the infinite universe. This comprises the reigns of both Ocean and Jove whose treasures run parallel to each other in a process of eternal vicissitude and change.[14]

It seems to me probable—although as far as I know the suggestion has never been made—that this final sequence of *Heroici furori* was suggested to Bruno by Balthasar de Beaujoyeulx's *Balet comique de la royne* presented at the French court of Henry III in Paris, in 1581, when Bruno was there.[15] The entertainment was more a masque in the English sense than a ballet, with a written text and illustrations that were published soon after the event. This means that Bruno could have studied the *Balet comique* in detail, even if his status in Paris was not such as to admit him to the court for the event itself. In either case, he would have found a masque based on the concept of a struggle between Circe and the higher virtues.[16] The entertainment began with the introduction of a mythological *tableau* figuring Circe in a sumptuous garden, with Pan just outside its confines playing on his pipes in a little wood. Beyond the garden and the castle of Circe could be seen in the distance the streets of a town intersecting at a focal point in line with the base of the throne from which the king watched the entertainment. The perspective of the scenography thus underlined, according to a long-established tradition, the centrality of the presence of the monarch within the world of the court drama. In the course of the long and complicated mythological action, interspersed at frequent intervals with music and dance, Circe managed to enchant the Naiads and with them Mercury, the messenger of the gods. To solve the crisis, Jupiter and Minerva appeared on the scene and, with the aid of Pan and his rough satyrs, defeated the enchantress with their divine powers, although allowing her to remain free within the precincts of her garden. Released from the enchantments of Circe, the Naiads then combined in an intricate geometrical dance during which they managed always to remain facing the figure of the king. The French queen then presented the king with a gold medal figuring a dolphin swimming in the sea, while the other ladies of the court presented their gentlemen with gifts that were all "things of the sea" decorated with representations of the sea nymphs. The sea imagery that dominated the central sequences of the entertainment had the function of reminding the king and his courtiers of their universal dimension by linking them to a process of perpetual generation and mutability: the great ocean of being. The dolphin, by leaping out of the sea, frees itself from the general process of mutability, showing its back in a momentary flash of higher being. Roy Strong sees in the dolphin image an allusion to the desired birth of a royal son and heir. Bruno himself, however, in an interesting discussion of some recurrent Renaissance emblems, sees the dolphin as representing philanthropy, or a benevolent love of man and the universe.[17] In the final sequences of the French court entertainment, Circe has been defeated by the royal presence, and her garden, purged of base mortality, appears in its full glory, flanked by high towers, and decorated with sparkling diamonds.

This mythical and emblematic entertainment of the French court was one of the most elaborate and discussed of the last decades of the sixteenth century, and it is highly probable that Bruno had it in mind when he composed the final sequences of the *Furori*. His use of his source, however, was subtly critical. Bruno's unique nymph remains herself a part of—indeed, a symbolic expression of—the world of corruptibility and mutability that the blind men rediscover in all its variety and vitality under her guidance. Like Bruno's Jupiter, in the *Spaccio*, she is herself involved in a universal process that is seen as the very process of life itself. She might dominate the sphere of history and time for an instant, and if she does that with intelligence, her contribution to history will be great, while the process of metempsychosis may allow her, through industry and wisdom, to rise even further in the universal scale of being. But she can never escape from the vicissitude and play of contraries that define a universe conceived now as the infinite expression of infinite plenitude. Bruno goes further. The songs of the seventh and eighth blind men conceive of the process of infinite vicissitude as essentially a revolutionary process. Just as day follows night, and knowledge progresses revealing what earlier was hidden, so the same vertiginous process of change and decay "suppresses the eminent and raises the lowly." This reelaboration, within a philosophical context of an infinitistic naturalism, of the traditional image of Fortune's wheel, imposes on the figure of the monarch an undeniably human dimension: if the monarch is a representation of divinity, so, to some degree, is everybody else, and indeed every aspect or atom of the infinite whole. Elizabeth I was no fool, and she appears to have understood very well what Bruno was getting at. In the only recorded remark that she made about him, she called him a man of no faith and no respect.[18]

All this would seem to suggest that Bruno had no role to play in the Stuart court masque, and that if he did end up there, it was through a process of misreading and misunderstanding. But I shall be arguing that this was not the case. One of the first masques in which his presence can, I believe, be traced is *Tethys Festival* by Samuel Daniel, and it so happens that Daniel was one of the few components of the Stuart court who had almost certainly known Bruno during his years in London and was in a particularly favorable position to have read with care, and understood, the terms of his philosophy. One of the few well-known and indisputable references to Bruno in England in contemporary English texts is to be found in the anonymous preface, signed N.W., to Daniel's translation from the Italian of an emblem book, *The Worthy Tract of Paulus Iovius*, published in London in 1585. Bruno's name is invoked here at a literary rather than a philosophical level: he is recalled as a scholar who, while speaking at Oxford, had defended the importance of translations. Daniel was an Oxford man, and although the reference to Bruno has to be attributed to the still unidentified N.W., its prominence seems to suggest that Bruno himself had played a part in stimulating the translation made by Daniel, who may well also have been present at Bruno's Oxford lectures. Later he would marry the sister of John Florio, the closest of Bruno's friends in London and, as his English–Italian dictionary witnesses, an enthusiastic reader of Bruno's works.[19] There are further biographical details that suggest that Daniel might have kept up a relationship with Bruno after his departure from England. Bruno left London in October 1585, in the retinue of the French ambassador, who was recalled to Paris for diplomatic reasons at that time. We know that Daniel too was in Paris from the end of 1585 until August 1586—a period that exactly coincides with Bruno's brief second stay in that city. Later Daniel traveled to Italy, and his knowledge of the language was undoubtedly sufficient for him to read Bruno's works with care.[20] Daniel's philosophical poem, *Musophilus*, containing a general defense of learning, which was published in London in 1599, is in the form of a dialogue between the poet

Musophilus and a figure called Philocosmus. The poem is dedicated to Fulke Greville, in whose house Bruno's Ash Wednesday supper, with its dramatic defense of the Copernican cosmology, is supposed to have taken place.[21] Daniel's poem is full of echoes of Bruno's works. It contains the idea of a heliocentric, infinite universe and propounds an ideal of learning familiar to the reader of the *Heroici furori*. Daniel's men of learning "set their bold *Plus ultra* far without / The pillars of those *Axioms* age propounds. / Discov'ring dayly more and more about, / In that immense and boundlesse Ocean / Of Natures riches, never yet found out."[22]

It is well to remember that at the end of the sixteenth century the concept of a boundless or an infinite universe, which contradicted in full the still officially accepted Aristotelian–Ptolemaic cosmos of the closed spheres, was closely connected with the name of Bruno. Scientists such as Kepler in Prague, or Thomas Harriot and the group around the ninth earl of Northumberland in England, all discussed the question of the infinity of the universe through an explicit reference to Bruno's cosmological dialogues.[23] For in works such as his *De l'infinito*, written and published in London in 1584, and his *De immenso et innumerabilibus*, published in Frankfurt in 1591, Bruno had been the first to expand Copernican heliocentricity in terms of an infinite universe containing an infinite number of worlds. The Lucretianism of these works is clear and explicit. Indeed the *De immenso*, which is largely a scientific poem, openly takes the *De rerum natura* as its literary as well as its philosophical model. The metaphor of the infinite universe as an unbounded ocean, although it could have been derived directly from Lucretius himself, is one that is found repeatedly in the cosmological debate of this period in references to, and readings of, Bruno. Someone so closely connected as Daniel was to Florio, who includes Bruno's *De l'infinito* among the texts used for the compilation of his dictionary, can be assumed to have known the modern Brunian derivation of this pregnant metaphor, with its subversive metaphysical implications. For if the natural universe itself is an infinite ocean of divine life and being, a crisis tends to occur with respect to the concept of a transcendental God.

Ocean imagery is the basis of Daniel's second mask, *Tethys' Festival*, in which the queen of King James I appeared as the figure of Tethys, queen of the ocean and wife of Neptune, and her ladies in the shapes of nymphs presiding over several rivers. This mask was presented on June 5, 1610, as part of the lengthy celebrations that marked the creation as prince of Wales of the young Prince Henry, who was soon to die before reaching maturity. Its text was published in the same year with an interesting preface by Daniel himself, who claimed that "shewes and spectacles of this nature are usually registered among the memorable acts of the time, being complements of state."[24] Later, however, Daniel admits to having received "rough censures" for his masque, although he fails to say what kind of censures they were. We do know, however, that he was not invited to compose another masque, and it was Ben Jonson who would become the official court poet, with Inigo Jones as his scenographer. Daniel's effort had clearly not been appreciated. In his preface, he excuses himself as being only a "poore inginer for shadowes" who frames "images of no result"—a remark indicating that Daniel saw his masque in terms of that fleeting, umbral concept of being to which Bruno had given powerful expression in his *De umbris idearum*.[25]

Daniel's spectacle is centered around water imagery, with Naiads and Tritons who sing to musical instruments a celebration of spring as the new prince is created. Tethys, the queen of the ocean, is associated with the "intelligence which moves the sphere of circling waves," and her principal attendant is a "lovely Nymph of stately Thames." The scarf that, on behalf of Tethys, is presented by Zephirus to the prince is clearly associated with the pervading imagery of water and waves, underlining the prince's participation in the sphere of mutability and vicissitude rather than his supernatural or divine attributes. The other gift, a precious, bejeweled sword, symbolizes his power to dominate the ocean of vicissitude with action and with thought—a dualism that is reminiscent of the two shining eyes of the nymph of the river Thames who opens Circe's magic vase and heals the blindness of the heroic searchers in the final sequences of Bruno's *Heroici furori*. In Daniel's masque, as in Bruno's dialogue, the oceanic sphere, when it is comprehended in its infinite lifegiving wonder, is one of "Love and Amitie," and the prince is exhorted not to attempt to move outside the pillars that define its boundaries. Apart from the anti-metaphysical implications of this warning, which are surprising enough in a court masque, there seem to be some unexpected anti-imperialistic emotions at play here. The prince is told not to envy the treasures and riches brought to Spain by her discovery and dominion of new continents, but to search for more certain riches at home: "For Nereus will by industry unfold / A chemicke secret, and turne fish to gold." The new science appears in Daniel's view to offer more than the new colonial adventures—a position that suggests a further influence of Bruno's Italian dialogues where the anti-imperialistic note is a powerful one. For Bruno was convinced that Europe could only take with her to the new world the corruption and disease of her own sick civilization.[26]

After the presentation of the gifts to the prince, Daniel and Inigo Jones mounted a richly elaborate sea scene dominated above by the opening of the heavens, which appeared as three circles of lights and glasses, one within another, and began to move circularly. Daniel's description of this part of the masque emphasizes the use of number symbolism, which has been acutely analyzed by John Pitcher in an interesting study of *Tethys'*

Festival.[27] Pitcher, however, fails to consider this elaborate scenic mechanism as an attempt to represent, through a play of multiple movements and reflections, Bruno's infinitistic cosmology. It was a central tenet of Bruno's cosmology that there is no sphere of fixed stars, but that all the celestial bodies move, even if most of them imperceptibly to the human eye—and also that all movements in the infinite universe are in some way linked together as reflections or echoes of each other.[28] This last idea was further expressed by Daniel and Jones where Tethys and her Nymphs emerge from the sea world and meander up riverbeds until they reach the tree of victory that links the earth to the revolving skies. The tree was a symbol frequently used in the court masques in Neoplatonic terms. In this case, the skies, into which the tree extends its highest branches, appear, as they always do in Neoplatonic philosophy, as the gateway to a fully transcendent and superior sphere of being. However, the tree could also be used as a Neoepicurean symbol, as Bruno used it in the same Italian dialogue in question here, the *Heroici furori.* There a gnarled and mature oak appears as a symbol of the Epicurean philosophy itself, which unites the natural universe at all levels in an organic and infinite vitality and harmony.[29] This natural harmony is expressed in Daniel's masque, as it is in the final pages of Bruno's *Furori,* by song and soft music accompanied by lutes. The words are formed of three lyrics that praise the rich variety of the natural universe in terms that distinctly recall the songs of Bruno's nine heroic lovers after their release from blindness by the unique nymph of the Thames. The theme of illumination that Bruno had expressed through the image of the gleam in the nymph's eyes is here underlined by a flash of sudden lightning that pervades the scene while the nymphs sing: "Feed apace, then, greedy eyes, / On the wonder you behold; / Take it sudden as it flies, / Though you take it not to hold: / When your eyes have done their part, / Thought must length it in the hart." The world of vicissitude is a world of shadows from which not even royal masquers can escape. In Daniel's masque, as in Bruno's philosophy, it is only memory and thought that can serve to lengthen the shadows into meaningful shapes and forms. Perhaps it is not surprising that Daniel received "rough censures" for his only attempt to write a court masque.

Leaving Daniel means addressing the problem of the long series of court masques composed by Ben Jonson between 1605 and 1630. The problem is a complex one. There certainly are signs that Jonson might have read, or at least heard about, Bruno's works, not only the *Candelaio,* but also the *Furori,* which seems to be echoed in more than one of his masques. But it should be borne in mind that this may have been due to Bruno's use of Neoplatonic imagery and masque-like formal techniques in the final sequence of his dialogue, and may not necessarily imply a takeover of his metaphysics or his political philosophy.

Bruno's presence in Jonson may possibly be traced back to his earliest masque, *The Masque of Blackness*, performed in 1605, some years before *Tethys' Festival*.[30] The structure of the masque bears some similarities to the final sequences of Bruno's *Furori*. A spectacular oceanic opening is centered on a group of vaguely Ethiopian nymphs who journey from Niger to the shores of the Thames to find cleansing illumination on her shores. The masque contains interesting philosophical implications, as the African origin of the nymphs is explicitly linked to an Egyptian and Hermetic concept of knowledge and nature. This is perceived in the masque to be a valid foundation of pristine knowledge, but ultimately insufficient, and in need of a higher form of intellectual illumination. The theme of illumination on the basis of a Greco-Roman concept of beauty symbolized by a group of British courtly nymphs is developed in a later but related masque, *The Masque of Beauty*, performed in 1608.[31] And we still find possible traces of the *Furori* in one of Jonson's later masques, *Love's Triumph Through Callipolis*, performed in 1631 shortly before his quarrel with Inigo Jones.[32] Here Jonson repeatedly uses the phrase "heroic love," and actually portrays Charles I as a specifically "heroic" lover whose proper place is the center of the temple of all beauty. In both cases, however, the possible reference to Bruno on the part of Jonson seems to me overlaid by more generally Neoplatonic influences that hardly allow the Nolan philosophy, in this case, to be considered as a major source, or even necessarily as a source at all.

Jonson was clearly little impressed by the post-Copernican cosmological speculations that were a foundation stone of Bruno's concept of universal vicissitude and reform. Indeed Jonson's masque of 1620 titled *News from the New World Discovered in the Moon* treats the new post-Aristotelian cosmos with elegant but satirical wit.[33] For Bruno's universe was based on the idea of a homogeneous material substance of infinite dimensions, in which constantly changing bodies assume fleeting forms through the agency of a world soul. This idea destroys at one blow the carefully graded hierarchies of universal being that were contemplated by the traditional cosmology and that created an unfathomable gulf between the world at the center and the celestial bodies above. In Bruno's universe, the celestial bodies could be seen as simple variations of the same substance that forms our own world, and space travel becomes at once a theoretical possibility. Bruno can thus be seen to be literally light-years away from both the metaphysical and the imperialistic implications of Jonson's masques. For Jonson constructs for the English court a sophisticated series of Neoplatonic fictions based on a cosmology that is not just earth-centered but England-centered. *The Masque of Beauty* ends with a song in which the island kingdom is exalted as the "fixèd" center of the universe. Its monarch, although surrounded by the ocean, "never wets / His hair therein." Outside and above the universal process of vicissitude, he is a perpetual and constant sun whose beam never sets. Surely D. J. Gordon, in his pioneering studies of these masques, was right to emphasize what he called Jonson's "outrageous compliments" to his king.[34] Jonson's monarch will never decay like Bruno's Jove, nor is there any Momus in his world to remind him of his limits or his mortality.[35]

With Jonson's departure as a writer of masques in 1631, Inigo Jones found himself creating the scenery for what turned out to be the final spectacular entertainments of the Stuart court. On a political level, the new monarch, Charles I, was trying to impose on England a prerogative of absolute power without Parliamentary control. On a religious level, he had just appointed Archbishop Laud to carry out his dreams of far-reaching, anti-Puritan religious reform. These were the last years of Charles's rule, before he led his court into catastrophe and his country into civil war. It is increasingly recognized that even the crystallized and formal rituals of the court masque began at this point to show unquiet signs of political dissidence, and to open cautiously out toward a contestation of the monarch they were supposed to be celebrating.[36]

Of particular significance in this context, by all accounts, was the masque of 1633–1634 called *The Triumph of Peace* created by the Inns of Court (or the community of London lawyers and judges) and written by John Shirley of Gray's Inn.[37] This is often called "The Lawyers' Masque," and it is recognized as a spectacle in which the lawyers were trying to edge Charles I toward a return to a regime of Parliamentary law and justice.[38] The occasion of this masque was the publication by William Prynne of his *Histrio-Mastix: The Players Scourge or Actors Tragedie*, dedicated to "His much Honoured Friends" of Lincolnes Inne and the four Innes of Court. Prynne's Puritanical and anti-monarchical attack on all forms of dramatic representation as works of the devil, or "sugared soppes of Satan," had infuriated the king, whose Star Chamber (the name then given to the British monarch's personal Court of Justice) had sentenced Prynne to life imprisonment, a heavy fine, and the loss of both his ears and his Oxford degree. The lawyers therefore found themselves in the uncomfortable position of having on one side to ingratiate themselves with their angered sovereign, while on the other wishing to make him aware of the barbarity of Star Chamber justice and the tyranny of personal rule.[39]

The lawyers showed considerable cunning and skill in using the complicated mythological machinery of the masque to carry their message. They did this through reference to a celestial reform in the heavens, or what they called "Jove's upper court," which they presented as having been invaded by corruption and vice. These have recently been quelled by the triumphant forces of Peace, Justice, and the Law, and have fled to the earth, where they seem (in what had already become a familiar topos in English Renaissance literature) to inhabit above all in Italy. There they are to be found in a "Forum or Piazza of Peace," where they form the subjects of a series of anti-masques that, under the comic surface, are surprisingly violent and threatening, striking an entirely new note in the formalized and ritualized world of courtly masquing. Peace, Law, and Justice prepare to clean up the piazza of the lower world as well, and to see that the menacing forces of vice are replaced by their equivalent virtues. These inevitably find their emblems in the figures of the English king and queen, who are celebrated as the representatives of divine and human justice in this world.

Although it has not, so far as I know, been recognized, there can be little doubt that Bruno's *Spaccio della bestia trionfante* was being used by the lawyers as a source for this idea of a universal reform carried out in the name of justice and the law. This is particularly clear in the first part of the masque, where the climax is reached with the appearance on a cloud of the three figures of Peace, Law, and Justice, with Law calling to Justice: "Descend and help us sing / The triumph of Jove's upper court abated / And all the deities translated." The presentation of the triumph of peace and justice in the lower world as a triumph of the absolute monarch was not completely foreign to the lawyers' source; as Bruno too had allowed the "diva Elizabetta" and Henri III of France to appear at his climax in these terms. However, as the lawyers may have been aware, Bruno's criticism of monarchy in the course of the *Spaccio* had been powerful and unambiguous. The constellation involved here is precisely that of the Lion, standing for monarchy: "a Lion," writes Bruno, "who brings in his wake the terrors of tyranny, fear and arrogance, a dangerous and odious authority, glorying in presumption and the pleasure of being feared rather than loved. This is a sphere of severity, cruelty, violence and suppression, tormented by the shadows of fear and suspicion, but into this heavenly space ascend Magnanimity, Generosity, Splendor, Nobility, Excellence, who administer to Justice and Mercy."[40]

Bruno's revolutionary view of vicissitude within an infinite universe was a long-term one. As the worlds rolled round in their unceasing dance, with time all the continents would become seas and vice versa, just as all those on high would become low and the low become high. Within his own time, however, Bruno was more concerned, in Machiavellian terms, to persuade the present princes to face political realities and reform their ways. He was prepared to celebrate them, praising the English queen as the "diva Elizabetta," and figuring the French Henri III "translated" into the constellation of the Tiara at the end of the *Spaccio*. Such ideas made him an ideal point of reference for the lawyers, whose concern at this point was not to depose their king but rather to persuade him to return to Parliamentary legality. It is generally conceded that they managed to unite a celebration of the Caroline court with some surprisingly explicit criticisms of its shortcomings. Bulstrode Whitelocke, in his memories of this masque, recorded in 1682 that one of the anti-masques, requesting a patent of monopoly for a new food for capons composed of raw carrots, "pleased the spectators the more, because by it an information was covertly given to the King, of the unfitness and ridiculousness of these projects against the Law."[41] The "Projects against the Law" were the royal monopolies that had been officially banned by Parliament in 1624 but that Noy (who was one of the lawyers involved in the masque) had been persuaded to reintroduce during the period of the royal prerogative. Orgel and Strong, in their essay on "Platonic Politics," recognize in the lawyers masque a daring and ingenious attempt to question the royal prerogative itself.

When Carew wrote *Coelum britannicum*,[42] which was presented at court five days after the lawyers' masque, he referred to, and developed in even more precise terms, the same Brunian source: the *Spaccio de la bestia trionfante*. Many readings of this masque find in Carew's refined and elegant poetry, united to some of Inigo Jones's most splendid and spectacular scenery, a last, triumphant restatement of that monarchical ideal that would be so rudely shattered by the Long Parliament and the civil war. This is the reading proposed by Orgel and Strong in their essay on "Platonic Politics."[43] They see *Coelum britannicum* as a counterstatement to the criticisms and political sniping of the lawyers' masque. Carew, in their opinion, presents "the King's own view of his place in the commonwealth" in a spectacle that they judge as "unquestionably" the greatest of the Stuart masques. As far as Bruno is concerned, Orgel and Strong, as we have seen, think that Bruno is essentially a mystic, in the perspective of the Hermetic reading of his philosophy suggested by Frances Yates. As they find few signs of this mysticism in the highly rationalized and, in their view, exquisitely Neoplatonic justification of absolute monarchy elaborated by Carew, they conclude that he was little attracted by Bruno. The *Spaccio* appears to Orgel and Strong to have been read by Carew only superficially, and without any real interest in Bruno's philosophy of reform.

I shall be developing readings both of Bruno and of Carew's masque of a rather different order. In my opinion, the reference to Bruno on the part of Carew is constant, and essential to the meaning of his spectacle. It goes much further than the borrowings already noticed and listed by Dunlap. Carew's use of the *Spaccio* is far more systematic and direct than any other reference to Bruno found in the Stuart masque, with the possible exception of Daniel's *Tethys' Festival*. Carew, however, gives us much more of a text than Daniel did. His work is in large part a direct translation of parts of Bruno's work, and when he is not translating literally, he is nearly always looking for appropriate translations into spectacular terms of the ideas promoted in his work by Bruno. These ideas were not, in my opinion, only or even primarily mystical. They were an effort to rethink, in the context of a post-Copernican and infinite universe, the ethical bases of a society that had lost its traditional hierarchical points of reference. A corollary of this process of rethinking was an effort to curtail the arrogance of corrupt and oppressive forms of power whose cosmic justification was looking more and more problematical. Carew turned to Bruno as a source not only for his fable but for the political meaning of his masque, which appears to me to have been written in line with the lively but indirect forms of criticism of absolute monarchy that had inspired the lawyers' masque only a few days earlier.

The opening of Carew's masque is centered on a witty dialogue between Mercury and Momus. This is Bruno's Mercury, sent from the "high senate of the gods" to refer the examples of corruption and decadence that have led to a reform of the heavens—a Mercury who, as in Bruno, stands specifically for memory as "the registration of acts" in the tables of time. Carew's touch is light and delicate. His Momus could well be taken as no more than a refined, court wit joking elegantly with Mercury about the fact that in every inn between earth and Olympus there is a book "where your present expedition is registered, your nine thousandth nine hundred ninety-ninth legation." Nine was a sacred number, associated with the Muses, knowledge, and wisdom. But the exchange is not a joke—it is a reminder to the king that time will remember his acts too, recording in the annals of universal history not only the shortcomings of the celestial Jove, but those of the earthly one as well.[44]

Bruno's Momus in the *Spaccio* was one of his most brilliant creations. Identified with the intellect in its search for meaningful action within the bewildering vicissitudes of an infinite universe, his strength lies in his lively skepticism and ironic irreverence for Jove's illusions of absolute power. Momus's skepticism is not just a critical attitude of the intellect as a voice of political dissidence, but recognition of the necessary limits of human action within the overwhelming vistas of infinite space and eternal time. In this new and frightening cosmological perspective, not even the celestial Jove can command more than a limited awe and respect, for he too is involved in the vortex of vicissitude, and in Bruno's cruel and devastating portrait of him, he has gray hair and decaying teeth. Carew takes over Bruno's Momus with brilliant verve, and makes of him a wittily disturbing voice of satirical dissidence. When Mercury chides him with the words: "Peace, railer, bridle your licentious tongue, / And let this presence (that is, the King) teach you modesty," Carew's Momus replies astonishingly: "Let it if it can."[45] This is really letting historical realities intrude into the rarified fictions of the court masque—just as the lawyers had done earlier, when they introduced the common people onto the stage, claiming that they had never seen a masque but were going to now.[46]

Kevin Sharpe notes that Carew's Momus imparts to *Coelum britannicum* a second, often undermining voice: "A tone of realism and cynicism pervades this masque. Momus is a warning that we should not reduce the masques, before we have carefully studied them, to monotone or monochrome. They may resonate with more tones and reflect more complex images than we have realized." This new realization makes it easier to see how Carew conducts his masque through a continued, direct reference to Bruno's text, adapting the thematic and rhetorical strategies of the original brilliantly to his own historical situation, and, as the lawyers had done before him, to the conventions of the masque. The anti-masque is once again used to represent "the monstrous shapes" of the vices that are being chased out of the ideally unsullied heaven of the English court. As in the lawyers' masque, the anti-masques assume a disturbing frequency and force, suggesting that the shadow in which they are represented may not be so easy to dissipate and destroy. Carew is further aware of the cosmological foundations of Bruno's philosophy that are introduced explicitly for the first time into the carefully stratified world of the masque by his Momus. Gone are the eight, nine, or ten revolving spheres of the Aristotelian–Ptolemaic cosmology, while the dimensions of the new post-Copernican universe stretch out to a new and exciting but also disturbing infinity. "Here is a total eclipse of the eighth sphere, which neither Booker, Allestre, nor your prognosticators, no, nor their great master Tycho were aware of."[47] And with what is again astonishing daring and licence, Momus proceeds to gossip about Jove's celestial reform with reference to what he calls "the eighth room of our celestial mansion, commonly called Star Chamber." The suggestion that the celestial reform might be contagious and lead to a "total eclipse" of the dreaded royal chamber of justice—one of the principal organs of Charles's absolute rule—is explicitly made by Momus, who literally warns the king that there may be those about who intend to "unfurnish and disarray our foresaid Star Chamber of all those ancient constellations that have for so many ages been sufficiently notorious, and to admit into their vacant places such persons only as shall be qualified with exemplar virtue and eminent desert."[48] It needed Carew's delicate touch for this to pass as a joke.

It has already been recognized by previous commentators that the long dialogue that follows between Plutus and Poenia takes over in a fairly direct form the episodes in Bruno's *Spaccio* in which Riches and Poverty both plead, without success, for a place in the heavens.[49] Many passages in this part of the masque are literal translation, and yet one of the most interesting aspects of Carew's use of his source material here is his critical attitude toward Bruno's discussion on a minor but intriguing point. Carew's Momus introduces the figure of Plutus by pointing out the dangers of cupidity and riches, just as Bruno had done at rather more length before him. Using a complex mixture of metaphors, Carew sees gold as a poison hidden in the bowels of the earth, an excrescence that men seek to their destruction: "this being the true Pandora's box, whence issued all those mischiefs that now fill the universe." In spite of the close reference to Bruno's original here, the image of Pandora's box is in fact introduced by Carew. It is not used by Bruno in the *Spaccio*, although it is used by him in another work, and in another context altogether. Bruno uses it in the Argument of *Heroici furori*, in a page of strong anti-Petrarchan polemic, addressed to Sir Philip Sidney, in which he sees the sighs of the courtly lover and sonneteer as "poisonous instruments of death issuing from a Pandora's box."[50] Carew is clearly not prepared to dismiss the courtly Petrarchan lover with such exuberant verve as Bruno does, and he distances himself from this theme by making Plutus the "true" Pandora's box. The point is an interesting one, suggesting as it does that "Carlomaria," the idealized Neopetrarchan union of the king and queen of English courtly love in the 1630s, was still a cherished convention, whereas the theme of luxury and overspending was one more open to satire and debate. Orgel and Strong's edition of *Coelum britannicum* gives detailed information on the high cost of the production, which caused some caustic comment even at the time. Carew, who was clearly himself a lover of courtly luxury, appears to be enjoying the ironies of the debate.

In another passage of this same debate between Poverty and Riches, Carew's Mercury takes leave of Poverty in terms that are an interesting comment on Bruno's apparent praise of her, and seem to refer to an English development of the discussion that had intervened between the *Spaccio* and the masque. In Bruno's dialogue, Poverty had not been awarded a place in the heavens, but nevertheless she had been praised handsomely by Jove as a friend of all those who are content to follow the laws and ways of nature, leaving the ambitious to the "poverty" of riches. In particular Jove had praised her for associating with philosophers, whose intimate meditations would be disturbed by the crowds and confusion attendant on wealth. Bruno's attitude is clearly ironical here: the passage is a wry comment on the poverty that accompanied him throughout his life. It would certainly be a mistake to read Jove's praise of poverty straight: it is a sermon—probably a memory of Bruno's monastical origins—that is being parodied in both its substance and its tone.[51] Why, after all, *should* the studious and the cultured be condemned to a "clerical" poverty? This very question is asked at the end of Jove's speech by Momus, who wonders, under his breath, what injustice of fate regularly leads to Riches staying away from those who could be claimed most to deserve her company. Jove, however, proves inflexible: unreasoningly (and, Bruno is suggesting, unreasonably) he decrees that fate will continue to be unjust on this point. Carew picks up the argument from there, adding to it what may well be echoes of Francis Bacon's open praise of riches and opulence in his *New Atlantis*, published in 1623. In Bacon too, there is no shame in praising riches, which he sees as the just and proper outcome for the men of learning of his new scientific society. Carew's Mercury appears to associate himself with this attitude by severely maligning Poverty as one who "Degradeth nature and benumeth sense / And Gorgon-like turns active men to stone." "We," continues Carew's Mercury (and it is difficult to decipher if this is a royal "we," or if it refers to the courtiers present, or to the moderns, or to the English specifically) "advance such virtues only as admit excess, / Brave bounteous acts, regal magnificence, / All-seeing prudence, magnanimity / That knows no bound."[52] Although Carew too may be indulging here in a note of irony at the expense of his luxury-loving royal master, it would clearly be a mistake to overestimate his critical attitude to the opulent world of courtly masques and ceremony.

The subsequent appearance of Fortune in Carew's scene recalls briefly Bruno's own discussion of the subject. In the *Spaccio*, as in the masque, this follows directly on the episode involving Poverty and Riches, who so clearly illustrate Fortune's caprices and her power.[53] This is recognized as far-reaching and awesome, particularly in the now infinite vastness of a universe involved in perpetual and bewildering processes of change. Fortune herself claims her right to a place in the heavens on the basis of a kind of rough justice depending on her blindness, which refuses to listen to specious arguments of favoritism or influence. Before Fortune all are equal: a good enough reason, she maintains, to justify her election to a celestial seat. However, both Bruno and Carew refuse her entry to the heavens. Bruno had claimed that Fortune has no real identity, insofar as what seems a caprice of Fortune to the limited mortal eye is really only part of a larger providential scheme. Carew's Mercury echoes this reasoning, inviting Fortune to vanish "And seek those idiots out / That thy fantastic god-head hath allowed." The real question at stake here is what concept of providence is involved, and what kind of reference to its powers is being invoked. The point is a delicate one on which Bruno suggested a solution likely to have proved congenial to Charles I, at least insofar as it denies what is seen both by Carew and by Bruno as a "lazy" Protestant fideistic solution: what is advocated is rather the ongoing search for knowledge of the intimate workings of the universal whole.[54] Knowledge of, and power over, the natural world will ultimately uncover the hidden workings of a higher providence, and save mankind from the blind caprices of an unjust fate. Carew once more faithfully translates Bruno's thought into his masque where Mercury dismisses Fortune's "vain aid" by claiming that "Wisdom, whose strong-built plots / Leave nought to hazard, mocks thy futile power."

Carew continues with his close reference to Bruno's celestial reform by introducing, immediately after the figure of Fortune, that of Hedone, or Pleasure. Carew defines her negatively as a siren who leads inquiring man away from his serious study of the course of things—a poison as subtly attractive as riches who displays "th'enameled outside and the honeyed verge / Of the fair cup where deadly poison lurks." This passage has been considered by nearly all Carew's critics as an original addition, at variance with Bruno's Epicurean tendencies. But this is not the case. It faithfully reproduces Bruno's episode of the constellation known as the Goblet, following him in the distinction, which recurs throughout his work, between a just and controlled principle of pleasure and severe condemnation of overindulgence in the material goods of life.[55] Historically, this distinction goes back to Epicurus himself, and forms the basis of the Epicurean movement as a serious school of philosophical thought. Later it would be picked up by Roman writers such as Cicero. By following Bruno in this classical distinction of a base hedonism from a controlled and sober principle of pleasure, Carew underlines the ethical and philosophical seriousness of his masque.

The reference to *Lo spaccio de la bestia trionfante* continues throughout Carew's masque. In insisting on this, I am once again going against received critical opinion that has always claimed that the final sequences of *Coelum britannicum*, which begin with a particularly fine scene by Inigo Jones depicting Stonehenge, where the royal masquers return to their origins and find renewed moral and historical vigor, have nothing to do with Bruno's text at all. It is certainly true that Bruno nowhere mentions Stonehenge. However, he does refer several times to the druids and their sun-worshipping religion, as well as their belief in metempsychosis, and he links their "prisca theologia" to that of Pythagoras and the ancient Egyptians.[56] The Stonehenge sequence in Carew's masque, in my opinion, represents a faithful "translation" into English terms of the long quotation from the Hermetic text of the *Asclepius* that Bruno introduces into the final dialogue of the *Spaccio*.[57] Not, of course, a literal translation, but a faithful transposition into English terms of that return to distant, uncorrupted origins that is the message proposed by Bruno through his quotation from the *Asclepius*. The return to Egyptian origins in Bruno signifies the reestablishing of a correct relationship between man and a universe in which the gods have not yet retreated into some impenetrable region in the skies. The Egyptians, as Bruno's Sophia says in her comment on the passage from the *Asclepius*, searched for divinity in the forms of beasts and plants, ascending to the divine through the secret and magic heart of things. Bruno sees this return to a most ancient origin in which the universe is unsullied and unspoiled, its magic resonances and harmonies intact, as a necessary foundation of any valid metaphysical or ethical philosophy for modern man. Carew follows him, making the Genius of his Stonehenge invoke for the king and queen the druid circle's "aged priests and crystal streams / To warm their hearts and waves in these bright beams."[58]

However, this philosophical primitivism, or "prisca theologia," was not, in my opinion, the final outcome of Bruno's philosophy—only the establishment of the foundations on which it must be built. In the final sequences of the *Furori*, the subject is developed by Bruno through the figure of Circe. The heroic lovers' passage through her kingdom, and their initiation into her magic, is a vital stage of their passage toward a revelation of the true harmonies of the universe. But Circe's magic leaves them blind, and it is only the two eyes of the unique nymph of the Thames who can restore their sight. This vitally important moment in the progress toward knowledge is called by Bruno himself one of "illumination." There is a recovery of an intellectual principle in the Greco-Roman philosophical tradition, at least in some of its manifestations: Pythagoras, the pre-Socratics, Epicurus, the Stoics, to some extent the moderate Skeptics, something of Plato, and even some aspects of the despised Aristotle. Once again, I believe that Carew found remarkably appropriate "translations" into his masque form of Bruno's ideas. From Stonehenge, the scene changes to "a new and pleasant prospect clean differing from all the other, the nearest part showing a delicious garden with several walks and parterras set round with low trees, and on the sides against these walks were fountains and grots, and in the furthest part a palace from whence went high walks upon arches, and above them open terraces planted with cypress trees, and all this together was composed of such ornaments as might express a princely villa."[59] This is a surprising description, especially if we pay sufficient attention to that "might." This is seventeenth-century Europe. But the princely villa, it seems, can no longer be taken for granted, and in any case, it is not the center of the scene, which is rather the garden. This is strikingly similar to the Argument of the second dialogue of *The Ash Wednesday Supper*, where Bruno describes the "picture" of his universe, using the metaphor of a painting. Here too there is a royal palace, but it is only one of a number of elements that make up the contemporary scene: "a royal palace here, a forest there, a glimpse of the sky above, and on one side the half of a rising sun."[60] Carew's garden possibly goes too far in leading us toward a rather too formalized, rationalistic landscape, clearly of French inspiration. Bruno's representation, fifty years earlier, gave a freer reign to nature. His greatest philosophical work, the *De immenso*, finishes with a description of gentle, fertile hills and vales, with the philosopher who is invoking the sound of their echo that reverberates over the landscape. This is not Egypt, nor the mountain of Circe, which were rather Bruno's point of departure. His landscape of arrival has been inhabited by the nymphs and the naiads for many centuries: it may make us think of the hills of Tuscany, for example, or indeed (which is where Bruno conducts us at the end of the *Furori*) the valley of the river Thames.

Carew's masque does not finish with the scene of the garden, which is only the terrestrial space in which the masquers enjoy their nightly revels. When these are over, Carew and Inigo Jones mount a final scene that represents the celestial outcome of the universal reform. The reference to Bruno's *Spaccio* remains a close one, particularly where the terrestrial and the celestial spheres are conceived of as parallel reflections of each other's treasures rather than separated spheres of being. Carew uses the same image as Bruno had done to represent this concept: that of the River Eridanus whose privilege it was "In heaven and earth to flow, / Above in streams of golden fire, / In silver waves below."[61] Carew then proceeds to apply to his English monarchs the terms of praise that Bruno had reserved for the French Henri III in the final pages of the *Spaccio*.[62]

These represent some of Bruno's most subtle and enigmatic pages. They formed the basis of Frances Yates's claim that Bruno was expressing a mystical cult of monarchy, but I think this is a misreading of the text.[63] The two central images used here are the altar, occupied by the centaur Chiron, and the constellation of the Crown, or the Tiara, which Bruno reserves for Henri III. Chiron, with his double nature, half man and half beast, is used by Bruno as a metaphor for Christ's double nature as half God and half man, so it does look as if Bruno is presenting the French king as also part human and part divine, as well as a defender of the faith. Bruno's Momus, however, ridicules this whole complex of ideas by saying that he will never believe that half a trouser and one sleeve add up to more than a whole trouser and two sleeves. That is to say, Momus conceives of the divine presence in the universe in terms of a total immersion or immanence, rather than as an emanation of a transcendental god. He is sternly invited to keep quiet, and to believe that profound mysteries are at work here that he should not inquire into. Momus agrees to believe in the idea of a pure, divine mind of which Chiron is half a representation, although making it clear that his belief is held only to please Jove—in other words, he has no choice. Jove, for his part, goes on to say that Chiron, or Christ, will be venerated as the priest of his altar because altars are necessary, and because Christ is the priest of the moment. He might even be the eternal priest, but this is not certain, and fate may have decreed otherwise. This is a Machiavellian concept of religion as a social and pragmatical necessity, and its mystical overtones are the means by which it is imposed on an ignorant populace.[64] Bruno's choice of Henri III for translation into the constellation of the Crown is also Machiavellian in tone. He is chosen not on a principle of divine right, but because he maintains his kingdom in peace and order. It is not even certain that he will in the end be assumed into the Crown—we are waiting on time, concludes Bruno's Jove, to see who will be most deserving of such a merit.

Carew was necessarily less reserved than this in his praise of Charles I and his queen. He does not allow his Momus to intrude, with his corrosive wit, into the final moments of the masque, with their inevitable royal apotheosis. And he does appear to allow the royal couple a timeless and motionless perfection that places them outside and above the process of vicissitude and change. But Carew gives with one hand only to take away with the other, for he abandons the traditional image of the sovereign as the sun, allowing him to figure only as a star, though a greater and more eminent one than the rest. By doing this, he maintains the new infinitistic cosmology as his universal background up to the end, for, as in Bruno's universe, every star has become a sun, with its own planets and satellites revolving around it. The revolving spheres no longer exist, and earth itself has become a negligible point in the boundless cosmos: a "Decrepit sphere grown dark and cold." It is the royal stars who glow with heat and light, shedding their resistless influence on "the uncertain tide" of human change. Carew concludes by claiming that propitious stars will crown every royal birth: "Whilst you rule them, and they the earth." This sounds perfectly respectable as an ending of a court masque. But if we bear in mind Carew's source, the praise says rather less than it appears to. For the whole point of Bruno's *Spaccio* was that it had eliminated one by one the talismanic signs of the zodiac, and with them the astrological concept that the stars do in fact rule the earth. In the final picture of Carew's masque, there are no stars left in the heavens but only a serene sky. "After which," the text tells us, "the masquers dance their last dance, and the curtain was let fall."

Serious scholars of those tense years of the 1630s, which lead up to the Long Parliament and civil war, rightly warn against the dangers of interpreting them in the light of hindsight. There are, however, signs that in referring to Bruno's dialogue in the context of a court masque, Carew was quite aware of what he was doing. Bruno's unique nymph at the end of the *Furori*, as we have seen, initiated the blind searchers after knowledge into a double sphere of being made up of both matter and mind, body and soul—both being inextricably linked as two aspects of a single truth through the image of the divine nymph's eyes. Bruno did not deny the existence of a higher plane of being, which he identified with the divine, although he refused to think of it as purely spiritual, and maintained that, in any case, it remained outside the field of human vision. Nevertheless in the pages of dedication to Prince Julius of Brunswick at the beginning of the Frankfurt trilogy, Bruno praises the prince as a Trismegistus: three times wise. And one of the aspects of his wisdom is the witnessing of divine truth on earth.

What appears to be a justification of the idea of divine right, however, inevitably becomes ambiguous in the course of Bruno's trilogy. For the sphere of the divine increasingly assumes the form of an infinite plenitude of which the whole infinite universe is the seal, shadow, or reflection, while the three spheres of Neoplatonic being merge into the three dimensions of Euclidian space. The mind participates in the divine through its reflections in the three-dimensional infinite universe, so that it is doubtful whether the monarch can be distinguished from other forms of universal life in any terms other than the particular glow with which his mind warms to the divine intimations of immortality. Once again Carew found remarkably apt translations of Bruno's concepts, for it seems inescapable that the monarchical principle, through reference to Bruno's philosophy, had become involved in a universe of relativity that rendered obsolete the traditional Neoplatonic schemes of courtly masquing. Carew appears to have been aware of Bruno's preoccupation with the number three: the trinity that had to be understood in new and demanding ways as the structural principle of the infinite universe itself. Carew's masque was not only given in 1633 but it was also presented on Shrove Tuesday of that year, which fell on February 18. The previous day, February 17, was the anniversary of Bruno's death at the stake thirty-three years earlier; the following day was Ash Wednesday, for which Bruno had written a *Supper* in which he had claimed that the divine is within us all, closer to us than our very selves.[65]

I have argued that it is misleading to dismiss Carew's reference to Bruno as marginal and slight. By following some of the central episodes of Bruno's celestial reform outlined in *Lo spaccio de la bestia trionfante*, Carew succeeded in presenting the action of a masque peculiarly suited to the tensions of the moment: spectacular mythological machinery was at hand to satisfy the art of Inigo Jones and the courtly taste for pageantry, while the presence of Momus offered a focal point of elegantly subversive wit. Through the reference to Bruno's infinite cosmology, Carew could subtly question the absolutism of Charles I while at the same time developing his masque within the terms of a coherent moral and metaphysical discourse of universal reform. It would be interesting to know whether the Venetian ambassador was aware of Carew's Italian source when he referred to *Coelum britannicum* as a "very stately and solemn" masque.[66] However that may be, the reference to Bruno, as I have tried to show in this chapter, was central to what Carew was trying to do. I have further attempted to show that Bruno's presence in the Stuart court masques was a constant one from the beginnings of the form in the reign of James I. It was not a neutral presence. It made itself felt above all in those masques that were posing in increasingly problematical terms the political statement of absolute monarchy that the form was presumed to assert.

If my argument is correct, the question of the originality of Daniel, the lawyers of the Inns of Court, and of Carew, in their attempt to oppose within the schemes of courtly masquing the increasing absolutism of the Stuart monarchy has to be addressed. The attempt by Orgel and Strong to eliminate Bruno from their reading of *Coelum britannicum* appears to be largely determined by an anxiety to leave Carew as the effective author, if not of the fable then at least of the concept of power being propounded through it. It is not clear, though, why, since the studies of D. J. Gordon, a generally conceded reference to Plato and the Florentine Neoplatonists as the intellectual basis on which the apology of the Stuart monarchy was founded is to be considered acceptable, and not in contradiction with the originality of Ben Jonson and his followers, while a reference to Bruno has to be so fastidiously avoided in order to save the originality of Carew. The fact is that none of those concerned with the making of the English masques, beginning with the consummate artist of the Stuart court, Inigo Jones, were political philosophers with a proposal of reform of their own to put forward in what were clearly also political messages. They were only showing themselves well informed, and abreast of the tide of the times, in making references to the major intellectual movements reaching them from Renaissance Europe.

It is clear that the name of Giordano Bruno, even then, was a more difficult one to digest than those of the revered Plato or his Florentine followers, Marsilio Ficino and Pico della Mirandola, with their close links with the Medici court. Bruno was, after all, a lone wanderer through Europe, an exile for most of his adult life. He had, furthermore, denied the doctrine of the Incarnation of Christ, and had been punished publicly and dramatically for doing so. His name was studiously avoided by most throughout the seventeenth century, even where his influence, as recent studies have shown, was undoubtedly felt, at times in decisive terms.[67] For Bruno had proposed a rethinking of the cosmic order that was making itself felt, if at times surreptitiously, in the most advanced areas of the new science. His cosmic vision had further been accompanied by a proposal for universal moral reform, based on an advanced concept of the rights of the individual, and of liberty of thought and expression, which had clear anti-absolutist implications.[68] On the other hand, Bruno was not a rationalist of the Enlightenment. His doctrine of universal vicissitude tended to deny the possibility of human reason to dominate a universe whose own laws obeyed an intellectual principle of infinite complexity and perfection. Bruno's attachment throughout his life to symbolic and emblematic forms of expression is intimately linked to his awareness of the limits of the human reason. His sense of the mysterious powers of images and words to combine to create meanings whose ultimate sense eludes the human mind linked his philosophy to the expression of poets and artists, whose inquiry he considered as equally valid with that of philosophers themselves.[69] Given that he himself used at times expressive techniques that he appears to have taken from the courtly masques of Renaissance Italy and France, he may well have seemed an ideal point of reference to those poets of the English court who were looking for an alternative philosophy of power to oppose to the absolutism they feared.

It is not surprising that the reference to Bruno being discussed here was not publicly declared. This reticence on the part of the English masque writers, which was shared with many others of the time, nevertheless creates some still unsolved problems in a study of his influence on the genre. Who brought him into the picture in the first place? Was it Inigo Jones himself, whose Italian was undoubtedly sufficient to permit him to approach Bruno's complicated and idiosyncratic texts? It is interesting to note that Jones was the scenographer for all the masques in which Bruno's influence is clear. It seems unlikely that he would have been kept in ignorance of the source being used, even if it was proposed by the poets in the first place. However, the library list of Jones's books, although it contains numerous Italian works in the original, has no book in it by Bruno.[70] In the present lack of documentary evidence, it would appear that it was Daniel, whose close links with John Florio are a documented fact, who acted as the way in for Bruno. It may well have been the explicitly admitted failure of Daniel's early masque that led to the long period of unsullied Neoplatonism under the vigilant eye of Ben Jonson. It was surely not by chance that Bruno surfaced again in the masquers' world during the unquiet years of the 1630s, when things began to be said, or at least murmured, in the court itself that helped to lead, only a few years later, to the dramatic crumbling of the crystal walls of absolutism so carefully constructed around their court by the Stuart king and queen.

NOTES

1. The identification of the *Spaccio* as Carew's source was first made by Robert Adamson in an article on Carew in the ninth edition of the *Encyclopedia Britannica* (1875–1889).

2. Dunlap Rhodes, ed., *The Poems of Thomas Carew*, Oxford, UK, Oxford University Press, 1949, 275–76.

3. See Yates (1964), 392–93.

4. Stephen Orgel and Roy Strong, eds., *Inigo Jones: The Theatre of the Stuart Court*, 2 vols., Berkeley–London, University of California Press, 1973, vol. I, 49–75.

5. The importance of Machiavelli's republicanism as expressed in his *Discorsi* as a major source for Bruno's idea of civil society in the *Spaccio* has been underlined by Ciliberto (1986), 176–78, and has since become a frequent *topos* in the critical discussion. See also Annabel Patterson, *Censorship and Interpretation: The Conditions of Writing and Reading in Early Modern Europe*, University of Wisconsin Press, 1991, and John Kerrigan, "Thomas Carew," *Proceedings of the British Academy* 74 (1988): 348.

6. See Joanne Altieri, "Carew's Momus: A Caroline Response to Platonic Politics," in *Journal of English and Germanic Philology* 88, no. 3 (July 1989): 332–43; 339, and the few relevant remarks in Ricci (1990b) and Ciliberto and Tirinnanzi (2002).

7. For Bruno's years in England, see Aquilecchia (1991) and Rowland (2008), 139–87. In agreement with these two commentators, I am not convinced that the documents presented have proved the thesis of Bruno as a spy advanced in Bossy (1991).

8. See Bruno (2002), vol. 1, 477–79.

9. Ibid., vol. 2, 171–96 and 487–500. For a comment on the dedicatory letter of the *Furori*, see Farley-Hills (1992) and chapter 5, "Petrarch, Sidney, Bruno," in this volume.

10. See Firpo (1993), 189.

11. Bruno (2000c), 11–12.

12. Ibid., 907. For a possible use of this passage by Shakespeare, see chapter 7, "Bruno and Shakespeare: *Hamlet*," in this volume.

13. See Yates (1943).

14. See Bruno (2002), vol. 2, 746–53. For a reading of these final pages of the *Furori* in terms of a prophecy rather than a masque (but the two readings are by no means antithetical), see chapter 6, "The Sense of an Ending in Bruno's *Heroic furori*," in this volume.

15. Baltasar De Beauioyeulx, *Balet comique de la royne*, Paris, 1582. Francis Yates refers briefly to the *Balet*, but only with reference to Bruno's earlier work on memory, the *Cantus Circaeus*, in Yates (1964), 202.

16. For a discussion of this ballet within the artistic and historical context of the court entertainments of Renaissance Europe, see Roy Strong, *Art and Power*, Suffolk, UK, Boydell Press, 1984, part II, chapter 3.

17. See Bruno (2002), vol. 2, 190. For Bruno's contrast between being seen as the "womb" or great ocean of eternal life, and the momentary glimpse of the

"back" as referring to the limited vision possible within the individual life span, see Canone (2003).

18. This remark is attributed to the queen by Julio Caesare La Galla in his *De Phoenomenis in orbe lunae*, Venice, 1612, where it is quoted in Greek. The text of this work was republished in A. Favaro's edition of Galileo's *Opere*, vol. III, Florence, Le Monnier, 1930. For this comment, see 352. It is known that Queen Elizabeth I possessed a volume of Bruno's works bound in black leather. The volume contained *La cena de le ceneri, Spaccio de la bestia trionfante, De la causa principio et uno,* and *De l'infinito universo et mondi.* For the importance and history of this volume, see Sturlese (1987), xxiv–xxv.

19. For the close relationship between Florio and Daniel, see Joan Rees, *Samuel Daniel: A Critical and Biographical Study*, Liverpool, UK, Liverpool University Press, 1964. For the linguistic dimension of the Bruno–Florio relationship, see Wyatt (2002). For an extended comment on the Bruno references in the letter signed N.W., see Gatti (2008c).

20. Mark Eccles, "Samuel Daniel in France and Italy," *Studies in Philology* XXXIV (1937): 148–67.

21. It has been pointed out that the itinerary followed by Bruno and his friends in dialogue II of *The Ash Wednesday Supper* indicates that the supper was not held in Fulke Greville's private house but in his chambers in the Royal Palace at Whitehall; see Bruno (2002), vol. 1, 468, note 8. On June 3, 1592, the Venetian inquisitors conducting the first phase of Bruno's trial asked him whether in any of his writings he had ever mentioned an Ash Wednesday supper, and if so what did he mean by it? Bruno replied that he had composed a book titled *The Ash Wednesday Supper* in five dialogues that investigated the movements of the earth, and that the dispute, held with some doctors, took place in England during a supper given on Ash Wednesday by the French ambassador whom he was serving and to whom it was dedicated (see Firpo [1993], 188). This account clearly suggests that the supper actually took place. Bruno at his trial may have placed it in the French ambassador's house, rather than Fulke Greville's chambers as the text clearly states, from a lapse of memory, or because it would have sounded to the inquisitors a more respectable location than the Protestant Greville's rooms.

22. Samuel Daniel, *Musophilus*, London, 1599, fol. CIIIr.

23. Gatti (1989), 56–57.

24. The text of Daniel's masque, complete with preface, was first republished in *The Progresses, Processions, and Magnificent Festivities of King James I*, ed. John Nichols, London, J. B. Nichols, 1828, 346–58. See also Orgel and Strong, *Inigo Jones*, op. cit., vol. 1, 191–96.

25. This is Bruno's first published work to have survived. It appeared in Paris in 1582 in reply to a request from the French king, Henri III, to know more about Bruno's art of memory. It is now in Bruno (2004), complete with Italian translation and comment.

26. See Ricci (1990a). Bruno's anti-imperialism may also have been known to the Shakespeare of *The Tempest*. A possible relationship in this sense is discussed in Tarantino (2002).

27. See John Pitcher, "'In those figures which they seeme': Samuel Daniel's *Tethy's Festival*," in *The Court Masque*, ed. David Lindley, Manchester, UK, Manchester University Press, 1984, 33–46. This reading of Daniel's masque questions the negative criticism of it in the works of Orgel and Strong; see in particular Roy Strong, *Henry, Prince of Wales and England's Lost Renaissance*, London, Thames and Hudson, 1986, 155–60.

28. The idea is repeated many times in Bruno's works. It finds one of its earliest expressions in the first dialogue of *The Ash Wednesday Supper*. See Bruno (2002), vol. 1, 454–55.

29. Ibid., vol. 2, 622–25.

30. Orgel and Strong, *Inigo Jones*, op. cit., vol. I, 89–93. For more on a possible Bruno–Jonson relationship, see chapter 8, "Bruno's *Candelaio* and Ben Jonson's *The Alchemist*," in this volume.

31. Orgel and Strong, op. cit., vol. I, 93–96.

32. Ibid., vol. 1, 405–7. For a possible influence of Bruno, see Graham Parry, *The Golden Age Restored: The Culture of the Stuart Court, 1603–42*, Manchester, UK, Manchester University Press, 1981, 186.

33. Orgel and Strong, *Inigo Jones*, op. cit., vol. I, 307–12.

34. See D. J. Gordon, *The Renaissance Imagination*, Berkeley–Los Angeles, University of California Press, 1975, 184.

35. At the most, Jonson's feelings about the inadequacies of the Stuart court seem to have taken the form of nostalgia for that of Elizabeth I; see Anne Barton, *Ben Jonson, Dramatist*, Cambridge, UK, Cambridge University Press, 1984, chapter 14, "Harking Back to Elizabeth: Jonson and Caroline Nostalgia."

36. Opinions diverge as to how fully the court masques reflected the increasing state of political crisis, as well as to which ones were more open to the signs of the crisis. For studies of the political situation as it was reflected in the court culture of the 1630s, see R. Malcolm Smuts, "The Political Failure of Stuart Court Patronage," in *Patronage in the Renaissance*, eds. Guy Lytle and Stephen Orgel, Princeton, NJ, Princeton University Press, 1981, 165–91. For a study of the last of the Stuart court masques, *Salmacidia Spolia*, as a specifically political statement reflecting the crisis of the times, see Martin Butler, "Politics and the Masque: *Salmacida Spolia*," in *Literature and the English Civil War*, eds. T. Healy and J. Sawday, Cambridge, UK, Cambridge University Press, 1990, 59–74.

37. England was still using the old Julian calendar according to which the year finished on the 25th of March. As the events we are concerned with took place in February, the year would have been considered 1633 by British contemporaries, but 1634 on the Continent and in the writings of modern scholars. For the text of Shirley's masque, see Orgel and Strong, *Inigo Jones*, op. cit., vol. 2, 546–53.

38. Orgel and Strong, *Inigo Jones*, op. cit., vol. 1, 49–75, recognize the critique of the court in this masque. For more detailed discussions, see Martin Butler, "Politics and the Masque: *The Triumph of Peace*," in *The Seventeenth Century* 2 (1987): 117–41, and Lawrence Venuti, "The Politics of Allusion: The Gentry and Shirley's *The Triumph of Peace*," *English Literary Renaissance* 16 (1986): 182–205.

39. See William Prynne, *Histrio-Mastix*, London, Thomas Thorpe, 1633, and, for the historical background to the publication of Prynne's book and his pun-

ishment, Charles Carlton, *Charles I: the Personal Monarch*, London, Routledge, 1983, chapter 10.

40. See Bruno (2002), vol. II, 192.

41. Orgel and Strong, *Inigo Jones,* op. cit., vol. 2, 542. Bulstrode's firsthand account of *The Triumph of Peace*, given in his *Memorial of the English Affairs*, London, J. Tonson, 1682, 18 ff., is republished in full in Orgel and Strong.

42. Orgel and Strong, *Inigo Jones,* op. cit., vol. 2, 570–80.

43. See also, in line with this reading, Raymond D. Anselment, "Thomas Carew and the 'Harmless Pastimes' of Caroline Peace," *Philological Quarterly* 62 (1983): 201–383, and E. Veevers, *Images of Love and Religion: Queen Henrietta Maria and Court Entertainment*, Cambridge, UK, Cambridge University Press, 1989.

44. Orgel and Strong, *Inigo Jones,* op. cit., vol. 2, 571, and for the Brunian passage Carew is using as his source, Bruno (2002), vol. 2, 247–49.

45. Orgel and Strong, *Inigo Jones,* op. cit., vol. 2, 571.

46. Ibid., vol. 2, 552.

47. Orgel and Strong, *Inigo Jones,* op. cit., vol. 2, 574. For Sharpe's comment, see Kevin Sharpe, *Criticism and Compliment: The Politics of Literature in the England of Charles I,* Cambridge, UK, Cambridge University Press, 1987, 197.

48. Orgel and Strong, *Inigo Jones,* op. cit., vol. 2, 574.

49. Ibid., 574–76, and for Carew's source for this scene, Bruno (2002), vol. 2, 273–78.

50. Bruno (2002), vol. 2, 489.

51. Ibid., 280–87.

52. Orgel and Strong, *Inigo Jones,* op. cit., vol. 2, 576.

53. Ibid., and in Bruno (2002), op. cit., vol. 2, 287–95.

54. The terms of Bruno's anti-Protestant polemic have been much stressed in recent years. See Ordine (2007) and Ciliberto (2007). For an attempt to understand why, in spite of his strong anti-Protestant stand, Bruno wrote and flourished above all in aggressively Protestant centers of culture such as London and Wittenberg, see Ignegno (1987) and Gatti (2002).

55. Orgel and Strong, *Inigo Jones,* op. cit., vol. 2, 576–77, and Bruno (2002), vol. 2, 396–97.

56. Bruno (2002), vol. 2, 452.

57. Ibid., vol. 2, 363–64. These pages have been at the center of much critical discussion of Bruno's philosophy. Yates (1964), 211–15, considered them the main foundation of her Hermetic interpretation. For a critical consideration of the Yatesian reading of these pages, see Ciliberto (1986).

58. Orgel and Strong, *Inigo Jones,* op. cit., vol. 2, 578. It should be remembered that Inigo Jones thought that Stonehenge was not a prehistoric monument, but that it was built by barbarian invaders at the end of the Roman occupation. His reconstruction of it, which was published posthumously, underlines what he sees as a classical regularity and harmony in the original plan and discusses it as a work of the Roman decadence. Thus, in what is already a concept of the Enlightenment, Jones sees the return to valid ancient origins as including a dimension of classical antiquity. See Inigo Jones, *The Most Notable Antiquity of Great Britain vulgarly called Stone-Heng on Salisbury Plain*, London, 1655.

59. Orgel and Strong, *Inigo Jones,* op. cit., vol. 2, 579.

60. Bruno (2002), vol. I, 435. See Michael P. Parker, "'To my Friend G.N. from Wrest': Carew's Secular Masque," in *Classic and Cavalier: Essays on Jonson and the Sons of Ben,* eds. Claude J. Summers and Larry Pebworth, Pittsburgh, PA, Pittsburgh University Press, 1982, 171–91, for an interesting critical discussion of Carew's poem *To my Friend G.N. from Wrest,* where the country house of Henry de Grey, eighth earl of Kent, appears to become a substitute for the rapidly declining court, already fighting against the puritan Scots to impose the religious policy of Charles I's religious policy. It is to Wrest, in this poem, rather than the Stuart court, that Carew transfers his desire for harmony and order in both a social and a natural context. It is interesting to note that in the subsequent events of the 1640s, which Carew would not live to see, Henry de Grey became a Parliament man, although he took no part in the trial and death of Charles I.

61. Orgel and Strong, *Inigo Jones,* op. cit., vol. 2, 578. For Bruno's treatment of the river Eridanus, see Bruno (2002), vol. 2, 384–85.

62. Ibid., 400, and Orgel and Strong, *Inigo Jones,* op. cit., vol. 2, 579–80.

63. See Yates (1964), 360–98.

64. The importance of Bruno's reading of Machiavelli for his religious and social thought has been clearly documented and underlined in recent years by Italian scholars such as Alfonso Ingegno and Michele Ciliberto. See in particular Ingegno (1985) and Ciliberto (1986).

65. It seems to me that these three dates are brought deliberately together even if it should be remembered that the first and the last correspond to the old-style Julian calendar and the second, the date of Bruno's execution in Rome, to the new Gregorian calendar.

66. See the *Calendar of State Papers Venetian (1632–36),* 190.

67. See the relevant pages in Ricci (1990b).

68. See Calogero (1963) and Gatti (1989), 17–19.

69. For Bruno's identification of true poets and true painters with true philosophers, see Bruno (2009), 120–21.

70. Published as appendix III in John Harris, ed., *The King's Arcadia: Inigo Jones and the Stuart Court,* London and Bradford, Arts Council of Great Britain, 1973, 217–18.

10

ROMANTICISM

BRUNO AND SAMUEL TAYLOR COLERIDGE

ON SEPTEMBER 16, 1798, TWO young poets left England for Germany. Only a few days previously, they had published together a small volume of verses destined to change the course of English literature: the *Lyrical Ballads*. However, the implications of these poems, which proposed a reevaluation of the life of the sentiments and the spirit of the individual in forceful and unadorned language, had not yet been fully appreciated. Both their departure from England and their arrival in Germany went almost unnoticed.

Once in Germany, William Wordsworth, whose central concerns were always of a primarily literary nature, was overcome with homesickness for his native landscape, leading him to live in retirement with his sister, devoted to poetical composition. Samuel Taylor Coleridge, on the other hand, actively pursued the cultural novelties that permeated the German society of the end of the eighteenth century, and accordingly he started out on an intense program of philosophical studies. He enrolled in the University of Göttingen, where he attended with enthusiasm the lectures of J. F. Blumenbach and the course of critical Biblical studies held by J. G. Eichhorn. It was in Göttingen that he began a reading of Kant that he would continue assiduously in the coming years.

The German journey finished with a trip to Helmsted, where Coleridge visited the famous University Library, which he found similar to those of the colleges of Oxford and Cambridge. As he wrote in a letter: "we rummaged old Manuscripts, and looked at some Libri Rarissimi for about an hour."[1] When he returned to England in July 1799, Coleridge could read German with ease, and he brought back with him a rich collection of philosophical texts, among which were works of Kant, Fichte, and Shelling, all of whom he would study in depth in the coming years. It is in these volumes that Coleridge left a series of marginalia denoting a profound interest in their contents that, in the first thirty years of the nineteenth century, would be made public on numerous occasions. Coleridge's reading of these works would help to bring Britain into the flow of the post-Kantian philosophical debate that characterized the European culture of the period.[2]

Among the philosophical texts brought back to Britain by Coleridge was the volume by Friedrich Heinrich Jacobi, *Über die Lehre des Spinoza in Briefen an Hernn Moses Mendelssohn*, in the second edition of 1789. This text carried in an appendix a synthesis, translated into German, of some significant pages of Bruno's most metaphysical work, *De la causa, principio et uno*, which Jacobi commented on in the context of his polemical discussion with Lessing and Mendelssohn concerning the pantheism of Spinoza. Jacobi's reading of Bruno's philosophy as an anticipation of the concept of an infinite substance proposed by Spinoza would have profound repercussions on the German philosophies of the following period, in particular on the transcendental idealism of Schelling that opened the new century.[3] Coleridge, however, responded independently to Jacobi's comments, writing on a page of his copy of this important volume:

> It is doubtful whether to Bruno or Jacob Behmen belongs the honour of daring to announce the *substantial* meaning of the (verbally by all Xtians) acknowledged Truth, that God hath the *Ground* of his own existence in himself and that all things were created out of the *Ground*.[4]

In this brief note, we can find some of the central characteristics of Coleridge's reading of Bruno: the link, which he will persistently repeat, between Bruno's metaphysics and the mysticism of Jacob Böhme; the divinization of the universal, infinite substance, considered as an extension of the divinity that in this way expresses its fullness and truth. The note can also be considered as an expression of the anxiety of a man of Christian faith to reconcile his Christianity with those tendencies present in the work of both Bruno and Spinoza toward a radical pantheism—tendencies that gave rise in Coleridge to deep tensions and at times to ambiguities. These tensions and ambiguities would increase with the years, as Coleridge gradually abandoned his youthful unitarianism to accept a full idea of the Trinity.

There is no doubt that Coleridge's attitude toward Bruno becomes more complex and contradictory as he gradually returns to the Anglican Church, becoming the major theorist of the so-called "Broad Church Movement," which aimed at mediating between differing theological and dogmatic positions by proposing the Bible as a unique authority not only in the field of metaphysics and morals, but also in that of daily life and even of politics. In 1816, Coleridge would publish his *Statesman's Manual*, in which he proposed the Bible as a necessary text of reference for the concerns of a modern state, including its politics and its economy. For this reason, Coleridge in his mature years could hardly respond to the thought of a freethinker such as Bruno with the same unconditional enthusiasm as he had shown on his youthful discovery of him in Germany. Even so, as one of the friends he met on that journey would testify in later years, Coleridge always attempted to present Bruno in a favorable light, even from a religious point of view.[5] In his *Philosophical Lectures* of 1819, which were delivered to a wide public in London, Coleridge dedicated a number of intense pages to Bruno, praising him for having proclaimed with such vigor the fundamental unity of the universe and the dignity of the human soul. He even went further, recognizing his profound personal debt to Bruno's thought, and if he felt it incumbent on himself to warn his audience of the dangers inherent in Bruno's pantheism, Coleridge nevertheless declared that Bruno was no atheist: "this man, though a pantheist, was religious."[6] There is an interesting note in this context written by Coleridge in his copy of N. F. Haym, *Biblioteca italiana, o sia notizia de' libri rari nella lingua italiana*, published in London by Tonson and Watts in 1726, where Bruno's *Spaccio de la bestia trionfante* is referred to as "Libro ateistico, ma rarissimo" [an atheist book, and very rare]. Coleridge notes: "ne rarissimo, ne ateistico. Catalogus... maxime imperfectus." [It is neither rare nor atheist. This catalogue is very imperfect.][7]

This note is a help in understanding how it was that when Coleridge published the *Statesman's Manual*, he could choose as an introductory motto to the entire volume a quotation, which he adapted, from Bruno's *De immenso*:

> I beg you, pay attention to these things, however they appear to you at first sight, in order that, though you perhaps may think me mad, you may at least discover the rational principles behind my madness.[8]

In spite of the Anglicanism and the Trinitarianism of the mature Coleridge, his reference to Bruno's works would remain a constant factor throughout his intellectual life. It is in a letter of these mature years, for example, that Coleridge mentions Bruno together with Cicero and Luther as original thinkers and men of genius who, for precisely that reason, were imperfectly understood and often criticized.[9] So it can be said that Bruno remained always, for Coleridge, even in his years of increasing Christian orthodoxy, a thinker of the first order: one of the most important historical sources, together with Proclus and Böhme, of "that philosophy which attempts to explain everything with an analysis of the consciousness, and to construct a world in the mind using the materials furnished by the mind itself."[10]

It is in the context of this original reading of Bruno's works on the part of Coleridge that it is necessary to explain the curious lack in his personal library, at least as far as we know it today, of Schelling's dialogue titled *Bruno* of 1802, which, together with the aforementioned pages of Jacobi and the later section on Bruno in Hegel's history of philosophy, constitutes one of the principal texts concerning Bruno to be produced during the Romantic period.[11] It is possible that this is simply a coincidence, given that in 1802 Coleridge was in a state of profound depression due to a spiritual and matrimonial crisis made more acute by his dependence on opium. It is in 1802 that Coleridge decides to leave Britain for Malta, from there he will travel to Sicily and then be obliged to flee through Rome, Florence, and Pisa to escape the Napoleonic armies. It could, furthermore, be argued that Coleridge had little need either of Schelling's dialogue or of Hegel's later pages on Bruno. Like them, he had read Jacobi at the end of the previous century, in Germany, and had already linked Bruno's works to the philosophy of nature and the transcendental idealism of Schelling before the publication in 1802 of the latter's dialogue on Bruno. The first references to Bruno's works in Coleridge are dated 1801, and they indicate an intense reading of the *De monade* and above all of the *De immenso*, the final work of his Latin masterpiece known as the Frankfurt trilogy, published in that town in 1591.

It is not known how Coleridge obtained a copy of the volume containing both these works—whether during his German visit or after his return to England.[12] What is clear from his *Notebooks*, however, is that he was profoundly struck by his reading of these two texts. The pages on them in the *Notebooks* bear witness to Coleridge's habit of reading furiously through a book that interested him and then coming back to it with a more reflective and critical spirit in later years. In the brief space between a Monday and a Tuesday, in April 1801, Coleridge notes that he read "two works of Giordano Bruno, printed in one book with one title-page." To this observation, Coleridge added a series of long quotations and some notes relating to the quoted texts.[13]

Coleridge, who was both a philosopher and a poet, was particularly struck by the opening ode of the volume in which Bruno repudiates the mythological figure of Dedalus to dedicate himself to an interior flight of the soul, in search of knowledge of the divine. Coleridge judged this ode to have been written with great dignity and elevation of spirit, and in coming years, he would return to this poem to translate it into English. He appears to have read his translation during his philosophical lectures in London, although it survives only in the form of some fragments noted down by one of his listeners. What is clear, however, from the *Notebooks*, is that Coleridge at first preferred the *De immenso* to the *De monade*, which he found too mathematical, "linear," Pythagorean, and intensely obscure. He claimed not to have seen the atomistic *De triplici minimo*, the first work of this trilogy that was published separately, and that he would only discover later among a group of Bruno's texts conserved in Malta, but he did not consider that he had lost a great deal, if it was anything like the *De monade*.

The *De immenso*, on the other hand, was judged by Coleridge, from the opening pages, to be of a quite different order:

... a very sublime enunciation of the dignity of the human Soul, according to the principles of Plato—(Compare Stolberg) and then affirms his own principles.[14]

This is the only comment. The other notes that Coleridge confides to his *Notebooks* are made up of quotations that show that he was already, at least a year before its publication, reading the same pages in Bruno that would inspire Shelling's dialogue. Particular attention is paid to the sixth book of the *De immenso*, and especially to those verses in which the absolute One appears as the fountain from which the Creator orders every living soul to appear, so that the world—the whole universe—can be considered a world of living things. In this way, an eternal process is set in motion according to which the One becomes a multiplicity, and the multiplicity of things returns to the One. The seventh book is also quoted, and in particular those verses in which Bruno is ironic about the idea that all human beings have a unique origin. Bruno's own opinion, notes Coleridge, is that if all human beings in the world should disappear, the soul of the world would produce them again. Nature is a perfect mother who feeds her creatures without requesting any reward.

In these early notes of April 1801, there are also signs of an interest on Coleridge's part for another aspect of Bruno's philosophy—that is, his philosophy of nature or his science. The *Notebook* for that year contains some observations found in the *De immenso* in praise of the Copernican theory, which Coleridge approves of, observing that it was Bruno and not Descartes who first denied the narrow confines of the Aristotelian–Ptolemaic universe. Then there are repeated quotations from those parts of book VI that refer to the microcosm/macrocosm analogy, particularly with respect to the circulation of the blood: "And throughout our body, the blood circulates and recirculates, as if throughout a world, an astral body, the moon."[15] Coleridge is also interested in those pages in which Bruno explains his idea of the vacuum as an empty space filled by the Divinity, or by his ideas on gravity as caused by movements of affinity and repulsion.

This group of passages on the more scientific aspects of Bruno's inquiry suggests that Coleridge was reading Bruno in the light of an English tradition of commentary as well as a German one. It is known that, in the final years of the eighteenth century, when he was still a Unitarian in his religious beliefs, Coleridge was in sympathy with figures such as Joseph Priestley and Erasmus Darwin who, although still materialists and rationalists in the tradition of the Enlightenment, were nevertheless beginning to develop a concept of matter as animated by vital powers, in a continuous state of evolution. Coleridge followed with much interest such new scientific developments as Priestley's studies of electricity, or the botanical studies of Erasmus Darwin, which were producing the first ideas concerning evolution. In a work such as Priestley's *Disquisitions on Matter and Spirit* of 1777, Coleridge would have found references to "the famous Jordano Bruno" seen as a precursor of Locke and Andrew Baxter in the conviction that all the powers of matter should be considered as the direct work of a divine power. Priestley's thesis that matter should not be thought of, in traditional terms, as inert substance, but rather as essentially pervaded by forces of attraction and repulsion, would lead Coleridge toward a fully vitalistic idea of nature that he could later reconcile with the idealism he derived from Schelling. So it is no surprise to find Bruno's name mentioned by Coleridge, as one of the first exponents of a dynamic philosophy in the physical sciences, in one of the most articulate pages of his natural philosophy. In *Aids to Reflection*, Coleridge repudiates as excessively abstract the Cartesian idea of nature as "a lifeless machine whirled about by the dust of its own grinding"—an idea that he claims had been of incalculable importance when moving bodies were considered as geometrical constructions or as subject to algebraic calculations, but that was clearly inadequate when it came to considering them as "a truth of fact." Coleridge reacts against Descartes's lifeless nature by invoking the names of Bruno and of Kant, and proposing the idea of a world "created and filled with productive forces by the almighty *Fiat*."[16]

It is impossible not to be struck by Coleridge's effort in these pages to reconcile Bruno with the German critical philosophy through an eloquent reference to the Bible. Insofar as he recognizes the Bible as the ultimate authority not only on spiritual and moral matters but also on philosophical ones, Coleridge feels justified in considering nature as a dynamic force, vibrating with obscure and vital energies, not only on the basis of preceding philosophies but also on the basis of the story of Genesis, as well as, more generally, the whole tone and text of the Scriptures. Coleridge had read Bruno's *The Ash Wednesday Supper*, to which he refers several times in his effort to affirm Bruno's seriousness as a philosopher. He claims that Bruno was unjustly accused of being a heretic, and that this is evident if it is remembered, as the *Supper* testifies, that in London he frequented the company of Christians as prestigious as Sir Philip Sidney or Sir Fulke Greville.[17] Nevertheless, Coleridge could hardly have missed the pages of the fourth dialogue of that work in which Bruno, anticipating Galileo and Francis Bacon, attempts to circumscribe the authority of the Scriptures to a moral sphere, necessary for the instruction of the masses but not of the natural philosopher, who has no need of such instruction to justify his conclusions about the natural world. In the light of Bruno's claim in this sense, it comes as no surprise to see Coleridge, in another place, taking his distance from Bruno's ideas about the creation of the world. For Bruno's belief in the eternity as well as the infinity of the universe could hardly be reconciled with that Biblical "story" of a creation in time that Coleridge himself considered as perfectly in line with common sense and the experience of the senses.

This reference is of great importance because Coleridge is associating Bruno's thought here with a discussion that will assume a prominent position in nineteenth-century culture: the discussion concerning the theory of evolution. In a page of his philosophical manuscripts, unfortunately undated but probably written quite late in his life, Coleridge assumes a decidedly critical attitude toward the new evolutionary ideas that were beginning to circulate at that time.[18] He is anticipating here the terms of the violent discussion that will break out in Victorian England after the publication of Charles Darwin's *On the Origin of Species* in 1859. For in this manuscript note, Coleridge already expresses a decided repudiation of all those theories that daringly attempt to work their way back toward the bestial larva from which mankind originated. Such an attempt, as Coleridge was already aware, tended to present mankind as the "gay image" of one or another species of monkey. Coleridge does not refer here, as he could have done, to *The Ash Wednesday Supper*, where Bruno himself had announced, in his *Proemiale epistola*, a concept of the fundamental and vital unity of an animated and evolving universe that would have "made apes roar with laughter"—a passage that earned him many a quotation toward the end of the century on the part of those historians of science who were attempting to trace the historical development of the evolution idea.[19] What Coleridge does do is to distinguish between two differing evolutionary concepts of the origins of humankind: one fundamentally rational that works back to the beginnings stage by stage, and another of an Epicurean derivation that can be expressed in the image of a gigantic and spontaneous birth, giving rise to innumerable new creatures that feed on the innumerable breasts protruding from the original clay. Coleridge expresses his dislike of both these theories, although he admits to preferring the second to the first, and it is to this Epicurean theory that he links the name of Bruno:

> A modern Philosopher and Poet, and in both characters a man of vigorous and original Genius, no Epicurean but in as ill odour with Divines as Epicurus himself (I mean the Philosopher from Nola, Giordano Bruno, whom the Idolators of Rome burnt for an Atheist in the year 1600), assigns the same origin to the human race and supports his opinion both in his Latin poems and the Prose annotations at great length. It is indeed a natural consequence of his Dogma, that the Earth is "etherogeneum integram animal."

After a series of quotations from the sixth and seventh books of his favorite Brunian work, the *De immenso*, Coleridge goes on to underline in this note how Bruno had revived the Epicurean concept of an infinite universe, justifying a priori his idea of the fixed stars as suns, each at the center of its own planetary system, on the basis of an interactive center/periphery process of revolution, seen as a primary law of matter.

The final part of this note consists of an attempt to distinguish the thought of Bruno from what Coleridge considers the undeniable atheism of Epicurus himself. The planetary souls that Bruno sees as vivifying nature, Coleridge claims, are everywhere considered as ministerial powers, and nature herself as the vicar or creature of a great *Opifex*.[20] If it is true that for Bruno, mankind is born from the womb of the universe and not directly from a Divine Creator, nevertheless his work is pervaded by a spiritual principle, and by the eloquence of a true piety and morality. In this sense, Coleridge claims, it vies with the work of that Fénelon who had so strenuously denied the Epicurean philosophy in the name of the "existence de Dieu."[21]

In another part of this same note, which constitutes one of the principal moments of his meditation on Bruno's philosophy, Coleridge defines Bruno's law of matter as a "Law of Polarity." In Coleridge's view, this is one of the ways in which Bruno anticipates modern ideas—in particular, those of the Romantic philosophy itself. This idea of a law of "polarity" has been recently at the center of critical attention toward Coleridge's own works—an attention that increasingly tends to privilege the philosopher with respect to the poet, and to emphasize the fundamental importance of his interest in logic.[22] It has already been underlined how Coleridge repeatedly claims to have learned such a logic from one of Bruno's works, the *De progressu et lampade venatoria logicorum*, published at Wittenberg in 1587.[23] In his *Notebook*, Coleridge claims to have seen this work while in Malta, among a group of ten other works of Bruno's that he found there. It is easy to see how readily Coleridge would have been struck by Bruno's idea of a "hunt" for absolute truth through use of the categories of *Differentia*, *Contrarietas*, and *Concordantia* that he derived from Raymond Lull.[24] As far as its particulars are concerned, however, Coleridge admits that he had read this text only hurriedly, and had forgotten much of it. Later, after his return to England, he tried in vain to find another copy. For this reason, it is important to stress that Bruno's logic of contraries would have been available to Coleridge from other works as well, such as the passages quoted by Jacobi from the *De la causa*, or various pages of the *De monade* or the *De immenso*, which seem to have been the texts of Bruno that Coleridge most constantly frequented.

Bruno's concept of the universe is clearly defined in chapter I of book V of the *De immenso*, titled "Everything is made up of the same elements, so that everything is in everything." Here Bruno expresses his idea of the universe as a mirror or seal of a perfect intelligence, whose imprint is not only evident *within* nature, but whose incorruptible order *becomes* the order that unites natural bodies. For Bruno, the order of bodies "follows" that of the divine intellect, as the footprint follows the foot and the shadow follows the body. The natural order, however, is not made up of static structures, but rather of powers, influxes, and forces in a state of continuous tension and vicissitude.[25]

From a note on Coleridge's copy of the works of Richard Baxter, it is clear that Baxter was, in Coleridge's opinion, a precursor of the idea of a fundamentally tripartite logical disposition of the mind, as well as of the universe that it perceives. Coleridge refers to the Kant of the *Critique of Pure Reason* and again to the Bruno of the *Logica venatrix veritatis*, whose contents he says, once again, that he was unable to remember clearly.[26] Immediately after these references, Coleridge makes an important mention of the Pythagorean *Tetractys*: "the eternal fount or source of nature, sacred to the contemplation of identity, and first in order of thought with respect to any kind of division." This memory of Pythagoras together with that of the logic of Raymond Lull suggests that Coleridge had in mind a series of "figures" that Bruno explicates in the *De monade*, a work of a clearly Pythagorean inspiration. There the eternal fount or source of nature is imagined as a circle that Bruno calls the "Ring of Apollo," because it symbolizes the eternal perfection of the divine intelligence. Within this circle, an equilateral triangle represents the triadic movement of logical thought. The first angle of the triangle indicates the apprehension of the whole; the second, the understanding of the simple elements in their dualistic tension; the third, the moment of synthesis, which Bruno calls the moment of discourse. Beside the Ring of Apollo, Bruno draws a triangle that contains within it three equal and contiguous circles, while the sides of the triangle act as tangents to the three circles. This is the triangle of the mind, which explicates the natural order as a logical principle of trichotomy; Bruno calls it the "Table of the Graces" and claims that it contains the secret of universal order. It is by passing through this triangle that the mind arrives at the seal of the *Tetractys*, or Sign of Four, which Bruno calls "Ocean." This seal teaches us that many things are consequential, because given a triangle it is possible to derive from it all forms of parallelogram. The "polar" logic of thesis and antithesis is resolved in the coincidence of contraries, or the principle of identity, and teaches us how to cross the ocean of being.[27]

In his page of comments on Baxter's book, Coleridge seems to follow Bruno's reasoning very closely. Nevertheless, we find in Coleridge a more insistent emphasis on a transcendental principle anterior to that "ocean" of being that, in Bruno, becomes the only sphere available to human knowledge. Coleridge, for his part, distinguishes very clearly between the *Tectratys*, which precedes division in the order of thought, and the *Trichotemia*, the Sign of Three, which is the universal form of division. The *Tectratys* is a *Prothesis*, and, insofar as it is anterior to the *Thesis*, it cannot be considered a part of it. In this way, we have a *Prothesis* ("eternal fount of nature") as an anterior unity, which becomes multiplicity in the "polar" terms of *Thesis* and *Antithesis*, to then resolve itself in the new unity of the synthesis (as in Bruno's figure of the *Three Graces*). This theme returns again and again, in many different contexts, in Coleridge's philosophical writings. Above all, the triadic structure of thought is amply illustrated in sections XIX and XX of his *Logic*, and if there is no mention of Bruno in this text, there can be no doubt, as we have already seen, that his "magic" of contraries was well known to Coleridge. The fact that the explicit references in Coleridge's *Logic* are to Kant and Schelling does not mean that Bruno was absent from his thought, as the importance of Bruno's "polar" multiplicity in the development of the new transcendental idealism had already been fully recognized by Schelling himself in his dialogue titled *Bruno*. It will be remembered again by Hegel, where he sees in Bruno a thinker for whom knowledge of the unity of form and matter meant "to study the opposing and repugnant terms of things, the Maximum and the Minimum."[28]

Coleridge thus moves in this aspect of his reference to Bruno along a path already traced by the German philosophers. A more original element can be found in a recent interpretation of his *Logic* in the sense of a revision of the transcendental philosophy of Kant that reveals the dependence of its categories on the structures of a transcendental grammar.[29] This reading sees the underlying idea of Coleridge's *Logic* as a claim that the forms of a transcendental grammar determine our perceptions and the structure of our thought. This is similar to the idea expressed by Bruno in his figure of the Rings of Apollo, according to which the triadic movement of the intellect comprises not only the moment of the apprehension of unity and the moment of the comprehension of the contrary elements in multiplicity, but also the moment of "discourse."

It is in the context of this relationship between the apprehension of unity and the moment of discourse that the problem arises of Bruno's influence on Coleridge's aesthetic inquiry. Already many years before the elaboration of his ideas about the artistic imagination in the *Biographia Literaria*, Coleridge had shown a marked tendency to refer to Bruno in an artistic context, linking his name to two of the major poets of the English tradition, Shakespeare and Milton. The note appears in the *Notebook* in which Coleridge wrote down his impressions during his journey to Malta, and it is linked to a series of observations on the value of the lives of the great poets, and on the idea of the journey as a search for the places rendered "sacred" by their story. In a series of notes dated April 19, 1804, Coleridge writes that it would be of no interest to him personally to know that Shakespeare had planted a myrtle in this place or that Milton had lain down in that particular field. On the other hand, he admits that he had been profoundly moved by the sight of the beach on which Bruno had probably sought refuge from an enraged priesthood. In the following note, Coleridge develops these thought on the lives of great men:

> ... a Shakespeare, a Milton, a Bruno, exist in the mind as *pure Action*, defecated of all that is material and passive.—And the great moments, that formed them—it is hard and an impiety against a voice within us, not to regard as predestined, and therefore things of Now and For Ever and which were Always. But it degrades this sacred Feeling, and it is to it what stupid Superstition is to enthusiastic Religion, when a man makes a Pilgrimage to see a great man's Shin Bone found unmouldered in his Coffin, etc.[30]

Here the memory of the spirit of the "great man" as pure Action, which may seem already to conjure up the spirit of Carlyle, has to be put in its proper context—that is, an artistic context, or a discourse on poetic genius, with Bruno appearing together with the names of Shakespeare and Milton.

When, in 1815–1816, Coleridge writes the *Biographia Literaria*, where he defines the artistic imagination as a faculty involved in a continuous struggle to "idealize" or to "unify" those fragmented perceptions that logic alone cannot recompose, Bruno is still present in his mind, even if he appears only briefly. In these pages, Coleridge lays the foundation for a new aesthetic that will have profound repercussions on the British literary and artistic tradition, which are far from being exhausted even today. He makes no secret of the fact that he is moving in the context of the new German philosophy, quoting from Kant, Fichte, and Schelling as thinkers who had already elaborated a new definition of the artistic imagination with respect to the mechanical and associationist theories that had characterized the period of the European Enlightenment. It is in this context of Coleridge's reference to the new German philosophy that the name of Bruno reappears as one of those who had already developed a dynamic concept of thought.[31] And if in the *Biographia Literaria* itself Bruno is not present in the parts of the book devoted to Coleridge's thoughts about the poetic or artistic imagination, it should not be forgotten that he had already figured prominently in the *Preliminary Essays* written and published shortly before the *Biographia*, which prepare the way for Coleridge's mature thought on the arts.[32] These essays are preceded by a quotation from Bruno's *De umbris idearum* published in Paris in 1582. The choice of quotation is a significant one and constitutes a crucial moment of reference to Bruno's philosophy on the part of Coleridge.

Its significance seems not to have been correctly estimated even by those commentators who already have (or ought to have) considered the relationship between Bruno's aesthetics and Coleridge's artistic imagination. There is, for example, no mention of Bruno at all in Thomas McFarland's important chapter on "The Origin and Significance of Coleridge's Theory of Secondary Imagination," while Giancarlo Maiorana, although generous in his estimation of the importance of Bruno for Coleridge's thought on this subject, discusses the relationship almost entirely in terms of Bruno's final Italian dialogue, the *Heroici furori*, written and published in London in 1585, which is never mentioned by Coleridge and which he seems not to have read.[33]

The *De umbris* figures among the group of Bruno texts that Coleridge found in Malta, and that he later had difficulty in finding again. For this reason, he quotes it from memory, with some minor differences with respect to the original, but without any changes of importance. "We may say that the sun, which remains eternally one and identical, appears with a different face according to different observers," Bruno had written in the opening speech of this dialogue, which is an introduction to his theme pronounced by Hermes.[34] The sun, although always one and identical, appears differently to different observers in different circumstances, and so what Hermes calls his "solar art" will necessarily give rise to a fragmented vision of the truth, always different according to different circumstances and observers. This absolute individuality of every single perception of the truth is the secret that Bruno reveals in his work, which is that same *De umbris idearum* that Hermes presents with many misgivings, anticipating that it will be misunderstood. Hermes fears above all what he calls the "armed bowmen"—that is to say, the grammarian pedants who manage to convince themselves that they are in possession of an unchanging and universal truth. The other figure involved in the discussion, Philothimus, replies that it is essential to find the courage to present new ideas, even if they are often neither respected nor understood: "If everybody was only afraid, and remained silent, nobody would ever attempt new works and nothing of any dignity would ever be achieved."[35] Philothimus continues by invoking the figures of those whose imaginations are especially brilliant and inspired: Mercuries sent to earth by the gods to lead peoples' minds back toward a divine illumination. They are figures whose arduous and exalted vision of the truth may remind us of the Magus-Poet of Coleridge's *Kubla Khan*, with his gleaming eyes and flowing hair:

Weave a circle round him thrice,
And close your eyes with holy dread,
For he on honey-dew hath fed,
And drunk the milk of Paradise.

Coleridge's *Kubla Khan* was published in 1816 at the insistence of Lord Byron, although it had been written several years earlier, in the summer of 1797 or 1798. In a famous note of introduction to the published poem, Coleridge declared that it was written in a dream, during which all the images rose up in his mind as things, without the mediation of rational or logical modes of discourse. The beautiful, writes Coleridge in his later *Preliminary Essays*, is above all "Multeity in Unity." This definition appears to echo the twenty-fourth "Intention" of Bruno's *De umbris*, where the reader is invited to consider how a multiple light produces multiple shadows from a single body, and how innumerable lights produce innumerable shadows, even if they do not appear to us in sensible form.[36] The solar art proposed by Bruno consists in an effort to fragment the deceitful unity of common perception in order to pursue within the shadows of diversity a higher and ultimately infinite unity— an art that is clearly in line with Coleridge's well-known distinction between a primary (or common) and a secondary (or artistic) imagination. Schelling too had opened his dialogue *Bruno* with a lengthy discussion of the beautiful, which he saw as leading toward Bruno's infinite unity. In the *Preliminary Essays*, Coleridge, for his part, makes no explicit mention of Bruno after the initial quotation from the *De umbris*, but the choice of quotation itself, made with the secure touch of a masterly philosopher-poet, is a significant indication of the way in which Bruno had already defined the sense of a solitary "solar" art. This idea of art clearly played an important role in Coleridge's own definition of the unifying powers of the artistic imagination.

In the context of the search for an infinite unity, Schelling had already noted several times in his *System des transzendentalen Idealismus* the importance for the new idealistic philosophy of the Christian mystical tradition. In the case of Coleridge, a particular importance can be assigned to his reading of the works of Jacob Böhme in the translation by William Law (1754)—four volumes in which, between February 1808 and March 1826, Coleridge left a dense series of manuscript notes that have only recently been published, and that seem destined to become one of the major documents of English Romanticism.[37] The notes begin with a series of observations on Böhme's life that highlight an episode of his infancy when he leaves his playfellows to go and explore a cave. Inside he finds a dish full of coins, but rather than pocketing them, he runs outside to tell the other children what he has found. When they go back into the cave together, he can no longer find the treasure, although some years later, he hears that a stranger to those parts, instructed in the magical arts, had found the coins and taken them away with him, only to die, later, a violent death. In his note on this episode, Coleridge wonders if it might not be an allegory through which Böhme wished to indicate a philosophical treasure of which he and his companions were unable to enter into possession, but which Bruno had revealed in his works. However that may be, the note clearly indicates that Coleridge intended to carry out his reading of Böhme in the light of Bruno's philosophy, and he confirms this intention by writing one after the other the date of Böhme's birth (1575) and death (1624), followed by the date of Bruno's death, which this time he mistakes slightly, giving it as 1601 rather than 1600.

The image of the cave is clearly significant. It constitutes a rewriting of the well-known Platonic myth according to which the light of the ideas shines outside the cave, leaving mankind chained in the shadowy interior to live a life of miserable exile. Böhme, however, like Bruno himself in Coleridge's opinion, wants to find the treasure of truth that remains hidden among the shadows by penetrating ever more deeply into the heart of the cave itself: their philosophies propose to search among the deepest shadows in order to discover the traces of infinity that they conceal. What they are both pursuing is thus, for Coleridge, the divine qualities that lie within every aspect of the universe. It is therefore no coincidence that it is in the course of his reflection on Böhme's concept of quality that we find an explicit reference to Bruno in Coleridge's marginalia. A quality is that aspect of every elemental power that expresses the specific energies of its species, although in the divinity itself there is an absolute synthesis of the struggle between contrary qualities. In the idea of this synthesis, Coleridge finds one of the most profound mysteries that the human mind is called on to contemplate: "Plato in *Parmenides* and Giordano Bruno passim have spoken many things well on this awful Mystery / the latter more clearly."

Such references indicate how Coleridge in his later years, although ever more intent on reconciling his idealistic transcendentalism with Christian doctrine, never repudiated Bruno's philosophy. On the contrary, his reading of Böhme offered him the context in which to return with a new enthusiasm to that One and infinite Maximum that Bruno had spoken of in the *De immenso*, first read by Coleridge between a Monday and a Tuesday of April 1801.

One of the major Coleridge scholars of our times, Thomas McFarland, has written of Coleridge's Bruno/Böhme relationship with considerable skepticism. He considers the presence of Böhme within Romantic culture in general, and Coleridge's in particular, of great philosophical importance, while that of Bruno appears to him as little more than a cultural fashion, largely based on an interest in his dramatic life and death. In McFarland's opinion, Bruno's works, especially as read by the Romantics, have no real philosophical consistency. This accusation is surely mistaken, although, as we will see, it could be moved against some of the commentators of the end of the nineteenth century. Exponents of the early Romantic period, such as Coleridge, however, had little information available to them about the exact terms of Bruno's life and death, for the first full-scale biographies would appear only later in the century. Figures such as Jacobi, Schelling, or Hegel, and indeed Coleridge himself, rarely refer to Bruno's life story. Their pages on Bruno show quite clearly that their interest is centered on his works, and that it is primarily philosophical. It is dedicated above all to those works, such as the *De la causa* or the *De immenso*, in which Bruno attempts to define the relationship between his infinitistic and vitalistic philosophy of nature and the metaphysical status of the first cause. Their reading of Bruno's philosophy is a new one, of remarkable speculative and historical significance. It is carried out on the basis of a direct confrontation with a selected number of Bruno's texts, in spite of the fact that they were still not readily available in modern editions.[38]

Unfortunately, in the case of Coleridge, the fragmentary nature of his philosophical discourse, little understood by his British contemporaries, led to only a few hints about Bruno's works reaching a public that would remain largely indifferent to Coleridge's enthusiasm. Only his faithful admirer, Thomas De Quincey, who followed Coleridge's example in studying the new German philosophy of the period, would reply to his invitation to read Bruno's works. His library contains a copy of the same volume, the *De monade* bound together with the *De immenso*, that was such favorite reading with Coleridge. In his copy of this volume, De Quincey wrote: "Bought this day, Wednesday, May 31st, 1809; brought home this evening between 8 and 9 o'clock."[39] Later, in a letter to the publisher Blackwood of 1830, De Quincey would say that he was "rich" in Bruno texts, and that he had actually read them. He proposed to the publisher that he write an essay on Bruno's philosophy, but the project seems to have gone no further.[40] As for the other friends and admirers of Coleridge, they appear to have remained quite untouched by his interest in Bruno. When the essayist, Charles Lamb, wrote to Coleridge on August 26, 1814, expressing lively criticism of the Reverend Julius Charles Hare, who had refused to lend Coleridge the volumes of his precious collection of Bruno texts, Lamb confided candidly (and confusedly) to Coleridge that he had never touched the books of "Bishop" Bruno.[41]

More surprisingly still, there appears to be a total lack of references to Bruno on the part of the other major poets of British Romanticism, in spite of the fact that many of them were far more at home with Italy and its language than Coleridge himself. Shelley's intense reading of Italian literature, for example, seems not to have touched on Bruno's texts, which could well have had much to say to such a Neoplatonic and revolutionary poet, with his lyrical pursuit of the ineffable beauty of an infinite One. Even Byron, who would surely have appreciated the Epicureanism of such a heretical proponent of free thought, appears to have lived for several years in the Mocenigo palace in Venice without becoming aware of the role played in the arrest and trial of Bruno by Juan Mocenigo, whose denunciation of Bruno to the Venetian Inquisition in May 1592, while Bruno was living in his palace, had led ultimately to the philosopher's death at the stake in Rome in 1600. It is true that there are two Mocenigo palaces in Venice involved in these two stories, an older one inhabited by Bruno and a more modern one inhabited by Byron. Two distinct branches of the family were involved. Nevertheless, the Mocenigo family and their Venetian palaces forge a close link between these two colorful and unorthodox rebels, and it is surprising that Byron appears to have had no knowledge whatever of Bruno's life or works.

It thus seems necessary to conclude that the story of Bruno and British high Romanticism involves almost exclusively Samuel Taylor Coleridge. His reading of Bruno can be claimed as an integral part of his remarkable intellectual biography, unfortunately to a large extent confided to his untidy and fragmented manuscripts and marginalia that are being systematically edited only today.

NOTES

1. The most important documents relating to Coleridge's German experience consist of the letters he wrote to his family and friends. For the letter quoted, see *Collected Letters of Samuel Taylor Coleridge*, ed. E. L. Griggs, Oxford, UK, Clarendon Press, 1956–1959, 4 vols.: I, 285. For a modern account of Coleridge in Germany, see W. Jackson Bate, *Coleridge*, London, Weidenfeld and Nicolson, 1969, 91–96.

2. Most of Coleridge's annotated texts are conserved in the British Library in London. The *Marginalia* are published as six separate parts of vol. 12 in *The Collected Coleridge*, eds. Heather Jackson and George Whalley, London–New York, Routledge and Princeton University Press, 1980–.

3. For an account of the reception of Bruno's works in European cultures from his own day to the Enlightenment, see Ricci (1990b). See Ricci (1991) for a study of Bruno in the Romantic period in France and Germany.

4. This note is published in *Marginalia*, in *The Collected Coleridge*, op. cit., vol. XII, part III, 1992, note 12.

5. See C. Carlyon, *Early Years and Late Reflections,* London, Whittaker and Co., 1836, 193–95.

6. See S. T. Coleridge, *Philosophical Lectures,* ed. K. Coburn, London–New York, Pilot Press, 1949, 323–27.

7. See *The Marginalia*, op. cit., II, 976.

8. See *The Statesman's Manual* in *Lay Sermons,* ed. R. J. White, London–New York, Routledge and Princeton University Press, 1972 (*Collected Coleridge*, 6), 3 and 112. Coleridge was particularly fond of this passage, which he copied into

his *Notebooks* (ed. K. Coburn, New York, Pantheon Books, 1957–1961, 2 vols.: I, note 927) and then quoted again in a letter to his brother George (see *Collected Letters*, op. cit., III, 133). He later quoted it once more in *The Friend*, ed. B. Rooke, London–New York, Routledge and Princeton University Press, 1969 (*Collected Coleridge*, 4), 2 vols.: II, 87 for the edition of 1809 and 125 for the edition of 1818. For the original passage, see the *De immenso* in Bruno (1879), I, i, 208, where it reads: "*Ad isthaec, quaeso vos, qualiacunque primo videantur aspectu, (si iniqui judicii titulum abhorretis) adtendite: ut qui vobis insanire videtur, saltem, quibus insaniat rationibus, cognoscatis.*"

9. See in *Letters*, op. cit., IV, a letter dated April, 1819.

10. Ibid., III, in a letter to Lady Beaumont of January 1810.

11. Schelling's *Bruno* appeared for the first time in 1802 published by Unger in Berlin. Hegel's lectures on the history of philosophy were delivered in Berlin starting in 1818. For his pages on Bruno, see the Introduction to this volume, note 26.

12. There was as yet no modern edition of the trilogy, which Coleridge must have been reading in a first edition. For the extant copies of first editions of this work, see the bibliography of Bruno first editions by Sturlese (1987), 118–31.

13. See *The Notebooks*, op. cit., I, notes 927–29.

14. See *The Notebooks*, op. cit., I, note 928. The Stolberg referred to was probably Friedrich Leopold Stolberg (1750–1819), also a poet and a philosopher, who had proposed a Christian interpretation of Platonism. Together with his brother Christian (1748–1821), he moved in the circle of Goethe, as well as of Klopstock, whom Coleridge had met while in Germany.

15. See *The Notebooks*, op. cit., I, note 928. For Bruno's possible anticipation of Harvey's discovery of the circulation of the blood, see Gregory (2002).

16. See *Aids to Reflection*, ed. John Beer, in *Collected Coleridge*, 1993, op. cit., vol. 9, 400.

17. See the letter to Henry Crabbe Robinson, *Collected Letters* IV, June 1817, in which Coleridge writes: "Giordano Bruno's ... Ash Wednesday Week [*sic*] contains a highly curious and interesting account of his adventures in London."

18. This is one of Coleridge's most extended and significant references to Bruno, which remained for many years hidden in his private manuscripts (BL Egerton 2801, 15–77). It was first published by A. Snyder, with the briefest of comments, in "Coleridge and Giordano Bruno," *Modern Language Notes* XLII (1927): 427–36.

19. For Bruno's laughing apes, see Bruno (1977), 68.

20. In a late note of January–February 1820, Coleridge seems to reconsider this point, postulating a transcendental potency in which three heterogeneous powers are necessary for it to act on bodies. In this context of thought, he notes, now with disapproval: "the Worlds *not* animantia as Giord. Bruno holds them." See *The Notebooks*, op. cit., IV, note 4639.

21. The reference is to François de Salignac de la Mothe Fénelon, *Oeuvres philosophiques*, part I, chapter LXXVIII: "Les suppositions des Epicuriens sont fausses et chimériques." First published in 1718. See the edition published in Amsterdam by Zacherie Chatelain in 1731.

22. See in particular T. H. Levere, *Poetry Realized in Nature: Samuel Taylor Coleridge and Early Nineteenth-Century Science*, Cambridge, UK, Cambridge

University Press, 1981 (in particular the section titled "Powers and Polarity," 108–21), and T. McFarland, *Romanticism and the Forms of Ruin*, Princeton, NJ, Princeton University Press, 1981 (in particular chapter 5, "A Complex Dialogue: Coleridge's Doctrine of Polarity and Its European Contexts"). Coleridge's *Logic* has been published in the *Collected Coleridge*, op. cit., vol. 13, ed. J. R. Jackson, 1981.

23. O. Barfield in *What Coleridge Thought*, Oxford, UK, Oxford University Press, 1971, was the first to underline the importance of Coleridge's doctrine of polar logic. His book contains an appendix that attempts to read Bruno's *De progressu et lampade venatoria logicorum* in this context, but without succeeding, as the author himself admits, in getting fully into focus this important link between Bruno and Coleridge. This may be due to the fact that Coleridge makes only a very generic and hurried reference to this particular work

24. The eleven Bruno texts that Coleridge found in Malta are listed in his 1804 *Notebook* as: "1. De umbris idearum [*De umbris idearum*, Parisiis 1582]— 2. Acrostismus [*Camoeracensis acrotismus*, Vitebergae 1588]—3. De progressu et lampade venatoria Logicorum [*De progressu et lampade venatoria logicorum*, Wittenberg 1587]—4. Artic. 160. Advers. Mathematicos [*Articuli centum et sexaginta adversus huius tempestatis mathematicos atque philosophus*, Pragae 1588]—5. Candelajo, Comedia [*Candelaio*, Parigi 1582]—6. Della Bestia [*Spaccio de la bestia trionfante*, Parigi (London) 1584]—7. La cena delle Ceneri [*La cena de le ceneri*, London 1584]—8. Dialoghi della ca. [The editor of the *Notebooks* interprets this incomplete title as referring to two rare works, *Dialogi duo de Fabricii Mordentis* and *Dialogi Idiota triumphans*, published in Paris in 1586. It seems more probable that they refer to the Italian dialogue *Cabala del cavallo pegaseo*, Parigi (London) 1585, given that the title is inserted among a group of Bruno's Italian works.]—9. Dell'infinito etc. [*De l'infinito universo et mondi*, Venezia (London) 1584]—10. De triplici minimo etc. [*De triplici minimo et mensura*, Francofurti 1591]—11. Explic. Triginta etc. [*Explicatio triginta sigillorum*, London 1583]." See *The Notebooks*, op. cit., II, note 2264.

25. For chapter 1 of book V of the *De immenso*, see Bruno (2000c), 704–14. For a comment on this aspect of Bruno's thought, see Granada (1994).

26. See Coleridge, *The Marginalia*, I, op. cit., p. 347. There is no work of Bruno's called the *Logica venatrix veritatis*, which is probably an ill-remembered title for the *De progressu et lampade venatoria logicorum*.

27. For these pages, see Bruno (2000c), 286–303.

28. For Hegel's pages on Bruno in his *History of Philosophy*, see note 26 in the introduction to this book.

29. See J. C. McKusick, *Coleridge's Philosophy of Language*, New Haven, CT, Yale University Press, 1986, in particular 119–48.

30. See *The Notebooks*, op. cit., II, note 2026.

31. In particular in chapter 9, titled "Philosophy as Science." See S. T. Coleridge, *Biographia literaria*, eds. J. Engell and W. Jackson Bate, Routledge and Princeton University Press, London–New York, Routledge and Princeton University Press, 1983, 145.

32. *Preliminary Essays: On the Principle of Sound Criticism concerning the Fine Arts, dediuced from those which animate and Guide the True Artist in*

the Productions of His works. The three essays that compose this work were first published in 1814 in *Felix Farley's Bristol Journal* and then republished in J. Cottle, *Early Recollections Chiefly Related to the Late Samuel Taylor Coleridge,* II, London, Longman and Rees, 1837.

33. For McFarland's chapter, see *Originality and Imagination,* Baltimore, MD, John Hopkins University Press, 1985, 90–119. See also Maiorana (1982).

34. See Bruno (2004), 16–21, and the comment by Nicoletta Tirrananzi at 387–88.

35. Bruno (2004), 20–23.

36. Coleridge's definition is found in the third essay in Cottle, op. cit., 221. For Bruno's 24th–26th Intentions, see Bruno (2004), 72–77, and the comment at 426–28.

37. See *The Marginalia,* op. cit., I, 553–696. It was Thomas De Quincey who gave Coleridge these volumes of Böhme, translated by Law, in February 1808. Some months afterward, De Quincey visited Coleridge and saw that the first volume was full of notes written in his own hand.

38. Coleridge, for his part, was extremely severe with those who read philosophical texts superficially. In a marginal note to the ninth volume of his copy of the *Geschichte der Philosophie* by Wilhelm Gottlieb Tennemann, published in Lipsia in 1798–1819, he writes: "It grieves me to say that this Volume is a mere Bookseller's Order executed in the true book-making style—In short, with the exception of the account of Pomponatius, it is a poor compilation from common books, and the article on Giordano Bruno especially heartless and superficial—a mere skim from one or two only of Bruno's writings—while his interesting attempts in Logic and Mnemonic are passed over altogether." This comment was first published by A. Snyder, op. cit., 436.

39. See Sturlese (1987), 125.

40. See H. Ainsworth, *Thomas De Quincey. A Biography,* Oxford, UK, Oxford University Press, 1936, 329.

41. *The Letters of Charles and Mary Lamb,* ed. E. W. Marrs Jr., vol. III, Cornell University Press, Ithaca, NY–London, Cornell University Press, 1978, 107–8. For J. C. Hare's collection of Bruno texts, at present held by Trinity College, Cambridge, UK, see the relevant entries in Sturlese (1987).

11

BRUNO AND THE VICTORIANS

GEORGE HENRY LEWES

WHEN J. C. SHAIRP PUBLISHED HIS *Studies in Poetry and Philosophy* in 1868, he included a section on Coleridge emphasizing the break that, under his influence, separated the Romantic and idealistic period of the beginning of the century from the culture of the Enlightenment. From the point of view of Shairp, which is also that of Coleridge, the Enlightenment was based on a utilitarian attitude that denoted an active but restricted and unimaginative intelligence, notably deprived of fantasy, profound sentiment, a sense of reverence, or spiritual sensibility. Shairp added that, in the Victorian England in which he was writing, there were clear symptoms of a renewed dominion of this rationalist spirit, after the temporary pause due to the Romantic interlude.[1] This was the moment of affirmation of the utilitarianism of Bentham and John Stuart Mill, joined to a lively interest in the positivism of Comte. The result was a strong revival of the rational spirit that was acting as the theoretical basis of the robust development of the positive sciences and the industrial revolution that characterized so much of the culture of Victorian England.[2] John Stuart Mill's own 1840 essay on the philosophy of Coleridge, which makes no mention of the latter's reading of Bruno, expresses a generous tribute to the "great awakening" operated by the new idealism. Nevertheless, that same idealism was judged by Mill as a traditional philosophy, enclosed within the bounds of a nebulous and abstract metaphysics, which his own school had abandoned in order to proceed in the footsteps of the rationalism of Locke and David Hartley.[3] A reference to the Nolan philosophy in such a context could only be of a very different sort from that found in the works of Coleridge.

It was no longer such a difficult task to find Bruno's works. His Italian dialogues written in London had been available since 1830 edited by Adolf Wagner, while the first complete biography, by Christian Bartholmèss, published in Paris in the 1840s, as well as being an intellectual biography greatly facilitated a knowledge of Bruno's dramatic and tormented life and death. Nevertheless, the Victorian culture contains only rare references to those very texts that had challenged the critical intelligence of Coleridge. None of the major minds of the period, including Carlyle, Ruskin, Arnold, and William Morris, many of whom were closely linked to the Italian culture both of the past and of their own time, appear even to have heard of the works of the Nolan philosopher. On a more general level, the prevailing attitude of a society that can be defined as rich and decorous in its middle-class virtue soon showed signs of suspicion toward a figure whose clearly unorthodox tendencies suggested he would better be kept at arm's length. Clement Carlyon, the friend with whom, half a century earlier, Coleridge had discovered new philosophies and new books in Germany, wrote in his *Early Years and Late Reflections* of 1836 that Coleridge had always talked in favorable terms of Bruno's thought. Nevertheless, Carlyon himself is above all intent on repeating Coleridge's own warnings of the dangers of Bruno's and Spinoza's pantheism. Later he refers to Bruno's *De monade* in order to claim that he is unable to understand how sober and reasonable men can prefer such "rubbish" to the words of the Bible.[4]

Victorian culture, then, may be considered as essentially unfavorable toward Bruno. The few references that can be found in its most distinguished authors are almost always negative. Already at the beginning of the century, Thomas Zouch, apparently unaware of Coleridge's interest in Bruno, is clearly ill at ease in his biography of Sir Philip Sidney of 1808 when he has to mention the link between his subject and the "celebrated atheist" of Nola—no less suspect, according to Zouch, for having been praised by a notorious freethinker of the enlightment such as John Toland. Later, in the 1830s, the same disparaging spirit animates Henry Hallam who, in his *Introduction to the Literature of Europe in the Fifteenth, Sixteenth and Seventeenth Centuries*, describes Bruno as a fragile philosophical "meteor," adding that the *De la causa* is full of vain and presumptuous fantasies.[5] A more complex attitude is to be found in the work of the distinguished historian of science William Whewell, Master of Trinity College Cambridge, who, in his extremely popular *History of the Inductive Sciences from the Earliest to the Present Times*, recognizes the significant role played by Bruno's *The Ash Wednesday Supper* in the diffusion of the Copernican theory. Whewell is not only concerned with Bruno's science. He also mentions the harsh criticism directed by Bruno toward the English society of his times. Whewell's extended reference to Bruno in a history of science that carried great authority remains an important tribute, not only because it assured that Bruno's name and thought continued to circulate, but because it afforded an indication of the cultural context in which Victorian England would be most willing to recognize his merits.[6] A good example of this nineteenth-century "scientific" Bruno is the copy of his *De monade* held by the Library of the University of London. Originally owned by Pierre Gassendi, it passed into the hands of the mathematician Augustus De Morgen (1806–1871), who wrote in it the following note: "Giordano Bruno, born about 1550, burned at Rome Feb.y 17, 1600. A vorticist before Descartes, an optimist before Leibniz, a Copernican before Galileo."[7]

This is the new cultural climate in which the importance of the contribution of George Henry Lewes needs to be assessed. Lewes can be considered the only English intellectual of some note who, at the height of the Victorian period, dedicated prolonged and serious attention to Bruno's philosophy. Today Lewes may not appear as a figure of the first order, although his voice remains of considerable interest. Active in the middle years of the nineteenth century, Lewes's activity as a journalist and writer was marked by diverse philosophical interests. Furthermore, his long relationship with George Eliot, the pseudonym used by the female author Mary Anne Evans, one of the most gifted and intellectual of the Victorian novelists, brought him into contact with some of the foremost minds of the time.[8] Lewes first took Bruno into consideration for his *History of Philosophy from Thales to Comte* of 1846, where Bruno's name occurs repeatedly in a chapter concerning the birth of the positive sciences.[9] Lewes underlines the importance of his theory of doubt, with which Bruno criticizes from time to time the dogmas of the philosophers and scientists of his own day. He considers Bruno a precursor of Descartes and what he calls the "evolutionary pantheism" of Spinoza. Lewes's Bruno is thus once again a rationalist and a materialist, valued above all for his contribution to science and to the cause of reason.

Both Lewes and George Eliot were deeply influenced by the works of Comte, but Lewes's positivism appears to attenuate when, in 1849, he reviews the edition of Bruno's works edited by Adolf Wagner, together with the biography of Bartholmèss.[10] Here Bruno is recognized as the thinker who "all of us have discovered" in the pages of Coleridge and the German philosophers, even if Lewes himself scolds the transcendental idealists for having made the Nolan into "a kind of poetical pantheist." Unfortunately Lewes fails to propose at this point an alternative interpretation, preferring to adopt the safer solution of a synthetic biography composed almost exclusively of passages taken verbatim out of Bartholmèss. In spite of this, however, his brief biography is worth a mention as the first reconstruction of Bruno's life story in English published after the book by Bartholmèss, and so based on a more complete set of documents with respect to the previous sketchy biographical accounts. Lewes mentions Bruno again in 1855 in the two volumes he dedicated to the life of Goethe, where he takes into consideration the notebooks of the great German in which, among other things, he found it written that "nothing absurd or impious" can be found in the works of Bruno. Lewes also dwells at some length on those passages of the *De la causa* that Goethe, like all his generation, had found in Jacobi, observing (although by then somewhat tardily) that they deserve a philosophical explanation in terms of the multiformity and the unity of substance.[11] Two years later, Lewes would include Bruno in his *Biographical History of Philosophy*, presenting him once again in terms of the new science, as a precursor of Francis Bacon as well as of the pantheism of Spinoza.[12]

These references make up a consistently lively series that undoubtedly helped to keep Bruno's name remembered in a society that was not much inclined to favor him. Nevertheless, in comparison with the reading of Coleridge, for example, it is clear that there has been a considerable falling away in vigor and depth of critical attention. In George Eliot's and Lewes's library, housed in the Dr. Williams Library in London, there is only one Bruno first edition, the *De imaginum, signorum et idearum compositione*—the last of Bruno's works on the art of memory. To judge from the state of the volume, Lewes never seems to have opened it. More surprisingly, he does not appear to have possessed the edition of the Italian dialogues edited by Wagner. It is, indeed, difficult to avoid the suspicion that his reading of Bruno was carried out principally, if not exclusively, on the basis of his much scored copy of Bartholmèss.[13] When in later life, in 1871, Lewes wrote a letter to Gilbert Hammerton advising him to undertake a series of readings for the composition of his book on *The Intellectual Life*, he indicated to him the biography of Bartholmèss and another by Domenico Berti that had appeared in the intervening years, but he made no reference to specific works of Bruno.[14] It seems probable that Lewes himself used these biographies in order to become familiar with the thought of Bruno, avoiding the effort of a direct study of texts that may have presented him with insurmountable linguistic as well as conceptual difficulties.

American Transcendentalism: Ralph Waldo Emerson

On the 5th of August, 1833, a young American, who had just crossed the Atlantic for that very purpose, visited Coleridge in his house in Highgate. Coleridge was by that time considered "the grand old man" of English idealism, an experience of which he had been the major intellectual representative and mouthpiece. Obviously, the terms in which that experience was judged depended on the point of view of the observer. John Stuart Mill, as has already been noted, claimed in his essay on Coleridge of 1840 that his own rationalist and utilitarian position, developed in the light of Locke and Hartley, was a radically progressive one, whereas Coleridge's abstract metaphysics placed severe limits on what was nevertheless, in Mill's opinion, a philosophical intelligence of rare perspicacity. Ralph

Waldo Emerson, on the other hand, thought that our globe reflects the transparency of a divine law, not a mass of facts, and he found his truth in the predominance of ideas.

In spite of the general affinity between their philosophical and religious positions, the meeting between Coleridge and Emerson was not a success, and it was Carlyle who would later become Emerson's closest British friend. According to the account furnished by Emerson himself, Coleridge, "the great man," insisted throughout the meeting on expressing his firm faith in the Trinity, obliging his guest to admit that, in spite of the pastoral duties conferred on him after his religious studies at Harvard, his own beliefs were closer to the same kind of Unitarianism that Coleridge had preached in his youth.[15] Coleridge, it seems, pronounced a lengthy monologue, as he was in the habit of doing in his later years, paying scanty attention to the young American who had come so far to visit him. It is most improbable that they spoke of Giordano Bruno, although we know from his *Notebook* that Emerson was already reading him in 1831, dwelling at length on the same opening ode to the volume containing the *De monade* and the *De immenso* that had been so warmly praised by Coleridge in 1801. In his *Blotting Book* of June 1831, Emerson copied out, in the original Latin, the final verses of the ode: *Si cum natura sapio et sub numine / Id vere plusquam satis est* (my knowledge derives from nature and the will of the gods, and that is enough for me).[16] The choice is a particularly interesting one in view of the fact that in 1836, after his return to the United States, Emerson would publish his first book made up of a group of writings with the title *Nature*.[17]

After some introductory pages that define nature in terms of the "colors" of the spirit that confer their sentiments and beauty on to natural things, Emerson continues with a discussion of "comfort," "beauty," "language," "discipline," "idealism," "spirit," and "prospects." The critics, however, agree in considering the high point of *Nature* the final pages, in which Emerson expresses his ultimate vision of natural harmony through the voice of a so-called orphic poet. Given the lack of any kind of indication on Emerson's part, the critics have discussed at length the possible identity of the poet who dominates the concluding pages of the essay, which are densely packed with literary quotations and philosophical echoes. Many names have been proposed, but no definite and certain source has so far been tracked down. The tendency nowadays is to consider the "orphic poet" a rhetorical device introduced in order to emphasize to greater effect the ideas on nature of Emerson himself.

There are numerous echoes of previous authors in these pages, and an echo of the works of Bruno is clearly present among them. The spirit builds itself a house, writes Emerson, and beyond its house lies the world, and beyond the world is the sky. When it leaves its house, the human spirit appears to Emerson to be involved in heroic acts of purification and understanding: "so shall the advancing spirit create its ornaments along its path, and carry with it the beauty it visits and the song which enchants it; it shall draw beautiful faces, warm hearts, wise discourse, and heroic acts around its way, until evil is no more seen." At the moment in which he enters into possession of his natural kingdom, man is seen by Emerson, in the final words of his essay, as a blind man full of wonder after regaining his sight. It is difficult at that point not to remember the final pages of Bruno's *Heroici furori*, where nine blind philosophers sing and dance in ecstatic joy after discovering that "double bliss: one due to the recovery of the long lost light, and the other due to the discovery of a new light, which alone can reveal the image of the greatest good on earth."[18]

The probability that what we find here is not just a casual similarity in images is suggested by the note written in Emerson's diary on December 8, 1834:

> Why not strengthen the hearts of the waiting lovers of the primal philosophy by an account of that fragmentary highest teaching which comes from the half (poetic) fabulous personages Heraclitus, Hermes Trismegistus, and Giordano Bruno, and Vyasa, and Plotinus, and Swedenborg? Curious now that I first collect their names they should all look so mythological.[19]

This reference to antique myth suggests that the text that Emerson had in mind when writing this note was the *Furori*, at least as far as Bruno's contribution to this final page of *Nature* is concerned. For in Bruno's text, numerous mythological figures combine to create the image of the *Furioso* in his search for new horizons of knowledge. Furthermore, in spite of his disappointing meeting with Coleridge, Emerson continued to speak of Coleridge with enthusiasm. Some days after the meeting, while talking to Alexander Ireland, he made a series of comments on the *Biographia Literaria* and *The Friend*, claiming that they contained: "many admirable passages for young thinkers, many valuable advices regarding the pursuit of truth and the right methods to be adopted in its investigation." In both the *Biographia* and *The Friend*, Emerson would have read quotations from Bruno's *De immenso* together with generally favorable comments on his philosophy.[20]

Emerson's meditation on the figure and thought of Bruno developed over the years to become a reflection that accompanied him at length, without ever finding adequate expression in his texts. The numerous notes on Bruno in his diaries, even after the publication of the essays on nature, continue right up till 1862, but during that time there is no single reference, at least in explicit terms, in his published works. A note on the scholar's courage of 1847 reads: "The Scholar's courage may be measured by his power to give an opinion of Aristotle, Bacon, Jordano

Bruno, Swedenborg, Fourier. If he has nothing to say to these systems let him not pretend to skill in reading."[21] Emerson, however, has little to say about what Bruno's system is, so that, although it is possible to speak of a Coleridgean reading of Bruno, it would be difficult to say as much of Emerson. Many of his notes, nevertheless, remain of considerable interest. For example, in the 1850s, and precisely in 1854, Emerson read Henry Hallam's *Introduction to the Literature of Europe in the Fifteenth, Sixteenth and Seventeenth Centuries* in a decidedly critical spirit. It has already been noted that Hallam mentioned Bruno in most negative terms, and Emerson complains of the insufficiency of his comments on a series of figures who, in his opinion, deserved treatment of a different kind. These are Bruno, Everard, Digby, Herbert of Cherbury, Böhme, Franciscus van Helmont, Henry More, Emanuel Swedenborg: "All these he passes, or names them for something else than their real merit, namely, their originality and faithful striving to write a line of the real history of the world."[22]

Emerson's interest in Bruno extended to his circle of friends and correspondents. Thanks to Rita Sturlese, his copy of the *De monade* together with the *De immenso* has recently come to light—a volume that, as has been noted, was previously of fundamental importance also for Coleridge's interpretation of Bruno. Emerson's *De immenso* is held at present by the Houghton Library of the University of Harvard, together with other rare books owned by him, and it carries the dedication: "R.W. Emerson from A.B. Alcott October 1842."[23] Later, in 1855, a note in Emerson's diary records his intention to send "Bruno" to Ch. D.B. Mills. On October 16, 1856, Mills writes to Emerson to tell him that he is about to send back the Bruno volumes.[24] Of particular significance here is the interest in Emerson's study of Bruno shown by Amos Bronson Alcott: a pioneer in the field of educational studies as well as being a key figure in that "American Transcendentalism" of which Emerson was one of the major representatives.

Emerson announced the objectives of the "American Transcendentalist" of New England in an essay of 1842: the very same year in which Alcott presented him with the *De monade* and the *De immenso*.[25] These "new views" of New England, writes Emerson, are essentially old ideas presented in a new and specifically American light. It is not his intention to link the transcendentalist idea to the German post-Kantian philosophy, but rather to propose it as a constant attitude of the human spirit that finds its truth not in facts, in history, and in the force of circumstance, as in the materialist scheme of things, but in the power of thought and will, in inspiration, in miracles, and in the cult of the individual. For Emerson, the enemy is the utilitarianism inspired by Jeremy Bentham that dominated at that time in America as well as in England, and that confused the bases of philosophy and of culture with "the foundations of a bank or of an office of currency exchange." In his struggle against this materialism, which Emerson thought of as gray and grim—"a tower of granite" that a breath of thought could destroy in an instant—he invokes a series of thinkers who have scattered through the human universe the sparks of an idealism that is perhaps to be considered folly, but nevertheless contains a profound sense of the divine. Rather than at the modern German philosophers, Emerson looks at Plato, to whom he dedicates an essay of 1850 titled *The Philosopher*. Here, in the opening page of the essay, he offers his reader his idea of the "Bible" of any true man of culture: a book that should contain the works of Boethius, Rabelais, Erasmus, Bruno, Locke, Rousseau, Alfieri, Coleridge. This is the only occasion on which Emerson mentions the name of Bruno in a text published during his lifetime.[26]

Paradoxically, the echoes of Emerson's reading of Bruno can be found most clearly in an essay that never explicitly mentions his name: the essay titled *Circles* that appeared in 1844. Here Emerson defines the natural world in terms of a system of circles, slightly eccentric with respect to the common center so that the surface appears to be somewhat slippery and the appearances deceitful. A true philosophy consists of the search for those affinities that hold the whole together according to a profound and secret law. In such a context, the laws of a completely rational science are no longer sufficient, because relationships such as that of cause and effect are nothing more than the opposite sides of a single truth. Later, this skeptical relativism, which carries reminders of Bruno, is transferred by Emerson to an ethical plane that underlines the equivalence and the indifference of every action: may not our crimes be conceived of as the animated stones with which we build our temples to divine truth? Such a question brings to mind the words written on the same subject by Bruno: "Things small and vile are often the seeds of greatness and excellence; stupidity and folly often provoke great councils, judgements and inventions. It is clear that errors and crimes often give rise to important rules of justice and goodness."[27] Such statements should not be taken as praise of criminal practice, or as an attempt to subvert ethical laws, as Emerson was clearly well aware. For he finishes his essay with a reminder of the principle that denotes a higher order of truth and justice lying behind the apparent chaos and contradictions of the shadowy world of vicissitude and phenomena.

Emerson's Bruno is thus a submerged memory, which never becomes the object of a complex and stratified reading, such as that of Coleridge. Nevertheless, the presence of Bruno's philosophy at the center of the American transcendental movement remains a significant phenomenon, indicating that Bruno's influence on American culture is of no small importance and deserving of further research. What also needs underlining here is the fact that Emerson's notes on Bruno between 1856 and 1862 outline an image of the Nolan philosopher as a martyr of free thought in terms similar to those that will ever more insistently, even obsessively, become intertwined with the unification of Italy: a figure expressing an idea of secular freedom of thought defined in heroic terms. This is the Bruno who was burned at the stake, after fighting against the arrogance of an oppressive power:

> There are men who as soon as they are born take a bee-line to the axe of the inquisitor, like Jordan Bruno / in France, the fagots for Vanini / in Italy, the fagots for Bruno /in England, the pillory for Defoe.[28]

And then again in 1862:

> And I summon you to regard with due honour those men who born in each evil age, as soon as they are born take a beeline to the rack of the inquisitor, the axe of the tyrant, like Jordano Bruno, Vanini, Huss, Paul, Jesus and Socrates.[29]

The reader may be perplexed to find the names of Bruno and Vanini linked here to that of a pre-Reformer such as Jan Hus, and even more of a Christian disciple such as Paul of Tarsus—men with whom Bruno had little to share. The line of free thought traced by Emerson is undoubtedly somewhat distant from the European tradition of libertine independence from the churches, such as the movement that will celebrate Bruno at the end of the 1880s by building the monument to him in Campo dei Fiori in Rome. Nevertheless, there is a full recognition on Emerson's part of Bruno's fight against all forms of tyranny, and of his standing as an authentic "hero" of an indomitable search for truth.

FIN DE SIÈCLE: BETWEEN POSITIVISM AND ESTHETICISM

Although Emerson and American transcendentalism were of some help in keeping Bruno's name in circulation during the middle years of the century, in Britain a deep silence reigns after Lewes's few remarks on his philosophy in the 1850s. It is only at the beginning of the 1870s that a book is published in Italy that reminds readers that a philosopher from Nola had written in London a cycle of six Italian dialogues that were, at that precise moment, at the center of attention in the secular and anti-clerical circles of the newly united Italian peninsular.

In March 1871, a brief but elegant biographical account of Bruno came out in *Frazer's Magazine*. It was based on the first Italian biography of Bruno, written by Domenico Berti and published in 1868, and made reference also to the Italian translation of Shelling's dialogue by the Marchioness Florenzi Waddington, published in Milan in 1844 and reissued in Florence in 1859 in an augmented and corrected edition.[30] The article in *Frazer's Magazine* was anonymous.[31] It is known, however, that its author was Isa Blagden, a friend of Robert and of the recently deceased Elizabeth Barrett Browning: all of them part of an Anglo-Italian circle of Florentine friends. The author was thus a particularly appropriate person to introduce into the English culture of the moment the terms of the intense debate that was already raging in the new Italy (united into a modern state with its capital in Rome only in 1870) around the name of Bruno. At that point, the debate was in its initial stages, but some years later it would lead to the publication of the new national edition of Bruno's Latin works, and to the erection of the monument in the Campo dei Fiori where Bruno had been burned at the stake.

Only a few days after the appearance of Blagden's article, *Macmillan's Magazine* published an essay of a very different tone, more centered on the philosophy but at the same time more inclined toward the presentation of a negative image of the philosopher. The essay was signed by Andrew Lang, and it expresses all the ambiguous indecision of a culture prepared to recognize the scientific contribution of Bruno's thought, but deeply suspicious of his history of heresy and rebellion.[32] Lang reminds his readers that Bruno announced before Bacon the superiority of the moderns over the ancients; that he anticipated the skeptical doubt of Descartes, and that he suggested important philosophical themes to both Spinoza and Leibniz. Nevertheless, he considers the *Spaccio de la bestia trionfante* a work too obscene to be even cited, and finds Bruno's attitude to religion a form of fanaticism and madness. Bruno's death is thus judged to have been inevitable, and his works, on a final analysis, superficial and frivolous.

The nineteenth-century battle over Bruno had begun in England as well as Italy. These two essays represent the beginning of a debate that would continue for the next twenty years and more. Following the events that, in Italy, had led to the formation of an international committee for the erection of the statue in Campo dei Fiori, publications in Britain on Bruno multiply rapidly, forming a mass of heterogeneous material that is often of limited scientific value. On the other hand, it makes of him one of the figures most discussed and commented on during the final decades of the century. Already in 1877, Annie Besant, in a brief pamphlet dedicated to Bruno that develops a rhetoric similar in emphasis to that of her Italian counterparts, can be found writing:

"Who was Bruno?" is a question now so often heard, that a brief answer to it may prove acceptable to those of our readers who know nothing of this grandest hero of Freethought, this man who lived and died so nobly that he carved his name for ever on the marble temple of Fame.[33]

Annie Besant, who mentions in her pamphlet the plans for the erection of a statue in honor of Bruno at Rome, would play an important role in the "Theosophical Society" founded in England at the end of the century by Helena Petrovna Blavatsky (usually known simply as Madame Blavatsky) who would disseminate a form of occultism mixed with diverse elements of oriental philosophy and mysticism. Although often obscure and not free from fraudulent claims, her thought would be of some influence in bringing about the gradual dissolution of that rigid Victorian Anglicanism that had been so hostile to Bruno's thought and fame.

At the opposite pole, that of the positive sciences, the reawakening of interest in Bruno's works takes the form of proposing his thought as an anticipation of some of the ideas at the center of the contemporary scientific debate—in particular, the theory of evolution. One of the most influential English members of the international committee for the erection of the monument in Campo dei Fiori was Herbert Spencer, whose work proposes the application of Darwin's evolutionary theory to the fields of sociology and anthropology. A consistent reference to Bruno cannot be found in Spencer himself, but his interest in the figure and history of the Nolan philosopher must have contributed toward the discussion of Bruno's natural philosophy in the context of the argument about evolution that raged in England in the second half of the nineteenth century.

It is precisely in the context of this discussion that John Tyndall makes an important reference to Bruno, after mentioning Copernicus and the astronomical revolution, in a paper read before the British Association in August 1874, later published in his widely read volume titled *Fragments of Science*.[34] Considered in the first place as among the principal proponents of a coherent Lucretian theory of atoms and the infinitude of the universe, Bruno appears to Tyndall above all as a philosopher who postulated the unfolding of matter in a gradual process of evolution. Nature does not imitate the technical capacities of man; she does not work according to a process imposed by an external artifice. Bruno's achievement, according to Tyndall, was to understand that nature operates according to her own inner powers and virtues, which develop ever new forms and manifestations.

The paper read by Tyndall that contains this reference to Bruno was subjected to a series of harsh criticisms on religious grounds that the author himself deplored in *Fragments of Science*. In spite of this, Bruno's "evolutionary" theory makes a further appearance in an essay by Thomas Whittaker, entirely dedicated to Bruno and published in *Mind* in 1884. This is the most seriously philosophical of the English comments on Bruno made in these years. It offers a lucid and succinct synthesis of the principal ideas developed in his works, above all in the Italian dialogues written and published in London. Whittaker insists on the philosophical seriousness of Bruno's works and maintains that, contrary to the prevailing opinion, his thought develops in coherent terms. This judgment is supported by a detailed discussion of both the cosmological-metaphysical dialogues and the moral dialogues. In the former, Whittaker is particularly interested by the metaphysical "status" of the first cause, as well as by the theme of vicissitude deriving from an imperfect relationship between matter and form. It is precisely in this crux of Bruno's thought that Whittaker finds the origins of a process of evolution, tending toward that perfect coincidence of matter and form that would reestablish the quiet of an infinite unity as a resolution of the shadows of natural vicissitude. Whittaker's essay sanctions Bruno within British culture as a serious philosophical precursor of the theory of evolution. Later, when Henry F. Osborne publishes in New York in 1894 his volume on *From the Greeks to Darwin, an Outline of the Development of the Evolution Idea*, he too refers to Bruno's works as presenting a philosophy that insists on the perfecting power of intelligence or form at work within the processes of nature.

In 1887, Thomas Whittaker published another article on Bruno in *Mind* titled "Giordano Bruno and His Times."[35] The essay discusses the reading of Bruno's works by the German scholar Moriz Carrière published in 1847.[36] Whittaker appreciates the fact that Bruno is considered by many the greatest philosophical mind of his time, whose work defines a principle of unity lying behind the universal whole. He also praises Carrière for having underlined how the later systems of Spinoza, Leibniz, and Hegel develop a series of ideas already present in Bruno's works. Carrière's mistake, in Whittaker's opinion, was to propose a reading of Bruno in the light of the mysticism of Böhme, making the theistic element in his work the preponderant one. Whittaker chides Carrière for having quoted numerous passages from Bruno's *De l'infinito* that support his reading, without mentioning the numerous alternative passages that posit the divinity within the world of matter and of vicissitude. It could be objected that the allegory of the cavern, annotated by Coleridge with a note remembering Bruno, suggests that this problem is a complex one also in Böhme. However that may be, it is interesting to see Whittaker reflecting philosophically on the theme of the status of the first cause in Bruno, for it was the problem around which so much of the philosophical discussion of his works in the nineteenth century tended to move.

The 1880s also witness an increasing number of biographical studies of Bruno that aim to inform the English public of the reasons that were at the basis of the project to erect a statue in Rome in memory of the Nolan philosopher. An attempt was made to explain the arguments that divided the two factions: those who defended the hostility of the church toward this project, and those who approved of Bruno's philosophy, in England as well as in Italy. There was no lack of emphatically negative judgments, both concerning his philosophy and his image as a hero of free thought, nor did such criticism come only from the Catholic side, intent on defending the Vatican after its defeat in the battle to prevent the erection of the statue and the ceremonies planned to accompany it.[37] It was perhaps inevitable that a Catholic review such as *The Month* should, through the services of M. T. Kelly, present the Nolan as a man "corrupted by the canker of pride, and of a heart that, formed for noble aspirations, chose rather to do evil than good."[38] More unexpected, and therefore of greater interest, was a series of articles published in reviews of a less specific ideology, but of a more or less conservative stance, which opposed the increasing "Brunomania" in the name of a generic taste for moderation and for the respect of civil and religious institutions. In 1878, an influential secular publication such as *The Quarterly Review* came out with an article signed by John Wilson titled "Giordano Bruno and Galileo Galilei" that developed the comparison strongly in favor of Galileo.[39] According to Wilson, it was precisely because Bruno was tactless enough to insist on the heretical aspects of the Copernican theory and to express all that was most "anarchical" and "irregular" in the philosophy of the Renaissance that Galileo had to face the anger of the Inquisition. Ten years later, only a few months before the inauguration of the monument, another prestigious review, *The Athenaeum*, published an anonymous essay titled "Giordano Bruno" that expresses an analogous sense of dismay in front of the spreading "Brunomania."[40] Although this article stops short of approving of Bruno's death at the stake, as Wilson had had no qualms in doing, it nevertheless underlines the nebulous and equivocal aspects of a political and religious debate in which Bruno's name was only too often used as an instrument that had little to do with the real terms of his philosophy.

The *Athenaeum* article is particularly critical of a heavily slanted book, David Levi's *Giordano Bruno o la religion del pensiero*, which it considers as above all an attack on the Catholic Church, and not a serious historical or critical reading of Bruno's thought. But the harshest attack of all comes from Puritan Scotland, and in particular from the *Scottish Review*, which complains bitterly that the name of Giordano Bruno has been "suddenly invested with an importance that it never formerly possessed, either in his own or in foreign countries."[41] This is an article that furnishes interesting information about the international committee that had been formed for the erection of the monument, although it presents that initiative in a strongly critical light. Clearly someone was reacting to Bruno's criticism of the foolish and pedantic grammarians of the Protestant Reformation, for the *Scottish Review* severely questions the wisdom of some ladies of good family who allow their names to be pronounced together with that of an author of "obscene and scurrilous works" such as Bruno's drama, *Candelaio*. It emphasizes how even many Italians were opposed to the erection of the monument, and how the whole initiative had already brought about student demonstrations "which have sometimes amounted to riots." In two long articles dedicated to Bruno, this review publishes for the first time in English a considerable part of the documents relating to Bruno's trial in Venice, with a marked tendency to side with the religious authorities, and to approve of Bruno's condemnation by the Inquisition.

Such criticisms find a partial justification in the poor quality of some of the Bruno commentary of the 1880s. It is enough to cite as an example the fanatical heretic Arthur B. Moss, whose *Waves of Freethought* was published in 1885.[42] Moss had written a book titled *Was Jesus an Imposter?* that he himself described as "the most blasphemous book of the epoch." In *Waves*, Bruno is presented as a passionate and hot-blooded Italian, a true son of the "volcanic" south. The portrait that follows is pure fiction: the philosophy is of no interest, and is dismissed as a mere "curiosity." All the author's attention is concentrated on that "honest, brave life and noble death," which Moss exalts with an uncontrolled enthusiasm that cannot have done much to favor a serious interest in Bruno's thought. The romanticized biography by Constance E. Plumtree, *A Tale of the Sixteenth Century*, published in 1884, is hardly better.[43] The story opens with an improbable Nolan supper that gathers around the same table Bruno, his parents, the Nolan Neopetrarchan poet Tansillo (one of Bruno's major sources, and probably a friend of his father's, but who had died before he was born), and the French ambassador in London, Michel de Castelnau, Lord of Mauvissière. Castelnau was the protector of Bruno in London, but he never seems to have visited Nola, let alone in the unlikely role as a guest of the relatively humble Bruno family. Plumtree recognizes the fictional character of this plot in the preface, admitting to having been tempted to tell the story of an imaginary meeting between Bruno and Shakespeare—a temptation that, however, was happily resisted. It must nevertheless be admitted that Plumtree's generally opaque pages treat Bruno's thought with occasional seriousness, especially his pantheism that the same author had already commented on in a *General Sketch of the History of Pantheism* published in 1878.[44] Nevertheless, books such as these by Moss and Plumtree could be of little use to the English reader in search of a serious answer to the question: "who was Giordano Bruno?"

It is against this background of sometimes violent and often frivolous discussion and argument around the name of Bruno that the major studies of the 1880s acquire their value as properly scholarly contributions. In 1886, John Addington Symonds dedicated ample space to Bruno in his monumental study of the *Renaissance in Italy*. In 1887, I. Frith published the first serious biography of Bruno in English and, in the same year, L. Williams came out with the first English translation of Bruno's *Heroici furori*.

The history of Renaissance culture published by Symonds was undoubtedly influenced by the interpretative categories put forward in 1860 by Jacob Burckhardt. Symonds' aim, however, was somewhat different. He intended to offer an ample panorama of the Renaissance in its various expressions and phases of development, based on a series of detailed studies of particular aspects and themes. Symonds presents the Nolan philosophy as emblematic of the late Renaissance interest in inquiry into the natural world. In such a context, Bruno's anti-dogmatism and his lack of a philosophical system appear to him as virtues and not failings.[45] It is, in his opinion, precisely in these aspects of his work that Bruno can be claimed as most modern, contrasting in a positive sense with the over-rigid systematizing of a Hegel, a Schopenhauer, or a Herbert Spencer. Rather than a philosophy, according to Symonds, Bruno created a dream of the human intelligence in a process of continual expansion, aided by a desire that was noble and not heretical to emulate the wisdom of God. In his book on Sir Philip Sidney, published in the same year as *Renaissance in Italy*, Symonds considers Bruno as the most penetrating, lucid, as well as the most unfortunate of the "martyrs for truth" of the late Renaissance.[46]

Frith's biography, which came out a year later than the volumes by Symonds, appears to derive from the interest in Bruno on the part of one of the principal publishers of the period, Nicolas Trübner. Of German descent, Trübner had published in 1880 an English translation of one of the chapters of the German biography by Brunnhofer. Trübner invited Frith to write the first complete English biography of Bruno, and the book came out after a careful revision by Moriz Carrière.[47] It remains today of considerable value, maintaining a significant position among the works of Bruno commentary in the English language. It is based on a detailed reading of the works, both Italian and Latin, as well as of a large part of the secondary material then available, in English, French, and German. It has the character of an intellectual biography that considers Bruno at the same time among the founders of the new science—a disciple of Copernicus and a precursor of Bacon—and one of the principal sources of German idealism and the Hegelian dialectic. An unusual amount of interest is shown for the period in the works on the Kabbala, the art of memory, and the logic of Raymond Lull. On the whole, the book maintains the promise of the opening page, on which Bruno is presented as a courageous and original thinker: "a man destined to mark out a new era in philosophy." The appendices offer the reader a notable quantity of documents: a list of the works of the philosopher and of the available modern editions; information about the Noroff code containing still unpublished works (later to pass to the State Library in Moscow); a list of lost works; and a bibliography of modern studies. Last but not least, the Latin letter of Schiopp, describing Bruno's burning at the stake, is given in transcription.

Less satisfactory is the other Bruno publication of 1887, the first English translation of the *Heroici furori* by L. Williams.[48] The introduction to this work, as Williams himself admits, derives largely from the book by David Levi, where the development of the intellect and the soul described in the *Furori* is situated in a metaphysical context of a pseudo-Pythagorean nature. Williams is interested in the mystical and secret religious elements that he finds in Bruno, as he makes clear in a prefatory note to the second volume. He puts these in relation to *The Secret Doctrine* of Blavatsky, where a universal agent active in all forms of life, called Od, Ob, or Aour, is associated with the number nine that leads to the deepest secrets of being. It is by means of this reference to the Pythagorean aspects of Blavatsky's esoterical doctrine that Williams interprets the final pages of the *Furori*, in which nine blind men rediscover the light of the previously known Pythagorean truth, or the greatest good on earth. It is hardly a surprise to learn that Williams became a member of the "Theosophical Society" founded by Blavatsky herself. His translation of the *Furori*, which is often heavy and turgid from a linguistic point of view, is strongly influenced by this mystical derivation, which interferes throughout with a satisfactory rendering of the philosophical complexity and poetical and thematic variety of Bruno's text.

On the whole, however, these three publications of the 1880s concur in presenting Bruno as a cultural and philosophical figure of the first order. When, in 1885, the British Museum published its catalogue of printed books, there was already a substantial collection composed of twenty volumes of Bruno biography and criticism, as well as all the modern editions of his works and a considerable number of first editions as well. At the end of that decade, the English reader was therefore in possession of a rich and varied literature on Bruno that provided the basis for an informed discussion concerning the principal historical event of those years: the inauguration on June 9, 1889, of the monument to Bruno in the Campo dei Fiori in Rome.

The meaning of this much debated celebration was explained to the English reader shortly before the event in an essay that appeared in the review *The Nineteenth Century*, titled "Giordano Bruno and New Italy." Written by an author who signed himself as Karl Blind, the essay presents the relevant documents in detail, with the principal arguments for and against Bruno: on the one hand, the declarations by the Vatican deploring the erection of the statue and the ceremonies that accompanied its inauguration, and on the other, the declarations of those who supported them, including Francesco Crispi, then prime minister of Italy.[49] The article does not present a neutral attitude to the question but comes out in support of the liberal enthusiasts, expressing surprise at the violence of the anti-Bruno reaction on the part of those who wished to see the suppression of a public commemoration. In the final part of the article, the author links the name of Bruno to that of Galileo, considering the Nolan philosopher as a worthy precursor of the great Tuscan scientist. The last words of the author on this subject are: "the struggle against Obscurantism has still to be carried on."

On the morning of Monday, June 10, 1889, all the principal English newspapers carried articles on the events of the preceding Sunday in Rome. A Reuter telegram, picked up by among others the *Times*, the *Daily Telegraph*, and the *Daily Chronicle*, expressed decided favor of the initiative by describing in colorful terms the gathering of 80,000 participants carrying 1,972 banners and flags, who moved in procession through a crowd that uttered "indescribable" shouts of victory and joy. A synthetic account of the speech made by the radical member of Parliament, Giovanni Bovio, underlined the benefits deriving to mankind from the "martyrdom" of Bruno. This was followed by a description of the celebration in the Campidoglio that, with a well-calculated political choice, linked Bruno's death to that of Garibaldi, which had taken place on June 2, 1882. The telegram insisted on the fact that throughout the triumphant day in Rome, public order had been rigorously maintained. The *Daily News*, which carried a particularly enthusiastic service by a special correspondent of its own, also underlined this aspect of the event, claiming that from the enormous crowd, no remark was heard that could have offended religion. On the other hand, it seems that what these newspapers considered a Roman triumph of free thought was not adequately reflected in the celebrations organized by the Italian community in London. In its column titled *London Day by Day*, the *Daily Telegraph* wrote that a London event was supposed to take place in the restaurant "Monaco," but that when the participants arrived, they found a notice saying that it was against English law and that the event had been cancelled. According to the newspaper, some of those present interpreted this notice as a religious protest against the celebration of a "heretic," but the *Telegraph* goes on to correct this impression. The event was forbidden because the restaurant had no license authorizing it to open on Sunday. Some doubts seem to have been raised as to whether this failure had been deliberately manipulated. What happened in Rome was one thing, but there were clearly some in London who were not in favor of a Bruno celebration at all. Some days later, on June 15, *The Saturday Review* published an article that was decidedly critical of the whole affair:

The celebration at Rome was not itself a very respectable proceeding. ... It was notoriously intended (and was disapproved by the best class of Italian Liberals as being intended) very much less as a testimonial to Bruno than as an insult to the Pope, and perhaps to religion.

A more favorable side of the English debate concerning the Bruno commemoration in the Campo dei Fiori is to be found in the direct participation of some of the major English poets and writers of the period, who supported the event. The most prominent name is that of Algernon Charles Swinburne, who honored the memory of Bruno not only by participating in the international committee that supported the erection of the statue, but also by publishing some eloquent verses on the day of its inauguration in the Campo dei Fiori. His ode appeared significantly in the pages of *The Atheneum*, which had so fiercely attacked Bruno the preceding year. Now a very different music floated out from those pages, with the ample rhythms of Swinburne's pen evoking the ashes of a fire rising in the limpid sky of a Rome finally redeemed:

> ... Rome redeemed at last
> From all the red pollution of thy past
> Acclaims the grave bright face that smiled of yore
> Even on the fire that caught it round and clomb
> To cast its ashes on the face of Rome.[50]

A few weeks later, in August of that same 1889, another elegant writer dedicated himself to commemorating Bruno, although in rather different terms. A biographical narrative of a character fully *fin de siècle*, titled "Giordano Bruno: Paris 1586," was published in *The Fortnightly Review*. Its author was Walter Pater, who in 1873 had published his *Studies in the History of the Renaissance* with its celebrated evocation of Leonardo's Gioconda, whose mysterious beauty penetrates even the secrets of death.[51] They were pages that Oscar Wilde had heralded as the sacred writing of a new era dedicated to the cult of beauty. Pater's account of the Parisian Bruno proposes an intoxicating, subtly subversive, and dangerously liberating ethical message:

> ... that doctrine—*l'antica filosofia italiana*—was in all its vigour there, a hardy growth out of the very heart of nature, interpreting itself to congenial minds with all the fulness of primitive utterance. A big thought! Yet suggesting, perhaps, from the first, in a still, small, immediately practical, voice, some possible modification of, a freer way of taking, certain moral precepts: say! a primitive morality, congruous with those larger primitive ideas, the larger survey, the earlier, more liberal air.

Here the ethical implications of Bruno's natural philosophy emerge as an essential moment of subversion of the rigidly pragmatic morality of

the already declining Victorian era, slightly deformed, perhaps, by an Epicurean note nearer to that of Nietzsche than to a truly classical or Renaissance inspiration. The reference to the principle of a unique universal law
that had echoed throughout the pages of Emerson when he too reflected
on the ethical implications of Bruno's natural philosophy has now disappeared. The coincidence of contraries, with its identification of the crucial
point of reconciliation as a "point of indifference," is almost imperceptibly converted to a doctrine that becomes ever more insidiously amoral:

> The difference of things, and above all, those distinctions which schoolmen
> and priests, old or new, Roman or Reformed, had invented for themselves,
> would be lost in the length or breadth of the philosophic survey; nothing in
> itself, either great or small, and matter, certainly, in all its various forms, not
> evil but divine. Could one choose or reject this or that?

It is from America that, a year later, a message arrives that is nearer
to the vigorously libertarian notes of Swinburne. On January 14, 1890,
the Philadelphia Contemporary Club dedicated a memorable evening to
"Giordano Bruno: Philosopher and Martyr."[52] The guest of honor should
have been Walt Whitman, prevented from being present due to ill health.
He nevertheless sent in the text of an eloquent opening speech in which
he paid homage to the sacrifice of all those martyrs of the old world—and
in particular to Bruno—who with their "mental courage" have gained
the remembrance and the gratitude of the new world. The evening was
organized around two speeches of considerable importance, the first by
Daniel G. Brinton and the second by Thomas Davidson, who, in 1886,
had already published for the Index Association of Boston an essay on
"Giordano Bruno and the Relation of his Philosophy to Free Thought."[53]
This essay had been praised by E. Mead as the best thing written on
Bruno in the English language: an important recognition made in an ar-
ticle that presented in English translation Hegel's pages on Bruno in his
History of Philosophy.[54]

In his Philadelphia speech, Davidson attempts an ambitious analysis of
the sources of Bruno's thought, its character, and its value. He emphasizes
above all Bruno's concept of the monad, or the atom as the center of force
and the primal factor in any process of evolution. The permanent value of
Bruno's thought resides, for Davidson, in the idea of the one and infinite
universe as intelligible, and of the first cause as both transcendental and
immanent. Bruno's universe is filled with a divine intelligence comprehensible to the human mind, which, free from dogmas and without the
interference of intermediaries, attempts heroically to discover and penetrate its secrets.

Brinton, on the other hand, concentrates on the natural philosophy
rather than on the metaphysics, in a speech that reflects the interest in

the Anglo-Saxon world for Bruno seen as the precursor of a number of modern theories in physics. Some of his remarks can be considered as predictable at that point, such as his considerations on Bruno's ideas about evolution. Brinton is nevertheless interesting on Bruno's theories about the imperfect sphericity of the earth and, in general, on his theory of assymetry as a fundamental characteristic of the material world—an aspect of Bruno's thought already hinted at by Emerson in his essay on *Circles*. Although Brinton offers an evaluation that is not quite correct concerning Bruno's supposed faith in observation and sense experience, he is very good on the importance of his doctrine of doubt, which so irritated the theologians and the churches. He is also interested by what he sees as the absence in Bruno of a sense of sin insofar as evil is considered as merely a lack of the good, just as cold is a lack of heat. Nevertheless, Brinton does not consider Bruno a pantheist but rather an idealist who recognizes the progress of thought through contraries in terms that make him into a worthy precursor of Hegel. Furthermore, Bruno's hatred of theological doctrine seems to Brinton to bring him close to some of the Protestant sects, and in particular to the Quakers. In conclusion, Brinton expresses his approval of the choice of the newly independent Italy to honor this hero of free thought. By celebrating Bruno, the new nation has declared its intellectual autonomy, and has seen a way for survival in pursuing a course of active and philanthropic virtue. Undoubtedly it is of great interest to see Bruno emerge from this American celebration as the philosopher of an active philanthropy not adverse to radical Protestant leanings, whereas Bruno himself (although less consistently than the critical tradition has tended to assume) tends to attack the Protestant Reformation for its supposed refusal of good works and for its passive pietism.

The historical optimism and the love of active virtue that underlie the Brunian celebration of Philadelphia were beginning to appear as notes of ingenuousness, far from the decadent spirit that was pervading England at the end of the century—an end of the century in which Bruno became above all the hero of Walter Pater. In the month of April 1894, only a few years after Pater's essay had been published in *The Fortnightly Review*, the first issue of *The Yellow Book* appeared, with its celebrated opening essay by Max Beerbohm, *In Defence of Cosmetics*, constituting an elegant and melodramatic announcement of the end of an era. "For behold! The Victorian era comes to its end and the day of sancta simplicitas is quite ended." It is surely no mere chance if a literary reminiscence that leads to Bruno is present in this first number of *The Yellow Book*. In a contribution by Richard Garnett titled "The Love Story of Luigi Tansillo,"[55] there is an eloquent translation of Tansillo's sonnet beginning "Poi che ho spiegat'ho l'ali al bel desio" in which the Icarus figure is given a positive connotation, and which Bruno had incorporated into his

Heroici furori.[56] Here the atmosphere invoked is precisely that subtle and intoxicating taste for evasion from a world codified in rigid moral and intellectual precepts, which had already pervaded Pater's essay on Bruno.

In 1896, almost as a conclusion of the century on a note of refined but unquiet aestheticism, a fragment of an unfinished novel by Walter Pater was published posthumously. Titled *Gaston Latour*, its predictably French hero is an emblematic figure of the European *fin de siècle*. He moves languidly and dreamily between the end of the Middle Ages and the beginning of the Renaissance, dominated by the magical Circe and her spells. The novel narrates Gaston's passionate readings of Ronsard's odes, illuminated by the brilliant light of a world of the senses that was being rediscovered in all the beauty of its sounds and images. It includes a meeting between the hero and Montaigne, who helps him to discover the complex variety of human nature. Finally Gaston arrives in an "Italianized" Paris, dominated by the fashions of a late and decadent Renaissance. It is here that he goes to listen to Bruno's lectures, which teach him how the imagination seeks a truth that it is possible to know only in partial forms, as a kind of prophetic intuition, but that nevertheless contains fragments of the divine.

These Brunian pages of Pater's novel reproduce the essay of 1889 with an extended addition that refers to Bruno's first work published in Paris in 1582, *De umbris idearum*.[57] It is the first and only time in which Pater refers to a particular work of Bruno's. Even here, as the author candidly admits, it is not so much a question of the "shadows" evoked by Bruno as of "ideas and shadows of ideas" that reflect the unquiet thoughts and youthful dreams of Pater's hero. Invoking Bruno, Gaston Latour finds himself part of an irresistible and fatal spell, unable to determine "the practical and appropriate limits" of that doctrine of indifference that seems to cancel out, or at least to render superfluous, moral laws or precepts.

So it was that in the final decades of the nineteenth century, Bruno in England became less a part of a properly philosophical discourse, to be assumed as a seductive figure, image, or narrative model in a primarily literary context. It was a development that would have important repercussions at the beginning of the twentieth century, when James Joyce introduced Brunian motives and linguistic modes into his modernist narrative. Even within the last years of the nineteenth century itself, however, the Bruno of the literature of the period is no longer confined to the minor level of a Moss or a Constance Plumtree, but appears as a part of the major literary production of the period. Conjured up by the pen of a Swinburne or a Pater, the figure of Bruno is called on to express all the intoxicating suggestiveness, all the eloquent and at times exhilarated anxieties of an aesthetic *fin de siècle*. A serious Bruno scholar today might want to consider this interpretation of the Nolan philosophy as an impetuous and unscrupulous assumption of the figure of Bruno—an overtly "interested" and insidious cultural exercise in its subtle evocation of the deceptive vicissitudes of a universe no longer pervaded by the lucid intelligence of a Brunian first and infinite cause. For this presence of Bruno in the literary works of the final part of the nineteenth century is seldom based on a close reading of his texts. Paradoxically the large quantity of Brunian material accumulated during the course of the century had finished by rendering the texts themselves superfluous, rather than by stimulating a closer study of their pages.

Such considerations, however, should not lead to underestimation of the importance of the major cultural discourses in which we find a presence of Bruno in the second part of the nineteenth century in Britain. Apart from the significant interest shown by English (and American) writers of the first order, it has been possible to refer to a series of critical studies of undoubtedly high quality, such as the essays of Whittaker or the biography by Frith. Nor should one ignore the significant development, with respect to the name and story of Bruno, of a discourse regarding liberty of thought and opinion: a concept that the native English tradition had already defined in the essays of a John Milton or a John Stuart Mill. The nineteenth century in England and in the United States, although not without difficulties and at times through harsh debates, incorporated Bruno into its cultural discourse at various levels, sometimes wildly misinterpreting him, but at other times with an open-minded and balanced appreciation of his contribution to the culture of the modern world. Above all, the Nolan philosopher is appreciated where he tends to become a precursor of modern science such as it will be developed by a Francis Bacon or a Galileo Galilei: a tradition in which the intellectuals of the period often identified themselves. Nevertheless, a continuous reference can be found throughout the century to that philosophical trend that, in the wake of the new German idealism, defined the first cause metaphysically in terms of a divine intelligence that pervades and operates in every material manifestation. So that if it is necessary to admit that nineteenth-century Britain had no figure such as that of John Toland, who had acted as the primary inspiration for the diffusion of Bruno's works in the culture of the European Enlightenment, nevertheless it showed itself capable, above all at the beginning and the end of the century, of reading Bruno in new and stimulating ways. And if it is impossible to deny that Bruno's thought arrives in England at the beginning of the century mediated by German idealism in the first place, and then by the textual and documentary studies of French, Italian, and German scholars as the century proceeded, it is undeniable that a serious study of Bruno as a philosopher and poet begins to take shape in the course of the century, in England as well as on the continent of Europe. The interest in Bruno shown by the English nineteenth-century intellectual was often more in the nature of an occasional curiosity for one or other aspect of his works than a detailed and continuous study of his thought as a constant point of reference. Nevertheless, in the course of the century, a Brunian culture was formed of considerable consistency and importance, characterized by a large variety of interests and concerns.

It is difficult to deny, however, that the most profound and prolonged reading of Bruno in nineteenth-century Britain was that of Coleridge in the early years of the century. It is surprising to note how faint and inconstant the echo of that reading was to be in the century to come. But this aspect of the story is more a question related to the tormented history of Coleridge's own works, which only in recent years have become available to the public in a long-awaited complete edition of both the published works and the private papers. This at last makes clear that Coleridge's reading of Bruno, which took place before the publication of the first modern editions of his works, was characterized above all by continuous difficulties and frustrations related to the problem of finding the texts, as well as to the paucity at that time of a tradition of secondary literature. On the other hand, Coleridge's study of Bruno was supported by his vast knowledge of both ancient and modern philosophical traditions. In Coleridge, the poet and the philosopher unite in a reading of Bruno that, in spite of the German influences, acquires an independent and autonomous dimension, representing for the British culture of the time a genuine and exciting moment of discovery of a new philosophy: a new idea of the creativity of the human spirit that recognizes Bruno as both one of its precursors and as one of its most eloquent exponents.

NOTES

1. J. C. Shairp, *Studies in Poetry and Philosophy*, Edinburgh, Edmonton and Douglas, 1868, 119–20.

2. For the cultural climate of Victorian England, see Basil Willey, *Nineteenth Century Studies*, London, Chatto and Windus, 1949, and more recently *Victorian Thinkers*, ed. A. L. Le Quesne, Oxford, UK, Oxford University Press, 1993.

3. See J. S. Mill, *Coleridge*, in *Collected Works*, X, University of Toronto Press and Routledge, 1969, 117–65.

4. C. Carlyon, *Early Years and Late Reflections*, op. cit., 195.

5. Henry Hallam, *Introduction to the Literature of Europe in the Fifteenth, Sixteenth and Seventeenth Centuries*, J. Murray, London 1876 (1st ed., 1837–1839): I, 322; II, 105–11.

6. William Whewell, *History of the Inductive Sciences from the Earliest to the Present Times*, London, J. W. Parker, 1837, 3 vols.; republished in 1847, 1857, 1859, 1866, and 1875. The comment on Bruno is in vol. 1, section 2.

7. See Sturlese (1987), 123.

8. For information about Lewes, see V. A. Dodd, *George Eliot: An Intellectual Life*, London, Macmillan, 1990, chapter 16, "George Henry Lewes to 1850," and chapter 17, "George Henry Lewes and Marian Evans (1850–1854)." Lewes's philosophical position is described at 233.

9. See G. H. Lewes, *The History of Philosophy from Thales to Comte*, 3rd ed., London, Longmans, Green and Co., 1867 (1st ed., London 1846), chapter III, section VI.

10. The review was published anonymously in the *British Quarterly Review* IX (1849): 540–63.

11. G. H. Lewes, *The Life and Works of Goethe*, London, D. Nutt, 1855, 2 vols.: I, 100–102.

12. G. H. Lewes, *A Biographical History of Philosophy*, London, J. W. Parker, 1857.

13. These texts are included in W. Baker, *The George Eliot–George Henry Lewes Library: An Annotated Catalogue of Their Books*, London, Garland Publishing, London 1977, notes 132 and 321. The *De imaginum compositione* is registered as note 321, where it is erroneously attributed to Leonardo Bruni.

14. See *The George Eliot Letters, IX*, ed. G. S. Haight, New Haven, CT, Yale University Press, 1978, 29–30.

15. See R. L. Rusk, *The Life of Ralph Waldo Emerson*, New York, C. Scribner's, 1949, 191–92.

16. Bruno (2000c), 252. See *The Journals and Miscellaneous Notebooks of Ralph Waldo Emerson*, eds. W. H. Gilman, and A. R. Ferguson, Cambridge, MA, Belknap Press of Harvard University Press, 1960–1982, 16 vols.: III (1963), *Blotting Book*, note III, June 1831, 264.

17. See R.W. Emerson, *Essays and Lectures*, ed. J. Porte, New York, The Library of America–Cambridge, UK, Cambridge University Press, 1983.

18. Bruno (2002), vol. 2, 747.

19. *The Journals and Miscellaneous Notebooks*, op. cit., IV (1964), Journal A, December 8, 1834, 355.

20. See A. Ireland, *In Memoriam: Ralph Waldo Emerson*, London, S. Marshall, 1882.

21. *The Journals and Miscellaneous Notebooks*, op. cit., X (1973), Journal AB, 1847, 28.

22. Ibid., XIII (1977), Journal 10, 1854, 300. The negative comment on Hallam is taken up again at 310.

23. See Sturlese (1987), 119.

24. *The Journals and Miscellaneous Notebooks*, op. cit., Journal NO, 1855, 393.

25. See "The Transcendentalist," in *Essays and Lectures*, op. cit., 191–210.

26. See "Plato; or The Philosopher," in *Essays and Lectures*, op. cit., 633–54.

27. See the *Proemiale epistola* to *La cena de le ceneri*, in Bruno (2002), vol. 1, 439.

28. *The Journals and Miscellaneous Notebooks*, op. cit., XIV (1978), Journal SO, 1856, 95.

29. Ibid., XV (1979), Journal WAR, 1862, 175.

30. See Berti (1868; 2nd ed., 1889) and F. W. Shelling (1844; 2nd ed., 1859).

31. "Giordano Bruno," *Fraser's Magazine* III (1871): 364–77.

32. See A. Lang, "Giordano Bruno," *Macmillan's Magazine* XXIII (1870–1871): 303–9.

33. A. Besant, *Giordano Bruno*, London, C. Watts, 1877.

34. See J. Tyndall, *Fragments of Science for Unscientific People*, 6th ed., Longmans and Co., London 1879, 2 vols.: II, 137–203.

35. Whittaker's two essays on Bruno were republished in *Essays and Notices Philosophical and Psychological*, London, T. Fisher, 1895, 61–94 and 249–66.

36. Carrière's interpretation of Bruno is to be found in his *Die philosophische Weltanschauung der Reformationszeit in ihrem Beziehungen zur Gegenwart*, Cotta, Stuttgart–Tübingen, 1847. For a brief comment on Carrière's reading of Bruno in the light of a Hegelian idealism, see Blum (1998), 69–70.

37. Lars Berggren has written a history of the Bruno monument project in *Giordano Bruno på Campo dei Fiori. Ett monumentprojekt i Rom (1876–1889)*, Lund, Artifex, Lund 1991. See also Berggren (2002).

38. M. T. Kelly, "Giordano Bruno," *The Month. A Catholic Magazine and Review* LXXV (1892): 527–40.

39. J. Wilson, "Giordano Bruno and Galileo Galilei," *The Quarterly Review* CLXV (1878): 362–93.

40. See *The Athenaeum* 3143 (January 21, 1888): 82.

41. This essay was published in two parts in *Scottish Review* XII (1888): 67–107 and 244–69.

42. A. B. Moss, *Waves of Freethought*, London, Watts and Co., London 1885.

43. C. E. Plumptre, *Giordano Bruno: A Tale of the Sixteenth Century*, London, Chapman & Hall, London 1884, 2 vols.

44. C. E. Plumptre, *General Sketch of the History of Pantheism*, London, S. Deacon, London 1878, 2 vols. In the first volume, chapter IV, 348–66, Bruno is described as handsome, gallant, ardent, and intrepid, with a terminology that will be repeated in the later biographical romance by the same author.

45. See J. A. Symonds, *Renaissance in Italy*, VII, London, Smith Elder, London 1866, 135–98.

46. J. A. Symonds, *Sir Philip Sidney*, London, Macmillan, London 1886, 170.

47. I Frith [pseudonym for I. Oppenheim], *Life of Giordano Bruno the Nolan*, rev. Prof. M. Carrière, London, Trübner & Co., London 1887.

48. *The Heroic Enthusiasts. Gli Eroici Furori, an ethical poem ...* , trans. L. Williams, with an introduction compiled chiefly from D. Levi's *Giordano Bruno o la religione del pensiero*, London, G. Redway, London 1887.

49. K. Blind, "Giordano Bruno and the New Italy," *The Nineteenth Century* XXVI (1889): 106–19.

50. "Giordano Bruno," *The Athenaeum*, 3216 (June 15, 1889): 758. Swinburne wrote other odes to Bruno, one of which was titled: *For the Feast of Giordano Bruno: Philosopher and Martyr*. See *Swinburne's Collected Poetical Works*, London, Heinemann, London 1924.

51. W. Pater, "Giordano Bruno," *The Fortnightly Review* XLVI (1889): 234–44.

52. The speeches delivered that evening were published in a pamphlet titled *Giordano Bruno: Philosopher and Martyr: Two Addresses*, Philadelphia, D. Mackay, 1890.

53. T. Davidson, *Giordano Bruno and the Relation of His Philosophy to Free Thought. A Lecture*, Boston, Index Association, 1886.

54. See *Journal of Sepculative Philosophy* XX (1886): 206–19.

55. See *The Yellow Book* I (1894): 235–49.

56. Bruno (1958), vol. 2, 999. (See also the note by Gentile on 999–1000).

57. W. Pater, *Gaston Latour*, London, Macmillan, 1910 (1st ed., 1896), 158–59.

PART 3

BRUNO'S PHILOSOPHY OF NATURE

12

BRUNO'S NATURAL PHILOSOPHY

VER SINCE BRUNO STARTED TO BE studied seriously as a key
figure in the European philosophical tradition, there has been uncertainty as to what kind of philosopher he was. John Toland proposed him to the more radical components of the Enlightenment culture
of his time as a fundamentally anti-hierarchical thinker, drawing out all
the most subversive implications of his post-Copernican, infinite cosmology, with its relativization of values, not only spatial but also social, political, historical, and religious.[1] But when Friedrich Heinrich Jacobi included some pregnant passages from one of Bruno's major philosophical
dialogues in Italian, *De la causa, principio et uno*, in the second edition
of his critique of the pantheism of Spinoza, *Über die Lehre des Spinoza in
Briefen an Herrn Moses Mendelssohn*, published in 1789, it was Bruno's
metaphysical inquiry that was being brought to the reader's attention,
and that, in defiance of Jacobi's disapproval of its pantheistic tendencies,
would become a strong influence in the following half century on the
post-Kantian idealists, not only in Germany.[2]

In the opening years of the nineteenth century, when in Germany Shelling was writing his dialogue titled *Bruno: or a Discourse on the Divine
and Natural Principles of Things*, Samuel Taylor Coleridge, in England,
started what would become a lifelong reading of Bruno's philosophy that
is remarkable both for its conceptual subtlety and for its width of vision.[3]
For, on one side, Coleridge admired the studies of electricity of Joseph
Priestly, whose *Disquisitions on Matter and Spirit* of 1777 refers to "the
famous Jordano Bruno" as a precursor of Locke and Andrew Baxter in
the conviction that all the vital powers of matter should be considered
the direct work of God, thus making of Bruno the first exponent of a
dynamic philosophy in the physical sciences. And it is in these terms, as
one of the first thinkers to develop a fully dynamic idea of the processes
of both being and thought, that Coleridge refers to Bruno in his most
famous work, the *Biographia Literaria* of 1816, in a chapter titled "Philosophy as Science." On the other hand, Coleridge linked Bruno closely to
the Christian mysticism of Jacob Böehme, and to an idea of the divinity
as an absolute synthesis of a cosmic struggle between contraries. Indeed,
Coleridge would go so far as to write in his marginal notes to Böehme's
works, read in the English translation by William Law: "Plato in *Par-*

menides and Giordano Bruno passim have spoken many things well on this aweful Mystery / the latter more clearly."

Throughout the nineteenth century, comment on Bruno ran along this double track.[4] His works regularly found a dignified niche in the most qualified histories of science of the period, such as the section on the diffusion of the Copernican theory in William Whewell's *History of the Inductive Sciences from the Earliest to the Present Times* of 1837, or John Tyndall's widely read *Fragments of Science for Unscientific People* of 1879. There was also much discussion throughout the century of the influence that Bruno's vitalistic theory of matter had exercized on the major scientific debate of the period, the theory of evolution, which culminated in the substantial reference to his natural philosophy by Henry F. Osborn in *From the Greeks to Darwin, an Outline of the Development of the Evolution Idea* published in New York 1894. It was Bruno's intrepid inquiry into the new scientific theories of the late Renaissance, such as the implications of the Copernican revolution or the newly revived atomism, heedless of the protests being raised by the European theologians on both sides of the religious divide, which was celebrated by American figures of note such as Thomas Davidson during a memorable evening dedicated to *Giordano Bruno: Philosopher and Martyr* by the Philadelphia Contemporary Club in 1890. But the other side of the picture was always present, if often in a subdued form. Emerson, for example, was reading Bruno as one of "the waiting lovers of the primal philosophy," or "that fragmentary highest teaching which comes from the half (poetic) fabulous personages Heraclitus and Hermes Trismegistus", although he kept such thoughts to his private notebooks and journals. In the same midcentury years, the militantly Catholic and anti-Hegelian philosopher Franz Jakob Clemens, in Germany, made an important comparison between the theory of the coincidence of opposites in Cusanus and in Bruno.[5] Although unfavorable to the Italian, accused of illegitimately transposing an absolute identity from God to the infinite universe, thus confusing the identity that characterizes the substance of God with that of the substance of His effects, Clemens was the first to study Cusanus as a major source of Bruno's metaphysics—a theme that continues to lie at the center of comment on his philosophy today.[6] Toward the end of the nineteenth century and the beginning of the twentieth, with the revival of spiritualistic, esoteric themes, often of oriental inspiration, that aimed at polemicizing with the dominant scientific positivism of the age, Bruno can be found permeating the ardently undisciplined thought of the theosophical societies of the period. For Annie Besant, he was *Theosophy's Apostle in the Sixteenth Century*, according to whom "man's true and primitive form is divinity; if he has the consciousness of his own divinity, if he realizes it, he may regain his primitive form, and raise himself to the highest heaven."[7]

The nineteenth-century commentators of Bruno's philosophy had no apparent difficulty in reconciling these two dimensions of his thought. Hegel in his lectures on the history of philosophy paid as much attention to Bruno's dialectical logic of contraries (or what Coleridge before him had called Bruno's "polar logic") as he did to his resolution of those contraries in an absolute monad or the identity of an indeterminate One.[8] Influenced undoubtedly by Hegel's reading of Bruno, Isabel Frith-Oppenheim, in the excellently researched first book-length intellectual biography of Bruno to appear in English, published in 1887, was as eager to claim Bruno as a pioneer of the early stages of the so-called scientific revolution as she was to underline the modernity of his idealism.[9] But with the beginning of the new century, a polarization of interpretations of Bruno's philosophy becomes clearly evident against a cultural background dominated by the reasons of an increasingly scientific and technological society, with anti-metaphysical and neopositivist philosophical foundations. The book on Bruno by J. Lewis McIntyre, published in 1903, follows the positivist and neorationalist readings of the major Italian commentators of the second part of the nineteenth century such as Felice Tocco and Domenico Berti, for whom the magical and spiritualistic elements in Bruno's thought appeared as fastidious frills or leftovers from a previous age.[10] Appreciated in Italy by the early-twentieth-century editor of Bruno's Italian dialogues, Giovanni Gentile, McIntyre's volume is clearly concerned to present Bruno as primarily a precursor of Francis Bacon's scientific method, just as Gentile himself, in his essays on Bruno's thought as the culminating moment of the philosophy of the Renaissance, will place him just before his chapter on Galileo.[11] When, in the central years of the twentieth century, a number of distinguished French commentators dedicate their attention to Bruno, it is the scientific components of his thought that are at the center of their attention.

Paul Henri Michel's seminal essay on Bruno's atomism of 1957, followed by his book on the cosmology of 1962, together with the extensive treatment of Bruno's thought by Alexandre Koyré, both in his *From the Closed World to the Infinite Universe* and in his *Etudes Galiléennes*, represent authentic milestones in the study of Bruno's works in the context of the natural philosophy of the late Renaissance.[12] And if it is true that Koyré considered what he thought of as Bruno's "residual animism," deriving from an earlier phase of medieval and Renaissance Neoplatonism, as excluding him from the modern world, he was nevertheless of the opinion that Bruno's cosmological picture, at once prophetic, rational, and poetic, had profoundly influenced both the philosophy and the science of the centuries to come—a conviction whose enduring importance has been underlined by Eugenio Garin in his volume of 1975 on *Renaissances and Revolutions: Cultural Movements from the Fourteenth to the*

Eighteenth Centuries.[13] This is the Bruno we find in the major publications in English of the middle years of the century such as Dorothea Singer's translation of and comment on the *De l'infinito universo et mondi* of 1950, as well as Paul Oscar Kristeller's section dedicated to Bruno in his *Eight Philosophers of the Italian Renaissance* (1964). It is also the Bruno of Hélène Vedrine's major philosophical study titled *La conception de la nature chez Giordano Bruno,* published in 1967, which remains an important point of reference for scholars concerned with Bruno's natural philosophy and science today.[14]

For the first sixty years or more of the twentieth century, then, it seemed as if the die had been cast finally in favor of a Bruno whose philosophy found its historical collocation as a prelude and prophecy of the scientific revolution of the later Renaissance, which was thought of as the origin of the modern world. It was precisely this interpretation of both Bruno and of the modern world that was questioned by the studies of Frances Yates, and particularly by her influential book, *Giordano Bruno and the Hermetic Tradition,* first published in 1964.[15] It is worth noticing that Yates herself made no mention, and indeed seemed quite unaware, of the nineteenth-century anticipations of her Hermetic reading of Bruno; rather, she interpreted that century entirely in the light of the scientific positivism that was its dominant if not only outcome. In this conviction, it became for her the reign of error itself, which had given rise to what she began to define as the "old" reading of Bruno, which had enclosed him within the scientific-technological organization of existence while disregarding the magical and Hermetic dimension of his thought expressed in his search for the divinity as the ineffable unity of being.

Undoubtedly the influence of the studies of Aby Warburg and his successors, with their alternative reading of Renaissance culture in the light of its search for primitive origins, or a *prisca theologia,* cannot be overvalued in a consideration of the Bruno proposed by Frances Yates, a distinguished member of the Warburg Institute in London with which she had begun an association as far back as 1941. Clearly the book by her Warburg colleague D. P. Walker on *Spiritual and Demonic Magic from Ficino to Campanella,* published in 1958, is present in the background.[16] The immediate source of this radically overturned reading of Bruno, however, as Yates explicitly indicates in the introduction to her book, was the contemporary study of the Renaissance in the light of the presence of the Hermetic texts translated from Greek into Latin by Marsilio Ficino at the request of Cosimo dei Medici in 1463—a previously unsuspected presence demonstrated in a seminal paper by Paul Oscar Kristeller of 1938. This had become the basis of a new study of the period in the light of its magical and Hermetic doctrine proposed by Eugenio Garin and his school of scholars in Florence.[17] Garin himself had not extended this reading of the Renaissance to Bruno, and indeed has repeatedly insisted, in spite of his admiration of Yates's work, on the necessity of making distinctions between the different ways in which the Hermetic texts permeated different periods and areas of Renaissance culture, casting some doubts on the justice of an unmitigated Hermetic and magical interpretation of Bruno.[18] However, the Yates thesis itself, both in the original book on Bruno and in her later works, belies such distinctions. If Bruno is differentiated from Ficino and his Neoplatonist reading of the Hermetic texts, it is only in the sense of a less cautious and more radical assumption on Bruno's part of the Hermetic doctrines, made even more anti-rationalistic and anti-scientific by the Kabbalistic and magical strands that were later introduced by Pico della Mirandola and Cornelius Agrippa. In this perspective, it is Bruno's science that becomes for Yates a leftover from a previous century that had, in her view, insisted on dressing him in clothes that were theirs rather than his. And if it was difficult to deny that he had been reading Copernicus in a cosmological context and Lucretius in an atomistic one, and that such readings had been the subject of serious attention both by Bruno's contemporaries and in the following centuries, Yates thought she could explain away, in a few sentences or even in a footnote, both the infinite universe and the atomistic theory of matter as emblematic images of the mysterious secrets of being.[19]

It is not necessary here to trace in detail the long and complicated *querelle* that followed the publication of Yates's book of 1964. Some general commentary on how the field of Bruno studies adapted itself to the dramatic swing of the pendulum that led from the scientific Bruno of the first half of the twentieth century to the Hermetic Bruno of the last decades is, however, desirable in order to define the sense in which his natural philosophy will be considered in this chapter. For there can be no doubt that Yates raised a valid point in claiming that large areas of Bruno's works, such as his many texts devoted to Lullian and mnemotechnical themes, which Yates herself would look into in more detail in her volume on *The Art of Memory* of 1966, had been ignored or even despised by previous commentators.[20] These texts are today considered by many to be more closely connected to logic or to rhetoric than to the magical arts that Yates so insistently underlined.[21] Nevertheless, Bruno's detailed knowledge of ancient, medieval, and Renaissance magic that depended conceptually on the ubiquitous presence at the heart of matter itself of a vital spirit or universal soul, is nowadays considered, largely thanks to Yates's studies, to be present as a major aspect of his works. It is a concept that Bruno tends to radicalize rather than reject, incorporating it into his theory of matter as a substitute for the traditional idea of form, which thus acts from inside the universal material substance as a kind of creative force, or yeast. At other times, Bruno posits a boundary line between the world of things or becoming and the eternal envelope of indeterminate being that is seen as the magic or indefinable moment at which the logic of contrary forces begins. There seems no reason why such speculative definitions of magic, which appear again and again in Bruno's published works, should necessarily invalidate the scientific endeavor, or the attempt to penetrate, and appropriate for the use of civilized society, the forces at work within the world of becoming. And in fact, many of the most valid studies of Bruno in the post-Yatesian era have been concerned with an attempt to understand in what ways his natural philosophy, unwaveringly emphasized by scholars such as Giovanni Aquilecchia, Hélène Vedrine, Ramon G. Mendoza, or Leen Spruit, among others, can be reconciled with his magic, his reading of the Kabbala and his frequent references to the Hermetic texts.[22] So that what appears to be the agenda for the coming century is a reading of Bruno's works in their completion that is able to account both for his science and for his magic, without becoming shipwrecked in the shallows of the either/or attitude that dominated the twentieth-century debate.

A development of the critical discussion along such lines is made even more necessary by the recent publication of the first volume to present in integral form, surrounded by a dense apparatus of commentary and notes, the unpublished manuscripts that Bruno left unfinished at his death.[23] Titled *Opere magiche*, this large and well-produced volume gives the confusing impression that all the manuscripts published in it are concerned with Bruno's thoughts on the magical arts, although this is not in fact the case. By far the longest, and undoubtedly the major work that Bruno himself never published, the *Lampas triginta statuarum*, actually contains few if any references to magic, as the editors of the new volume admit in their notes to the text.[24] It is rather a reelaboration of Bruno's ontological considerations, already developed in his philosophical dialogue in Italian, the *De la causa, principio et uno*, written and published in London in 1584, on the relation of the apparently fragmented world of becoming and of things to the original principle of unified being—one of Bruno's most constant and characteristic themes, as Coleridge rightly claimed.[25] Other works, such as the *Theses de magia* or the *Medicina Lulliana*, appear to be little more than compendiums of notes of reading on those subjects, as the detailed quotations from Bruno's sources that are one of the major characteristics of this valuable volume make clear. This leaves the four brief works on magic, *De magia mathematica*, *De magia naturali*, *De vinculis in genere*, and *De rerum principiis* that Yates already knew, although only in the reduced form in which they were published in the nineteenth century in the third volume of Bruno's collected Latin works.[26] Curiously, however, as Michele Ciliberto notes in his introduction to the new volume, she made little use of these final, unpublished texts on magic, in spite of the fact that they indicate a definite interest on Bruno's part, in the final months before his arrest and imprisonment on the part of the Roman Catholic Inquisition, in the possible uses of magical techniques as a means of achieving a new dominion within the world of time and nature.

As we have already seen, the technical details of Bruno's natural philosophy have been the subject of a number of major studies during the twentieth century, and are too well known to need repeating here. Rather there is a need to restate the relationship he establishes between the two distinct philosophical poles between which his ontology constantly moves—of being and becoming, of permanence and time—and the sense in which he contemplates a new scientific activity in the context of a constant reference to eternal principles, or divine truths. This appears, indeed, in a general way, to be more and more clearly understood as the major characteristic of the so-called scientific revolution of the late European Renaissance, which today, after the discussion that has in the last decades involved the science of Isaac Newton with relation to his recently discovered papers on alchemy and his massive Biblical studies, can no longer be discussed in terms of a science versus religion interpretative scheme. In the case of Newton, this new realization has given rise to numerous differing emphases on the relative importance of his religion with respect to his science, or to the traditional inquiries in which he was still deeply involved, such as alchemy, with its cult of secrecy and its recognition of magical or occult qualities, and the modern scientific undertaking, based on shared and repeatable experiments, projected into the public domain. But in spite of some extreme positions to the contrary, such as that expressed by Betty Jo Teeter Dobbs in a much discussed paper of 1993, the consensus of the most qualified Newton scholars appears to be determined by their desire to preserve his position as the major figure of the early modern scientific experience, while at the same time recognizing the deeply felt need that his private papers—largely unpublished at his death, and for centuries ignored by Newton scholars—clearly express to relate his science to a dimension beyond logic and reason, which clearly involves an element of faith.[27]

The reference to Newton is not to be considered irrelevant here, as the purpose of this paper is to propose just such a synthesis as the basis of a new discussion of Bruno's philosophical endeavor. Indeed it is Bruno himself who spells out the meaning of his philosophy in these terms in the work that will be proposed here for comment and analysis—that is, the work titled *Lampas triginta statuarum*, or *The Lamp of the Thirty Statues*, which Bruno also left unpublished at his death. As we have seen, this work, first published in the third volume of the nineteenth-century edition of Bruno's Latin works, has recently appeared in the new volume of the posthumous manuscripts, together with an Italian translation and detailed comment and notes. Bruno is concerned here with precisely that relationship between eternal truths and the world of becoming that appears to have been a constant preoccupation of the new scientists up to and including Newton himself. Indeed, in a section of this work titled *The Field of Minerva, or Knowledge*, Bruno spells out with particular clarity his thought on such a relationship.

The *Lampas* shows a marked desire on Bruno's part to contain his very complex ontology within a coherent system of discourse. The lamps and the statues to which they refer, although they can be considered as magically endowed with their original light, should be seen primarily as the files in which Bruno stores his distinctions relating to the various grades of being. Each statue is itself divided into thirty subfiles, the *Field of Minerva* being no exception to this rule.[28] Knowledge, according to Bruno in these thirty sections, derives from an inner light in the mind that illuminates us as to the conclusions that we may draw from the first principles. These principles, or eternal truths, are not themselves the domain of reason but rather of faith. This is of two sorts: what Bruno calls a "well-regulated" faith, characterized as a simple recognition of the necessity of the first principles themselves, and what he considers an overexcited or perverse faith, based on the superstition of false prophets. This last remark clearly refers to revealed religions, and includes a reference to Bruno's long-standing anti-Christian polemic. The first principles themselves are not known by the mind, except insofar as it reasons a number of conclusions from them. It is these conclusions that constitute what Bruno calls "science," or knowledge, which he defines in a later paragraph in suggestive terms that, although based on ancient sources among which Aristotle's *Analitics* are specially mentioned, clearly project his idea of science into the modern world.[29]

Science is related to our powers of judgment, and it involves both sense experience and a process of reasoning. The results of the logic of such science must be articulated in some sort of discourse, which becomes a shared experience. The logical process defined must be repeatable: it requires a second examination that controls and verifies its exactitude. Only this process of verification guards the new scientific truth against the lies of impostors. Bruno thinks of geometry as an essential example of such a science. But another kind of science derives from the necessity of matter as much as from the necessity of form, and these two elements can concur together to constitute the "garment" of a new form of knowledge. This is knowledge as form, or the knowledge of knowledge, or the matter of intellectual truth.

These sections dedicated to Minerva, which include also references to far more traditional forms of knowledge, precede a long final section of the work dedicated to Venus and Cupid, who are seen as the forces of concord or harmony that bring sense and meaning into an otherwise confusing world.[30] Confusion, for Bruno, has to be clearly distinguished from Chaos, which, on the contrary, together with the abyss and privation, constitutes—in Anaxagorean terms together with what appears to be a clear reference to the *Liber chaos* of Ramon Lull—the first of the first principles themselves.[31] Chaos, the Abyss, and privation, which Bruno calls Ancient Night, are nothing or everything—substance in a state of complete indetermination—and as such, "they are so far from being accidents of things that they are, on the contrary, the principles according to which the accidents come into being, are related to one another, and enter into relation with substance."[32] For this to be the case, however, there must be a "superior" triad, which Bruno denominates the Universal Apollo, or a universal spirit or light, which brings the state of indeterminate privation into an ordered whole as a universe of individual entities. These are perceived by the mind as realities through the senses, but their order is "modelled by the artifice of fantasy and imagination" that, as the ancient philosophers understood, was in its proper function not a faculty designed to confuse the truths of reason, but rather to illustrate them, to explicate their order and maintain such order in the memory. The imagination is thus, for Bruno, an intimate part of the scientific activity.[33]

It is only in what Bruno calls this "universal perspective" that the world of nature can be understood as a third order of things, which, precisely because it is illuminated by the light of the one Apollo, or divine Monad, can also be considered as the good.[34] Science, therefore, for Bruno, is closely related to ethics: an understanding of the correct order of things in the natural world leads to a correct perception of what is virtue and what is vice. The essence of the scientific endeavor itself, however, is founded on the notion of quantity, and as such, Bruno presents it as the field of Ocean.[35] Here we find the attributes of magnitude in all its characteristics, which lead to a perception of the universe as a physical entity founded on the concepts of multiplicity and number. Within these concepts, Bruno emphasizes in particular the notion of addition, for it is through ideas such as increase, expansion, aggregation, and completion that the mind is able to conceive of a universe that is not susceptible of any further increment of any kind. That is to say, precisely the post-Copernican universe that is eternal in time and infinite in space that Bruno had presented to his readers in his Italian philosophical dialogues written and published in London in 1584 and that, probably at the same time as he was writing the *Thirty Statues*, he defined for the last time in his Latin masterpiece *De immenso et infigurabili*, published as the last work of his so-called Frankfurt trilogy in 1591.

The section of the *Thirty Statues* dedicated to "The Field of Ocean" thus reaches its conclusion with a paragraph titled "The Universe: The World," which is the way in which magnitude becomes quantitative and corporeal in its explication as physical reality.[36] And the physical universe is founded for Bruno on one specific quality, which is heat: a clear reference to that "giudiciosissimo Telesio" whom Bruno had already praised with unusual vigor in the second of his Italian philosophical dialogues written and published in London in 1584, *The Cause, Principle, and One.* Telesius, in his major work *De rerum natura,* first published in two books in Naples in 1570, had cautiously attempted to replace the Aristotelian physics by a new physical dualism based on a universal dialectical contrast between heat and cold: a contrast that Bruno had incorporated as a fundamental one within his own far more radically post-Copernican, infinite cosmology, using it as an explanation of the movements of the stars and planets of his infinite number of solar systems around their central suns.[37] In the *Thirty Statues,* moreover, heat is not only considered as a universal, lifegiving quality, but also as one that can be subjected to measurement as a quantity—a fact that guarantees for Bruno the possibility of a rational inquiry into natural things.[38] For the concept of size is to be considered "absolute" among all other physical realities. Size defines perfectly the particular and individual entities, but at the same time, it creates a similitude with that which is beyond size, or the infinite. The nature of the universe in its totality as infinite is thus confirmed by the mind as true.

A later section, which is dedicated to motion within the universe or the physical world and is titled "The Field of the Earth, or of Potency," appears at first sight to be founded on an Aristotelian concept of potentiality and to lead back to the traditional idea of matter as the passive element subjected to the potentiality of its specific form, which contains within it the impetus toward a motion defined by its individual nature and ends.[39] However, such an impression is mistaken, as Bruno reaffirms here his complete reversal of the Aristotelian equation, making matter into the active substance that underlies an infinite world of finite objects and contains within it the total potentiality of all forms—a potentiality that precedes the single form with its acts of motion, and on which all motion logically depends. Conceptually, this reversal of the Aristotelian relationship between matter and form, in the context of motion, can be considered as supplying the speculative foundation of a quantitative, universal law of motion, such as will eventually be formulated by Newton's law of gravity. However, Bruno here draws back well before defining such a possibility with any clarity, apparently resolving the question of motion in the more traditional terms of impetus theory. Thus he can still write that the impetus that leads to the motion of an individual thing derives from the internal principle of that thing's propensity toward a certain end, even if for Bruno such a propensity depends only in a secondary sense on the potential present in what is for him the accidental and impermanent nature of the individual form. Even Aristotle's substantial forms, which are the forms of a species rather than an individual of that species, partake in what is, for Bruno, a universal mutability. For they too ultimately depend on the universal potentiality present throughout the infinite substance, which precedes the single form and the single act of motion. It is on this infinite, universal potentiality that the single motion ultimately depends.

This line of reasoning reaches its logical conclusion in the following section titled "The Field of Juno," who is considered the mediator between the individual things of which the infinite universe is composed.[40] Here the idea of universal laws of physics is explicitly defined. Bruno sees the concept of laws in physics as intimately linked to the idea of civil laws, which are generally binding, such as treaties or oaths "or other things of that kind." The field of physics also contemplates a situation in which superior principles exercise their influence on inferior ones according to some law or prescription. These natural laws constitute the necessary intermediary principles without which nothing can happen in the world of things. Bruno thinks of them as knots, or chains, or even forms of glue that guarantee that all things are linked together in some universal formulation, which, like an oath, is repeatable and generally respected—a public and not a private or secret act. Here it is Plato rather than Aristotle who is called in as witness, insofar as his idea of a universal *spiritus* or world soul, which reflects throughout the physical universe the divine light of Apollo, is more agreeable to the idea of universal physical laws than Aristotle's concept of individual souls or forms, or even of substantial forms, whose motions are laws only unto themselves or, at the most, unto their species. Universal physical laws, moreover, are, according to Bruno, an assurance of the central role of human kind within the universe, for they are part of the human intellectual horizon. The eternal truths, or first principles, are explicated within the infinite world of things in terms of universally intelligible laws of physics capable of comprehension within the intellectual horizon of the human mind.

This scheme of natural philosophy extracted from the *Thirty Statues* represents an aspect only of that work, which is concerned in the first place to define the complex of first principles on which the world or universe depends. Even when Bruno does reach the final part of his work dedicated to the universe itself and the laws on which its movements depend, he has as much to say on the civic and ethical implications of a scientific inquiry into the natural world as he does on the more specifically logical or intellectual character of such an exercise. Nor can it be claimed that his thought on the inquiry into the natural world is always as clearly projected toward an endeavor that would rapidly become the modern scientific enterprise as the pages presented and commented on here would give to suppose. At times he is evidently concerned with more traditional philosophical concepts and inquiries, for which he found available an already established vocabulary. For example, the concept of links between every aspect of being in the universe, which in the *Thirty Statues* is seen as the necessary rational idea justifying the possibility of universal, natural laws, could be seen from the quite different perspective of the unknowable nature of the links themselves. In their occult essential nature, the links, or *vinculi*, will later become the center of the magical techniques of persuasion and dominion investigated in the work *De vinculis in genere*, also left unpublished at his death, where Bruno is attempting to inquire into areas of psychological tension that defeat the understanding of the conscious mind.[41] In many other parts of his *oeuvre*, however, both in the *Thirty Statues* and in a large number of his major published works, Bruno insists rather, in terms that Francis Bacon would surely have appreciated, on the necessity of turning the attention of the active, inquiring intellect toward physical realities, too long ignored, for attempts to comprehend the first principles that underlie the physical world are necessarily arduous and may finish up by engaging with phantasms whose truth remains in question. So it is Thetis, standing for the natural causes of things, the lover of the universal laws of matter that dictate the multiple and ever-changing forms of the natural world, who turns into the new "tiger" whose forces the human mind must now attempt to know and tame.

Bruno's considerations on Thetis, as the material substance that underlies all the specific formations that make up the natural world, are of particular interest in indicating at the same time both the possibility and the inevitable limits of an inquiry into natural causes. They may thus be used as the conclusive remarks to this paper. For Thetis, who is associated with the ocean in its infinite Protean capacity for metamorphosis, is by no means easy to "catch" or to define in definite and certain terms. Indeed, Bruno sees her as the wife of many husbands, none of whom can truly be said to possess her. Although her forms can be pursued by reason (*"subiectum ratione formabile"*), she resists all attempts at such dominion, and renders the hunt for her secrets difficult and problematic. Bruno portrays her as riding on a dolphin, whose back only at times appears as well defined above the moving waters of becoming. Thetis, after all, is the daughter of both the sky and the earth, and so should not be seen as crude or lowly matter, but rather as divinely inspired. Insofar as she represents nature, she is the object of natural philosophy, but insofar as she represents God, she is the object of theology, or metaphysics, or a philosophy of religion. The natural philosopher would be unwise to think that it is possible to penetrate her essence. All that can be hoped for is to understand something "around and about" her ways and habits.

When she assumes specific forms, such as a horse or a tree, then investigation through the senses, reason, and intellect into and about them ("*circa quod*") is in order, and to be encouraged. But it should not be assumed that the knowledge gained will ever reveal the ultimate secrets of the thing in itself. For this reason Bruno, in this section, pictures the human intellect as a sunflower, unwearily turned toward the ultimate truth of things, but destined to find joy mixed always with suffering, without ever reaching the final goal.[42] As I have already pointed out elsewhere, this crisis epistemology renders Bruno's inquiry into natural things more consonant with modern, post-relativity science than with the rational optimism of the mechanical sciences of the seventeenth and eighteenth centuries.[43]

NOTES

1. For Toland's relationship to Bruno, see Margaret C. Jacob, *The Radical Enlightenment: Pantheists, Freemasons and Republicans,* London, Allen and Unwin, 1981, and Ricci (1990b), 239–330.

2. See Ricci (1991), 431–65.

3. Hilary Gatti, "Coleridge's Reading of Giordano Bruno," *The Wordsworth Circle* 27, no. 3 (1996): 136–45, and chapter 10, "Romanticism: Bruno and Samuel Taylor Coleridge," in this volume.

4. For Bruno's nineteenth-century reputation, in England and elsewhere, see Canone (1998a).

5. Blum (1998).

6. See, for example, Mancini (2000), 245–74.

7. Besant (1913).

8. Hegel (1833). For a critical appreciation of Hegel's comment on Bruno's philosophy, see the *Introduzione* to Canone (1998a).

9. Frith (1887). Frith's book was published under a decidedly German influence by the London publisher of German origin, Nicolas Trübner. The book was controlled and revised by the German-born Moriz Carrière, a Hegelian as well as being himself a Bruno scholar.

10. McIntyre (1903).

11. See Gentile's frequent references to McIntyre's book in his notes to Bruno (1958).

12. Michel (1957) and [1962] (1973). Also Koyré (1957), and (1966).

13. E. Garin, *Rinascite e rivoluzioni: movimenti culturali dal XIV al XVIII secolo*, op. cit.

14. Singer (1950), Kristeller (1964), and Vedrine (1967).

15. Yates (1964).

16. Walker (1958).

17. Kristeller first noted the remarkable circulation of Hermetic texts during the Renaissance in the paper "Marsilio Ficino e Ludovico Lazzarelli: contributo alla diffusione delle idee ermetiche nel rinascimento," *Annali della R. Scuola Normale Superiore di Pisa*, Lettere, Storia e Filosofia, 2nd series, 7 (1938): 237–62,

republished in *Studies in Renaissance Thought and Letters,* Rome, Edizioni di Storia e Letteratura, 1956, 221–47. Garin's studies of the Renaissance in terms of its magical and Hermetic doctrines can be found above all in *Medioevo e rinascimento,* Florence, Sansoni, 1954, and *La cultura filosofica del Rinascimento italiano,* Florence, Sansoni, 1961.

18. See on this subject Garin's more recent *Ermetismo del Rinascimento,* Rome, Edizioni Riuniti, 1988.

19. See, for example, the famous page on which Bruno's Copernicanism is dismissed as "a Hermetic seal hiding potent divine mysteries of which he has penetrated the secret," in Yates (1964), 241, or the footnote 1 to 265 that contains all that Yates has to say on Bruno's neo-Lucretian atomism.

20. Yates (1966), in particular chapters 11–14.

21. See as examples Rossi [1960] (2000) and Ellero (2005).

22. See in particular Aquilecchia (1991) and (1993), Vedrine (1967) and (1993), Mendoza (1995), and Spruit (1988).

23. Bruno (2000b).

24. "Nella *Lampas* non emergono prove di un marcato interesse per la magia," Bruno (2000b), 1491.

25. See chapter 10, "Romanticism: Bruno and Samuel Taylor Coleridge," in this volume.

26. See Bruno (1879–1891), vol. III.

27. Dobbs's paper was delivered at the annual meeting of the History of Science Society and has been published as "Newton as Final Cause and First Mover" in *Rethinking the Scientific Revolution,* ed. Margaret Osler, Cambridge, UK, Cambridge University Press, 2000, 25–39. It was at once contradicted, above all by Newton's biographer, Richard S. Westfall, in the paper "The Scientific Revolution Reasserted," published in the same volume at 43–55. Also of much interest on the theological outlook that underlies the structure of Newton's *Principia* is the paper by J. E. McGuire, "The Fate of the Date: The Theology of Newton's *Principia* Revisited," published in the same volume at 271–95. For a more general treatment of the relation of the new science to its theological premises, see Brian P. Copenhaver, "Natural Magic, Hermeticism, and Occultism," in *Reappraisals of the Scientific Revolution,* eds. David C. Lindberg and Robert S. Westman, Cambridge, UK, Cambridge University Press, 1990, 261–301.

28. The thirty sections relating to the *Field of Minerva* are in Bruno (2000b), 1227–39.

29. See sections XXIV and XXV of the thirty considerations dedicated to *Minerva's Ladder, or the Disposition towards Knowledge,* which immediately follows the *Field of Minerva.* In Bruno (2000b), 1246–47.

30. For the sections dedicated first to Venus and then to Cupid, see ibid., 1248–77.

31. For Lull's *Liber chaos,* written about 1275, see Frances Yates, "Essays on the Art of Ramon Lull: Ramon Lull and John Scotus Erigena," now in *Lull and Bruno: Collected Essays,* vol. I, London, Routledge, 1982, 95–98. It is significant that Yates, in the introduction to her essays on Lull, states quite unambiguously that "The Lullian artist is not a magus" and that his arts are logical not magical.

She underlines how Lull traces the passage from Chaos, through Bonitas, to the elements of the natural world.

32. See the twenty-sixth consideration on Chaos in Bruno (2000b), 956–57, and the introduction, "Beginning as Negation in the Italian Dialogues of Giordano Bruno," in this volume.

33. These considerations on the imagination close a section of the *Lampas* dedicated to *The multiple forms of investigation*. See Bruno (2000b), 940–41. The relation between Bruno's science and the imagination has been treated at length in De Bernart (1986).

34. For the investigation into nature as the third order of being in the *Thirty Statues*, see Bruno (2000b), 1164–1295. The remaining part of the work is dedicated to technical instructions as to how to develop the mnemonic techniques that will allow the statues to be used as a mental filing system in which to store the wealth of data supplied by an infinite universe.

35. Bruno (2000b), 1180–83.

36. Ibid., 1184–87.

37. For Bruno's praise of Telesius, see Bruno (2000a), 166–67. For a critical comment on Bruno's important relationship with the work of Telesius, see Aquilecchia (1993a), 293–310.

38. For an attempt on Bruno's part to visualize such a measurement, see chapter 2, "Bruno's Copernican Diagrams," in this volume.

39. See Bruno (2000b), 1192–99.

40. Ibid., 1198–1209.

41. The *De vinculis in genere* is published in ibid., 413–584.

42. For the long and beautiful section of the *Lampas* titled *The Statue of Thetis, or Substance*, see ibid., 1122–43.

43. See Gatti (1999).

13

BRUNO'S USE OF THE BIBLE IN HIS

ITALIAN PHILOSOPHICAL DIALOGUES

THIS CHAPTER ORIGINATED WITH THE realization that during the composition of his philosophical works, Giordano Bruno made a constant and expert use of numerous Biblical texts. This may seem surprising, at first sight, in a philosopher noted above all, in his own days and in ours, for his heretical opinions with respect to the fundamental doctrines of both the Hebrew and the Christian religions. Nonetheless, Bruno's Biblical references do not appear to have a merely rhetorical or ornamental function, nor do they express a purely ironical or satirical attitude toward the Biblical texts, although they are certainly eccentric with respect to traditional readings of the Bible, sometimes with disconcerting results. It is perhaps more appropriate to speak of a habit of constantly "rewriting" the Biblical word, which involves the texts of the Old as well as the New Testament.[1]

I shall be concerned here only with Bruno's Italian works, written and published in England during his years in the French Embassy in London as a Gentleman Attendant on the Ambassador between 1583 and 1585. This does not imply that I consider Bruno's use of the Bible to be in any way less important in his Latin works—on the contrary, a study by Nicoletta Tirinnanzi of his use of the *Song of Solomon* in the *De umbris idearum*, written and published in Paris in 1582, shows how imprudent it is to ignore the Biblical presence in the works Bruno wrote in Latin.[2] For in the Latin works as well, Biblical texts are used by Bruno as authentic sources for many of his philosophical arguments. It can, however, be argued that his use of the Scriptures became particularly insistent in the philosophical dialogues in Italian composed in London where, above all due to his well-known relationship with Sir Philip Sidney and the group of intellectuals surrounding the earl of Leicester, it appears probable that Bruno was influenced by the new Biblical hermeneutics developed in the course of the sixteenth century by the Protestant Reformers.[3]

The attempt to study systematically the presence of Biblical texts in Bruno's Italian dialogues led me immediately to the realization that the number and variety of references were far greater than I had expected. It appeared necessary to base the study on some kind of objective data.

This led to the preparation of a table indicating the texts cited explicitly in the six London dialogues, using, for each text, the apparatus of notes in the most recent and authoritative editions of these works at that time. It is probable that many more Biblical references are present in these dialogues than have so far been indicated by Bruno scholars, but the aim of this table was simply, for the moment at least, to give a clear and ordered indication of what has already been noticed. The table is not published here, but the quantitative results obtained are used in the course of my argument. The result was in many respects surprising, and indicative of more far-reaching implications than it is possible to follow up in this brief chapter.[4]

Here only one or two points of particular interest will be taken into consideration. For example, it was not unexpected to find over twenty references to the Book of Genesis, in which recent commentators such as Robert Alter have found the creation of the world presented in a "properly technological" language.[5] It is true that Bruno's infinite and eternal cosmology denies the very idea of a creation in time. Nevertheless, Bruno often uses images of the creation taken from Genesis to define the sense and structure of his own infinite universe, seen as logically dependent on a first principle and cause. It is a universe that he also describes with at least seven references to the Book of Job that, in Bruno's opinion, Moses had added to his books on the law "as if it were a sacrament." This use of the "books of Moses" in the context of his cosmological arguments has already been the subject of a certain amount of comment on the part of Bruno's commentators.[6]

There has been much less, indeed no substantial comment on the remarkable number of references to the Psalms: at least 29 quotations as against 24 from Genesis and 7 from Job. The Psalms are the Biblical book most frequently used by Bruno in his Italian works. It is necessary in this context to remember the general comment on the Holy Scriptures developed by Bruno in the fourth dialogue of *The Ash Wednesday Supper*, the first of his Italian dialogues written in London, where psalms are compared to the frenzies of the poets. In other words, Bruno is considering them not so much as an expression of the Divinity, but as a poetic creation that expresses a human desire for God: "I would have everyone respect the words of divine men ... and I would have everyone respect the inspiration of poets who have spoken of such things with superior vision."[7]

It was precisely in this sense that the Psalms of David were a subject of attention in England at that time. Together with the sonnets of Petrarch, they were being taken as models for an exercise, both spiritual and metrical, which was leading to the development of a new literary genre in English: the Elizabethan lyric. Today a considerable amount of research is available into the metrical rewriting of all the Biblical psalms on the part of Sir Philip Sidney and his sister, the countess of Pembroke, who both underlined the intensely interior spirituality of David in the light of the comments on the Psalms of both Calvin and Theodore Beza. Calvin's comment on the Psalms had been published in English together with a prose version of all the psalms themselves by A. Golding in 1571, and Beza's comment together with a different English translation of the texts by A. Gilby in 1581.[8] It was in this context that Sidney emphasized, in his *Defence of Poetry*, that the original Hebrew form of the Psalms was already metrically structured. He saw them as the moment in which the poetical word and the search for God became one: "... even the name psalms will speak for me, which being interpreted, is nothing but songs; then, that it is fully written in meter, as all learned hebricians agree, although the rules be not yet fully found." David, Sidney wrote, "showeth himself a passionate lover of that unspeakable and everlasting beauty to be seen by the eyes of the mind."[9] It seems more than probable that the particular interest shown by Bruno, while in London, in the Psalms of David is linked to the fact that an important group of metrical psalms in English was composed by Sidney in 1584–1585. Those were precisely the years in which Bruno was in England, and in which he wrote his Italian dialogues, two of which he dedicated to Sir Philip Sidney himself.[10]

As far as Bruno's references to the New Testament are concerned, a quantitative count indicates his decided preference for the Gospel according to St. Matthew, which he quotes at least 17 times, compared with 7 quotations from Luke, 3 from Mark, and only 2 from John. Undoubtedly John's mystical perception of the Trinity would have appeared uncongenial to Bruno, who admitted to the judges of the Inquisition during his long trial that he had nurtured doubts about the Trinity ever since his flight from the Dominican monastery in Naples.[11] It is more difficult to understand the reasons for the clear preference for Matthew, which may have been due to what recent literary commentators of this Gospel, such as Frank Kermode, have called "the idea of transformation" that informs it. That is to say, the new message to which the Gospel bears witness is seen by the Jew Matthew as a "transformation" of the old message, or the religion of the Jews. The new story, Kermode writes, must be different and disturbing with respect to the old one, while at the same time it must respect the facts that the old story narrates, for precisely this respect constitutes the most profound source of its extraordinary novelty.[12] According to this reading, the Jew Matthew, more than the other writers of the Gospels, conceived of his work as a rewriting of the Old Testament, which he thus brought to its triumphant conclusion. It seems likely that a philosopher such as Bruno, who thought of himself as rewriting the history of the universe without wishing to contradict the doctrines of earliest antiquity, reacted with particular interest to this aspect of the Gospel of St. Matthew.

As far as Bruno's attitude to the *Antiqua vera filosofia* ("the true philosophy of the Ancients") is concerned, Miguel Granada has noticed how Bruno turns upside down the historical theses proposed by Ficino, Pico della Mirandola, and the other Neoplatonic philosophers of the Renaissance, who had indicated a chronological and doctrinal primacy of the Hebrew people and Moses. Bruno, on the contrary, strongly indicated a chronological and doctrinal primacy of the ancient Egyptians and Caldeans, "recognizing at best that the wisdom of the Jews expressed in the books of Moses and of Wisdom in the Old Testament could be seen as part of a *prisca theologia* but only in a second and derivative stage, as in the case of the Greeks."[13] This distinction made by Granada certainly has some theoretical justification for it, although it appears to take too little account of Bruno's awareness of the particular significance that the Biblical word had assumed over the numerous centuries of European history in which it had been considered as the Divine Word itself. For Bruno, as an ex-preacher, knew that the Old and New Testaments had played an absolutely primary role within the life of the community, while the Hermetic texts of the ancient Egyptians, even after the celebrated translation into Latin by Ficino, were still read and appreciated only by a restricted few. It was precisely this centrality of the Biblical word within the life of the community that had been reasserted with force by the Protestant Reformation, and that Bruno attempts to come to terms with in the opening page of the fourth dialogue of *The Ash Wednesday Supper*. It is a page in which the participants in the dialogue (already anticipating problems that will become paramount in the case of Galileo) reflect explicitly on the overall sense of the Scriptures for a modern society, obliged to face up to a post-Copernican cosmological discourse that denies the Biblical preference for a geocentric earth:

> THEOPHILUS. Yes, tell me.
> SMITHO. It's because the divine scriptures (whose meaning should always be recommended as deriving from superior minds which are unable to err) in many places affirm and suppose the contrary.[14]

It has already been noticed, in dealing with the psalms of David, how Bruno considered the word of the Bible as equivalent to the frenzy of the poets. His attitude is that the interpretation of the Biblical texts should no longer be subject to a series of dogmas considered as objectively true, but that what counts is the truth of the inspiration that underlies the text, or the personal search for God that it expresses. This idea of the Biblical word as human, and therefore subject to historical considerations, had originated in the preceding century with humanists such as Lorenzo

Valla, and had been further developed by Erasmus of Rotterdam at the beginning of the sixteenth century and later by some of the Protestant Reformers.[15] For his part, Bruno offers a formulation of the idea in unusually radical terms, considering the traditional mode or modes of Biblical exegesis as completely exhausted and irrelevant. Theophilus, who is Bruno's mouthpiece in *The Ash Wednesday Supper*, claims that: "the same Scripture is in the hands of Jews, Christians and Mahometans, which are sects so different from and so contrary to each other that they propose innumerable and completely contrasting readings; for all of them find there the truth which they desire." On the other hand, it is precisely this hermeneutical pluralism that appears to his English counterpart in the dialogue, Smitho, as a form of cultural and historical progress, for it opens up the possibility of reconciling the Biblical word, become uncertain and variable, with the new post-Copernican cosmology that is the subject of *The Ash Wednesday Supper*. Accordingly, Smitho concludes this part of the fourth dialogue of that work with a reference to what he calls the Biblical "metaphors," which he too sees as having become subject to a plurality of interpretations and meanings. According to Smitho, however, "for that very reason, they can now lie at peace with our philosophy."[16]

For Bruno, then, the Biblical word cannot be enclosed within any one specific or dogmatic line of interpretation, but rather appertains to many diverse traditions and so remains open to various and differing reformulations. It is in the light of this daringly unorthodox reading of the Bible that it is necessary to approach those passages in Bruno's texts where he takes a Biblical quotation and bends it to bring it into agreement with his own philosophical doctrine. The rest of this chapter will be dedicated to an attempt to illustrate Bruno's method of rewriting the Bible in the light of his own philosophy by considering, as a telling example of his method, some pages that are particularly significant in such a context. The pages concerned come from the introductory *Argomento ed allegoria del quinto dialogo de "Gli eroici furori"* ("The Argument and Allegory of the Fifth Dialogue of the *Heroic Frenzies*"), in which Bruno defines, at the very beginning of the last of the six philosophical dialogues written and published in London in 1585, the sense of its ending. Here the nine philosophers, blinded by Circe, travel through Europe in a search for illumination until they arrive on the banks of the river Thames, where Bruno was writing these Italian works. Once again, it is important to remember his link with the Sidney circle, especially as the *Furori* is dedicated to Sidney himself.[17]

Let us begin with Bruno's use here of the word "allegory," which in the sixteenth century still had a particular meaning linked to the traditional methods of Biblical exegesis. During the Middle Ages, four principal senses of Biblical interpretation had been distinguished, or four levels of meaning of the Biblical word. These have been studied by Henri de Lubac, who claims that allegory—which was considered traditionally the second level of Biblical interpretation—represents the sense of faith. For St. Augustine, it was to be considered a "dispensationis mysterium" that allows the reader to penetrate a profound spiritual region of the soul.[18] According to Origen, moreover, the difference between ancient myth and Christian allegory consists of the fact that the ancients enclosed the secrets of their theology in a series of fictions, while Christian allegory is not fiction but rather a level of meaning of the Christian story more real and profound than the naturalistic one.[19]

It appears clear that the Brunian allegory that closes the *Furori* is written in the knowledge of this religious sense of the word, closely connected with traditional Biblical hermeneutics. Furthermore, it need not surprise us if Bruno reduces the four senses of medieval Biblical hermeneutics to two—that is, the story and the allegory only. For he fails to distinguish them from the tropological—which is to say, the moral—or the anagogical—which is to say the eschatological—levels of meaning. This reduction carried with it the authority of St. Paul himself, who had raised the subject in the second letter to the Corinthians, chapter X. There St. Paul refers to a single level of deeper meaning of the words of faith—that is allegory, which assumes greater importance than the story itself. In the light of this passage in Corinthians II, it had become common practice among the interpreters of the Protestant Reformation to reduce the four traditional senses of the Biblical word to two—that is, the story and the allegory only—for Luther himself had claimed that the four traditional levels of Biblical meaning were too many, as they "tear Christ's mantel into four pieces." Sometimes even the allegory was considered by the Reformers as superfluous, or only the "froth" of the true Biblical word. Luther, however, tended to admit the allegory where the Biblical text itself appeared to justify it, above all when it was a case of reading the Old Testament in the light of the New.[20] Nevertheless, in the last analysis, the meaning of the Biblical word, in Luther's opinion, was one and unique. Calvin too insisted on a Biblical hermeneutics dominated by the idea of *sancta simplicitas*, although he allowed a metaphorical sense of the Biblical text insofar as it is to be considered the word of God himself "clothed" in terms that make it comprehensible to the human mind. "As for me," wrote Calvin in his comment on a delicate passage in St. Paul's Epistle to the Galations, chapter IV, 22:

je confesse bien que l'Escriture est una fontaine de toute sapience, très abondante, e qui ne se peut espuiser: mais je nie che la richesse et abondance d'icelle consiste en diversité de sens, lesquels il soit licite à chacun de forger à sa poste. Sçachons donc que le vray et naturel sens de l'Escriture, c'est celuy qui est simple et nayf.

[I confess that the Scriptures are a fountain of true wisdom, whose waters are abundant, and never dry up, but I deny that their riches and abundance derive from a plurality of meanings, which everybody is justified in interpreting as they wish. It must be understood that the true and natural sense of the Scritures is the one that is most simple and naive.] [21]

Although Bruno would not have agreed with Calvin that the sense of the Scriptures is unilaterally Christian, he also searched in the pages of the Bible for a simple and natural sense. It is one of the several ways in which the culture of the Protestant Reformation influenced him, in line with his admiration for the figure of Sir Philip Sidney, author of the sonnets to Astrophel and Stella, to whom this dialogue is dedicated.

Remembering Sidney, it is not surprising to note that the fifth dialogue of the *Heroici furori* is dominated by two women, who however, as the pages describing the *Argomento* specify, have only the task of narrating the story while "a more masculine mind" (obviously Bruno's own) clarifies its underlying sense. [22] This allegorical sense is outlined synthetically in the pages of the *Argomento* that deal with the final sections of the work. There Bruno tells his reader that the nine blind men of the final dialogue, which the common mind of the time will want to liken to the nine spheres of the traditional cosmology, exemplify the order and the diversity of all things within an absolute unity of being. Above them are ordered their intelligences, which, according to a similar kind of analogy, depend on a first and unique intelligence. Such a definition of the meaning of the allegory of the nine blind philosophers clearly appears here as Neoplatonic in origin, with overtly Christian implications. An attentive reading, however, will notice the significance of the fact that the nine blind men are likened to the nine traditional cosmic spheres only by the common imagination of the time, while the same common imagination constructs the analogy between their intelligences and a first and unique intelligence of a fully transcendental kind. Such analogies for Bruno himself are no more than modes of speech, with a value that has become purely semantic. This is a way of saying that only the common imagination, anchored still to the idea of nine cosmic spheres, outside of which exists the sphere of the divine, continues to believe in this double structure of being. For Bruno himself, being is identified in a unique substance, defined by both its extensive and its intensive infinity, which is itself divine. If any form of dualism is recognized as valid in this passage, it is only in the way in which this unique universe can be "read" according to two different levels of being, the material and the spiritual. These correspond to the naturalistic story and the allegorical interpretation, even if, in the end, these two levels of discourse have to be united to achieve that unique meaning of the Biblical word that the Reformers considered to be the Word of God, but that in Bruno has become the human word informing a purely philosophical discourse.

At this point, Bruno continues by identifying, in historical terms, those who had, in the first place, suggested to the minds of common men the kind of double structure of being that Bruno himself repudiates. They are the Kabbalists, the Chaldeans, the Magicians, the Platonists and the Christian theologians—all of them figures that Bruno refers to rigorously as "them," taking his distance from their ideas even if he had once himself traveled along such paths. Following a simple train of thought, which Bruno himself presumably at this point considers a simplification, "they" base their reasoning on the unique nature of their God. Bruno reminds them, however, that their God is also a trinity, so that "according to the laws of reflection and squaring," He becomes, in the moment in which He submits himself to fate and to mutability, nine—like the nine spheres, the nine muses, and, at least at the beginning of their journey, the nine blind men of Bruno's text. Others (but they are always "them," "they") speak of a kind of "conversion" so that that which is above fate is converted or transformed, placing itself under the influence of fate, time, and mutability. The sense of this conversion is illustrated by some quotations, one of which is taken from a Biblical source, chapter XX of the Apocalypse, where it is said that the evil dragon will be kept in chains for a thousand years, after which it will be freed. It seems probable that Bruno had in mind here the Aristotelian idea of conversion, which had made a comeback in the scholastic Middle Ages as the foundation of a theology that Bruno himself believed to be profoundly mistaken. This interpretation of the quotation from the Apocalypse seems confirmed by the manuscript papers titled *De infinitis* written at the beginning of the seventeenth century by the English scientist Thomas Harriot, who was part of the household of the ninth earl of Northumberland: an ardent reader of Bruno, and in particular of his *Heroici furori*. Referring precisely to this passage from the Apocalypse, Harriot writes in the margin of one of his manuscripts: "Aristotle. The devil that was bound for a thousand years and after let loose to deceave the people in the four quarters of the earthe."[23]

Still referring to this "conversion" as part of a hypothesis put forward by "them," Bruno continues by making a reference to an idea that he thinks he found in Plotinus (but the editors of the *Furori* have had difficulty in identifying the page), according to which "this conversion does not involve everything, or take place always, but only once." Bruno then refers to another and, in his opinion, more congenial idea, to be found in Origen, which claims that the conversion is eternal. (In this case as well, the passage has not been identified.) Possibly there is some confusion here on Bruno's part, as the idea of a conversion "which happens only once" seems to be associated with the "prime mover" of Aristotelian fame, who puts the whole created universe into motion only once, at the beginning of time. On the other hand, the idea of an eternally ongoing process of conversion can be found in both Plotinus and Origen.[24] In any case, Origen is clearly an important figure in this context, and it would seem important to identify correctly the page being referred to, particularly as Bruno calls Origen in this passage "the greatest philosopher" among the theologians. Undoubtedly the text being referred to is book III, chapter V, section 3 of the *De principiis*, where Origen claims that our world has had a beginning, but that after it, there will be another world like this one, just as he believes that there were others before it. It is precisely this eternal vicissitude of an eternal universe that guarantees for Origen the everlasting conversion of that which "lies above fate." Origen illustrates this idea with two Biblical references, one of which is taken from the book of Isaiah, chapter 66, verse 22, where it is said: "For as the new heavens and the new earth, which I will make, shall remain before me, saith the Lord, so shall your seed and your name remain." The other reference is to *Ecclesiastes*, chapter 1, verses 9–10. This latter reference is of particular interest, as these same verses from *Ecclesiastes* were frequently quoted by Bruno: "The thing that has been, it is that which shall be; and that which is done is that which shall be done: and there is no new thing under the sun."

It is only after this reference to Origen that Bruno intervenes with a direct comment of his own composed of a double consideration of great importance for an elucidation of his own thought. In the first place, he surprises his reader by praising the theologians for their insistence on a fully transcendental dimension of being above and beyond the reign of fate and mutability. For Bruno, however, such a transcendent concept of the divine is above all a useful concept at a social and ethical level, where it makes sense to fortify the moral foundations of society by speaking of an otherwordly afterlife where good deeds will be rewarded and bad ones punished. So what we have here is a concept of the political origins of religion, which Bruno certainly absorbed from authors well known to him such as Averroes or Machiavelli. At the same time, it would be an error to undervalue the Protestant context in which Bruno wrote his Italian dialogues, of which the *Furori* is the final example. The commentators of the Protestant Reformation of Catholic origin often tend to underline how the Reformation represented a return to an almost medieval conception of the dominion of the divine word with respect to the human one. On the other hand, commentators from the Protestant cultures themselves often prefer to notice how the fragmentation of traditional modes of religious discourse brought with it a new taste for dispute, with often surprising results, among which was a new "politicisation" of the idea of religion.[25] Such an idea was far from being repudiated by the circle around Sir Philip Sidney, to whom Bruno dedicated the *Furori*. On the contrary, the biography of Sidney written by Sir Fulke Greville after the death of his friend clearly delineates a political project cultivated by Sidney and his circle, who wished to use the increasing power and prestige of Elizabethan England to oppose the dominion of Spanish forces in continental Europe. They proposed to achieve this by calling the reformed religion to their aid in a design that clearly had a primarily political aim.[26] In this first explicit comment on the Biblical references that he introduces into his philosophical "allegory," then, Bruno appears to be in line with Sidney, to whom he dedicates his work.

Bruno's second comment on his own "allegory," on the other hand, praises with an explicitness of which Sidney could hardly have approved, those few true men of wisdom who have understood the fictitious nature of this transcendental dimension of being. The nine blind men who close Bruno's Italian dialogues are involved in a unique, eternal, and infinite process of mutability, dependent on a first cause that is its metaphysical but no longer fully transcendental principle of unity. Within this eternal process of mutability, Bruno recognizes a material as well as a spiritual dimension of being. At the beginning of their journey, the nine blind travelers are "enclosed within Circe's arms," which signifies "all-embracing matter" and is the cause of their blindness. Then they succeed in finding the nymph of the desert who opens every seal, discovers all secrets with "her twin splendour of the good and the true," and gives them back their sight by "sprinkling them with the waters of purification." Nevertheless, the illumination comes about, together with the end of the journey of which it is the symbol, without a break in the continuity of the narrative—the fundamental concept behind the more spiritual image of the nymph who gives them back their sight being that of the *sommo bene in terra*, "the greatest good on earth."[27]

It is precisely through a series of Biblical references that the sense of this final passage of the dialogue assumes its full significance. The "healing waters" that the new nymph sprinkles over the blind men clearly have a Biblical origin, and can be compared with the waters of baptism—even if the context in which they flow has been completely overturned. For the vase in which the healing waters are contained was given to the blind men by Circe herself—even if she was unable to open it and to gain access to them. This means that the spiritual dimension of being, to which Bruno gives the traditional name of soul, is now to be found within the folds of matter—a concept that later Bruno will develop in the context of an animistic form of atomism.[28] The final outcome of Bruno's idea of conversion can thus be defined as a reduction of the processes of being to an eternal interchange of shadow and light, matter and spirit, that takes place within a unique and infinite substance. This implies a corresponding negation of the transcendent, at least as a dimension knowable by the human mind. Bruno had referred in the first dialogue of the *De l'infinito universo et mondi* to an "inaccessible divine face." In this he can be likened to the ironic author of the Biblical book of Ecclesiastes, Qohelet, who, as recent commentators such as Harold Fisch and von Rad have underlined, does not doubt the existence or the power of God, but rather differs from the other Biblical authors in his conviction that it is not given to us to know or to comprehend directly the modes of His omnipotence.[29] All possibility of a dialogue with a superior being disappears. All that can be said is that His voice can be heard in every part, in every movement, within the natural world. Bruno had already defined his universe as infinite, and of an atomistic nature, which might well give rise to the expectation that all references to Biblical texts would disappear. On the contrary, the very sense of that universe, in these final pages of the *Furori*, is illustrated with great emphasis by Bruno precisely through a series of Biblical quotations, which reach their apex when he elucidates the sense of his infinite process of revolution or vicissitude with the words: "the inferior waters are equal to the superior ones."

The reference is to the second day of creation according to the account in the Book of Genesis, where, after the creation of light, the waters are separated from the waters by the creation of a solid wall called by God "the firmament." According to von Rad, "we must imagine the creative acts quite realistically as separation." The word "firmament" translates the Hebrew *Rāgia*, or "that which is firmly hammered, stamped. This heavenly bell, which is brought into the waters of chaos, forms ... a separating wall between the waters beneath and above."[30] The passage has constituted a crux for the learned commentators of Biblical texts. St. Augustine wondered whether the waters above the firmament were different from those that remained below.[31] The Hebrew *Genesis Rabbah* speculates as to whether the waters were divided in half, or whether those remaining above the firmament were greater in quantity than those below. And then, what do "above" and "below" mean exactly in this case? And how do the waters above the firmament remain in place, given their enormous weight? This final question had given rise to some particularly refined thoughts on the part of the commentators, such as the opinion of the Rabbi Tauhuma, in the *Genesis Rabbah*, that the waters remain suspended by the power of a word.[32] Furthermore, it was traditional to consider the waters above the firmament as a clear and unsullied region of the spirit that mediated between the natural universe and the perfection of a transcendent, divine intelligence. Origen, for example, considered the waters above the firmament to be a metaphor for the angelic spirits.[33] Other medieval commentators, on the other hand, considered the waters above the firmament as the origin of the flood, caused by their penetration through the gaps in the firmament. This more naturalistic interpretation would be radically developed in the course of the sixteenth century.[34]

Bruno was thus intervening in a discussion that had become particularly lively and sophisticated. As Giovanni Aquilecchia has noted, Bruno had already offered a new interpretation of this passage of the Book of Genesis in the fourth dialogue of *The Ash Wednesday Supper*, where the waters under the firmament are considered to be the waters on our own globe, and the waters above the firmament those on other globes. In this way, Bruno considers them both as parts of his newly infinite natural universe.[35] The quotation of the same passage in the Argument and Allegory of the fifth dialogue of the *Furori* remains faithful to this line of interpretation, while adding a marked spiritual dimension to the more radical materialism of the interpretation offered in the *Supper*. Now two principles run through the infinite universe in a play of perpetual alternation, one represented by the inferior waters of Circe that cause blindness, and the other represented by the waters of the nymph of the desert that illuminate—that is to say, one of the body and the other of the soul. A dualism of being, material and spiritual, is thus reproposed within the one infinite universe, according to which the waters above the firmament become healing waters and the nine blind men, now illuminated by the "double splendour of the good and the true" become "nine intelligences, nine muses." The central idea thus remains, even more emphatically than in *The Ash Wednesday Supper*, that of the impossibility of separating the superior waters from the inferior ones, "because the end limit of the superior waters is the same as the beginning limit of the inferior ones"; because there must be no "separation or vacuum" between them; because "everybody in unison celebrates the mighty and magnificent vicissitude which renders the inferior waters equal to the superior."

The conclusion that may be drawn from this study of the dense series of Biblical quotations to be found in the Argument and Allegory of the fifth dialogue of the *Heroic Frenzies* is that the Bible appears in Bruno's philosophical works as a text profoundly rooted in the European consciousness, but open to a series of radical rewritings. That is to say, the Bible is approached as a flexible container of traditional stories and images that illuminate the sense of both the material world and the world of the soul. Bruno, writing in London in 1585, knew that the canonical interpretations of the medieval ecclesiastical tradition had already been fragmented by the invitation on the part of the reformers to read those words with new eyes—that is to say, from a critical and innovative stance. Bruno himself accepted that invitation with a fervor that the Reformers themselves would hardly have been able to accept, as he finished up by taking the Biblical word outside the confines of Christianity itself. The Bible thus becomes a part of his own philosophical discourse. It is integrated into his own vision of an infinite universe with an effect that is both traumatic and disturbing. The meanings and interpretations of the past centuries are criticized, but so are the new interpretations of the Reformers. It is the new natural philosophy of an infinite and eternal universe inhabited by an infinite number of ever-changing worlds that becomes, for Bruno, the newly sacred word.

NOTES

1. Bruno cannot, of course, be considered the only writer in the European tradition to have "rewritten" in his own terms some of the major Biblical stories. See on this subject, P. Boitani, *The Bible and Its Rewritings*, Oxford, UK, Oxford University Press, 1999.

2. Tirinnanzi (1996–1997).

3. For Bruno's relationship with Leicester and his circle, see Ciliberto (1991), 29–195.

4. The table of Biblical citations referred to here may be consulted in the original Italian version of this essay in *La filosofia di Giordano Bruno: Problemi ermeneutici e storiografici*, ed. Eugenio Canone, Florence, Olschki, 2003, 199–216. Bruno would certainly have been reading the Bible in one of the many versions of the Latin vulgate, although his quotations are often in the form of a synthesis of the passage in Italian. It is probable that he knew one or other of the Italian versions of the Bible as well. For a history of the Italian Bibles of this period, see E. Barbieri, *La Bibbia italiana del Quattrocento e del Cinquecento*, Milan, Edizioni Bibliografica, 1992.

5. See *The Literary Guide to the Bible*, eds. R. Alter and F. Kermode, London, Collins, 1987, 33.

6. See, for example, the section "The Prophet Moses" in De Léon Jones (1997), 137–45.

7. See Bruno (1977), 179–80.

8. For the psalms translated by Sidney and his sister, see *The Sidney Psalms*, ed. R. E. Pritchard, Manchester, UK, Carcanet Press, 1992. For a study of Sidney's knowledge of the Hebrew comments on the psalms, see S. Weiner, "Sidney and the Rabbis: A Note on the Psalms of David and Renaissance Hebraica," in *Sir Philip Sidney's Achievements*, eds. M.J.B. Allen et al., New York, AMS Press, 1990, 157–62.

9. In Sidney's *An Apology for Poetry*, ed. G. Shepherd, Manchester, UK, Manchester University Press, 1973, 99.

10. For the dating of the composition of Sidney's translations of the psalms, see the introduction to J.C.A. Rathmell, *The Psalms of Sir Philip Sidney and the Countess of Pembroke*, New York, New York University Press, 1963, xi–xxxii. Rathmell writes in these pages: "The Sidneian psalms were metrically subtle, and constantly bring out the underlying allegorical sense." On Sidney's psalms, see

also the relevant pages in R. Zim, *English Metrical Psalms*, Cambridge, UK, Cambridge University Press, 1987.

11. "A glance at the biblical text indices of major patristic treatises reveals how massively the Gospel of John figured in the formation of classical trinitarianism. References to it typically overwhelm those to other biblical books"; see C. Plantinga Jr., "The Fourth Gospel as Trinitarian Source Then and Now," in *Biblical Hermeneutics in Historical Perspective*, eds. M. S. Burrows and P. Rorem, Grand Rapids, MI, William B. Eerdmans Publishing, 1991, 303–21. Bruno, on the other hand, admitted candidly to the inquisitors in the third Venetian hearing of his trial on June 2, 1592: "I have had some doubts about the name of the person of the Son, and the Holy Spirit, as I do not think of these two persons as distinct from the Father." See Firpo (1993), 170.

12. See Kermode's pages on the Gospel according to Matthew in *The Literary Guide to the Bible*, op. cit., 387–401.

13. See M. A. Granada, "Giordano Bruno e l'interpretazione della tradizione filosofica," in *L'interpretazione nei secoli XVI e XVII*, eds. G. Canziani and Y. C. Zarka, Milan, Franco Angeli, 1993, pp. 59–82.

14. My translation. See also Bruno (1977), 177–82.

15. For a still valid history of Biblical hermeneutics in the sixteenth century, see S. Berger, *La bible au sixième siècle: étude sur les origines de la critique biblique*, Paris, Berger-Levrault, 1879.

16. For a consideration of Bruno's attitude toward the "Biblical metaphors," see Ciliberto (1986), 107–15. For the more general problem of the contrast between Biblical texts and the new cosmology, see R. Fabris, *Galileo Galilei e gli orientamenti esegetici del suo tempo*, Rome, Pontificia Academia Scientarum, 1986.

17. These pages close the long introduction to the *Heroici furori*, dedicated to Sir Philip Sidney. See Bruno (2002), vol. 2, 509–21. For a comment underlining Bruno's use of the expressive modules of prophecy in the final pages of the main text of the *Furori*, see chapter 6, "The Sense of an Ending in Bruno's *Heroici furori*," in this volume.

18. H. du Lubac, *Exégèse médiévale: les quatres sens de l'Ecriture*, Paris, Aubier, 1959. The passage from Saint Augustine, taken from the *Factum audivamus: mysterium requiramus. In Jo.*, is quoted by Lubac in chapter VIII of his text titled: "L'allegorie, sens de foi."

19. On Origen in particular, and more generally on the implictions of the new Protestant Biblical hermeneutics for the development of a more modern natural science, see Peter Harrison, *The Bible, Protestantism and the Rise of Natural Science*, Cambridge, UK, Cambridge University Press, 1998.

20. Berger in *La Bible au seizième siècle*, op. cit., quotes from the passage where Luther claims that he always repudiated allegorical readings of the Bible except where the text itself suggested such readings, or where they were necessary to explain the Old Testament in the light of the New.

21. This well-known passage is quoted by Berger, op. cit., in his chapter on Calvin, 115–27. For Bruno's knowledge of Calvin's theology, see Ingegno (1987).

22. The two women, called Laodomia and Giulia, have been identified by Bruno's biographer, Vincenzo Spampanato, as two female cousins, for one of

whom, in his youth, Bruno seems to have nurtured an unrequited love. See Spampanato (1921), 64, note 3.

23. For Harriot's comment, see BL ADD. MS 6782, fol. 374v. The word "Aristotle" has been torn and is almost illegible—nevertheless, the strong attack against Aristotle's philosophy, and particularly his cosmology, that pervades these papers seems to confirm a reading already suggested by Harriot scholars. The ninth earl of Northumberland's library, to which Harriot had access, contained an important collection of Bruno texts, which included the *Heroici furori*, annotated by Northumberland himself. See the chapter "The Northumberland Texts" in Gatti (1989), 35–48.

24. For the philosophical definition of Aristotle's "prime mover," see *Physics, book VIII*, section 6. In the *Enneads*, VI 3, 13, Plotinus writes of the descent of the souls from the intelligible sphere toward the natural world that there is a correct moment for all of them, so that when they arrive, they can be said to descend, as if called by a herald, to penetrate the body that is appropriate for them. For this reason, it has been claimed that such a movement of descent is brought about by magic, or by a certain kind of irresistible attraction. Even in a single being, the ordering of the living world is thus brought about. "At the right moment, nature moves and generates everything that is . . ."

25. On this subject, see R. Weimann, "Discord and Identity: Religion 'Politicised,'" in *Authority and Representation in Early Modern Discourse*, ed. D. Hillman, Baltimore–London, John Hopkins University Press, 1996, 53–67.

26. See F. Greville, *The Life of the Renowned Sir Philip Sidney*, New York, Scholars' Facsimiles and Reprints, 1984.

27. For this passage, see Bruno (2002), vol. 2, 747.

28. For Bruno's atomism, see Michel (1957), and chapter 3, "Bruno and the New Atomism," in this volume.

29. See Bruno (2002), vol. 2, 43. See also G. von Rad, *Wisdom in Israel*, Nashville–New York, Abingdon Press, 1972, and H. Fisch, "Qohelet: A Hebrew Ironist," in *Poetry with a Purpose: Biblical Poetics and Interpretation*, Bloomington–Indianapolis, Indiana University Press, 1988, 158–78.

30. See *Genesis: A Commentary*, London, SCM Press, 1972 (translated from the original German *Das erste Buch Moses*). See also Psalms CV (CIV), ii, and Job XXXVII, xviii. According to C. Westermann in *Genesi: commentario*, Casal Monferrato, Piemme, 1995, 96: "in the so-called Sacerdotal Code (P) [which contains the proto-history of *Genesis*], the accent is placed on the representation of the world in its totality, on the categories of time and space, on living beings as the all divided into species: without denying the possibility of a disintegration of the whole." My translation.

31. See *St. Augustine on Genesis*, ed. R. J. Teake, S.J., Washington, DC, Catholic University of America Press, 1991, 165.

32. See Parasheh 4: Genesis I, 6–8, in *Genesis Rabbah: The Judaic Commentary to the Book of Genesis*, vol. I, ed. J. Neusner, Atlanta, GA, Scholars Press, 1985, vol. I, 37–44.

33. Quoted in B. Nardi, "Lo discorrere di Dio sovra quest'acque," in *Nel mondo di Dante*, Rome, Edizioni di Storia e Letteratura, 1994, 307–13.

34. The naturalistic interpretations of these verses have been discussed by E. Grant in *Planets, Stars and Orbs. The Medieval Cosmos, 1200–1687*, Cambridge, UK, Cambridge University Press, 1994.

35. As Aquilecchia points out, Bruno erroneously considers the word *firmamento* to stand for air, within which he thinks of the inferior waters, which are those of our own globe, as being distinguished and divided from the superior waters, which are those of other globes. See Bruno (2002), vol. 1, 526–27 and note 16.

14

SCIENCE AND MAGIC

THE RESOLUTION OF CONTRARIES

THIS CHAPTER ATTEMPTS TO MAKE A contribution to a discussion that has been developing for some decades, but that seems far from being exhausted. It becomes particularly relevent in the light of the recent book by this author that reproposes Bruno's thought as concerned, in many of its most central moments, with properly scientific and even technological subjects, in a modern sense of those words.[1] Such a reading of Bruno's thought creates a problem with respect to an approach such as that of Frances Yates, which claims not only to find in his works a radical culmination of the magical, Hermetic, Neoplatonic, and Kabbalistic themes already developed in previous Renaissance culture, but also to exorcise from a modern reading of his thought any serious interest for scientific arguments or logic: "Bruno is not at all in the line of the advance of mathematical and mechanical science," writes Yates.[2]

The line of argument developed in this chapter will be that it is erroneous to attempt to answer Yates's thesis by eliminating or neutralizing the magical aspects of a work in which such themes are clearly present. Rather, it is necessary to follow with particular attention the process of "deconstruction" (at the risk of flogging a useful modernist term) that the magical themes in Bruno's work are subjected to once they come up against the fundamental aspect of his natural philosophy: the infinity of the universe. In the attempt to develop such an approach, it will be necessary to insist on the *positive* function of the ambivalence and ambiguity that open up within Bruno's treatment of traditional magical doctrines.

Faced by the newly infinite dimensions of Bruno's universal, divinely animated substance, many of the central tenets of the Renaissance discourse on magic, which by the end of the sixteenth century were beginning to become repetitive and stereotyped, lose the axis around which they had rotated for centuries. The universal infinity proposed by Bruno tends to erode the traditional meanings of magical doctrine at precisely that point in which it denies the concept of limit with respect to the natural world. In this context, it is worthwhile to quote once again Bruno's frequently quoted description of his walk through the spaces of the new cosmos, no longer enclosed by any limiting celestial spheres:

Here is one who has flown through the air, discovered the sky, found himself among the stars, passed beyond the margins of the world, banished to oblivion fantasies such as the walls built of the first, ninth, tenth, and all the other possible spheres imagined by vain mathematicians or blind and banal philosophers.[3]

This walk through cosmic space is conceived of by Bruno as without limits and without end. Only the limited forces of the finite body and mind of the philosopher, or new scientific inquirer imprisoned in the co-ordinates of an ineluctible space-time, can constitute limits to our knowledge of the infinite world. The rays of Bruno's infinite number of stars or suns destroy, in an instant, the traditional idea of correspondences between a crystalline empyrean sky and the sky of the elemental world under the moon, for now the infinite space has become perfectly homogeneous, composed throughout of an infinite number of solar systems surrounded by a subtle ether. The idea of a hierarchy of being becomes problematical. Accordingly, changes become necessary in the crucial concept of contractions of the divine into the world of vicissitude, traditionally seen by the astrologers and magicians as taking place through planetary intermediaries and influences.[4] In the third of his Italian dialogues written and published in London, De l'infinito, universo, et mondi (On the Infinite Universe and Worlds), Bruno writes that in the light of an infinite universe, everything becomes relative, and there is no longer any "above" or "below." The result is evident: "that admirable order and ladder of nature" that had supported the traditional Aristotelian–Ptolemaic cosmology had been no more than a "pleasant dream."[5]

It is no longer necessary to penetrate the secrets of an infinite and divine mind that, according to the Christian theological and philosophical tradition, had expressed itself in a finite and limited act of creation. For Bruno, the expression of the divine in the world is infinite and total, in all its parts.[6] The secrets and mysteries, which are the stock in trade of the magician, become part of the here and now. At least virtually, they are within the reach of everybody; even if few, in Bruno's opinion, have seen the light and understood the terms of the new research to be undertaken. The new science will study a newly infinite universe made up of an infinite number of finite bodies—it will inquire into the forces and energies that animate it, its centuries-old evolution, the minimum particles that underlie its manifold changes and modifications, and the modes of their integration and disintegration within the infinite variety of accidents that make up the phenomenological world.

Within this new prospect, new words, which alone can give rise to new definitions of reality, become an urgent necessity, as Bruno writes explicitly in a much quoted page at the beginning of the Frankfurt trilogy.[7] Nevertheless, the old words need not disappear. Inevitably, however, they will become slippery, elide, fragment, open themselves up to new meanings and contents. Words, for Bruno, no longer live according to precise definitions sanctioned by the tradition. With respect to a linguistic ideology that will be developed by the Academy of the Crusca in Florence whose voluminous dictionary will attempt, at the beginning of the seventeenth century, to render canonical the traditional meanings of the Italian vocabulary, Bruno develops a linguistic theory and practice that is profoundly innovative and heretical.[8] In his philosophy, the meanings of words are enriched with unexpected ambivalences, for they are inscribed within the semantic vacuums between the atomistic minimums of an infinite and constantly evolving whole. Particularly significant in this context is the comment found in his work *De imaginum compositione*, where he is considering the poetical and mnemonic images found in the ancient Chaldean texts. Here Bruno claims that he has been inspired by ancient Egypt and by nature, both of which have become his guides toward a realization that it is impossible to stand still—impossible, that is, to go on repeating the concepts and images (the written language) of the Greeks, as certain ignorant Latinists do. Rather, it is necessary to search continually for new guiding principles around which the parts of sentences will compose themselves anew, giving rise to original forms of linguistic discourse that more directly express the processes of the natural world.[9]

Among the many limits that become blurred in Bruno's infinite universe, it is necessary to include that between words and things themselves, between a world that is now without end and without circumference, infinitely full of infinite forms of life, and the alphabets that attempt imperfectly to describe them: numbers, words, images, mnemonic tables, and rotating mnemonic wheels. The definition, always shadowy and ambiguous, becomes confused with the reality of the thing in itself, while the newly infinite universe recedes to incommensurable distances with respect to the limits of the inquiring human mind. It is Bruno himself who underlines this concept with a lively image in a crucial passage of *The Ash Wednesday Supper*:

> The problem is that our painter is unable to examine the portrait by taking advantage of those spaces and distances which are normally available to the masters of that art. The canvas on which the picture is painted remains too close to his face and eyes, and it is impossible for him to take a few steps backward or to one or the other side without the risk of jumping into the void like the famous defender of Troy.[10]

Here Bruno is attempting to come to grips with one of the problems that continues to assail scientists today—that is, how it is possible to obtain a clear picture of real things when the inquiring mind remains inside the picture itself, as an intimate part of the image being formulated. The problem clearly becomes infinitely complex when the reality being observed itself assumes infinite dimensions. Bruno attempts to solve this problem by insisting, as Eugenio Canone has underlined, that our knowledge is not concerned with the infinite object in its entirety, but only with the visible *back* ("dorso") of the object, with a speck of an infinitely extended universe. This means that we never see the truth in its entirety, face to face. That would mean penetrating into the depths of the *womb* ("grembo"), the infinite ocean that remains the dark and mysterious source of eternal life.[11]

It is precisely our limited possibility of conceiving, in the light of a finite reason, the whole picture of an infinite universe, and above all the intimate principle of unity that governs its mutations, rendering it ordered and coherent, that opens up a space again for a discourse concerning magic. Nevertheless, it is necessary, as always, to follow with particular attention the way in which Bruno redefines this word, like all the other words that he inherits from a philosophical and cosmological tradition of a radically different kind. It seems difficult to deny that the first phase of Bruno's philosophical speculation, and particularly the years of his English experience, are above all concerned with supplying a definition of a new natural philosophy. An exception should be made for the *Sigillus sigillorum*, the text of most importance in a consideration of magic in this first phase of Bruno's thought, published in London soon after his arrival there in 1583. In this work, magic is a major presence, listed by Bruno as one of the *rectores actuum*, together with art, mathesis, and love.[12] Nevertheless, the emphasis changes in the Italian philosophical dialogues that Bruno began to publish the following year in London. It is noteworthy that the entries in the *Lessico bruniano* relating to the Italian works contains 14 whole pages devoted to the entries *natura, naturale, naturalità, naturalitade, naturalmente, naturante*, while the entries *magia, magicamente, magico*, occupy barely one and a half pages.[13] Furthermore, the most part of these few occurrences of the word "magic" and its corollaries relate to the celebrated apology of Egyptian magic in the third dialogue of the *Spaccio de la bestia trionfante (The Expulsion of the Triumphant Beast)*. This apology is pronounced by the figure of Isis, who is obviously concerned with defending the Egyptian idea of a cosmos full of vibrating and mysterious life, with magical resonances, which she finds even within the humble things of nature such as onions and croccodiles. Bruno himself, however, in the introductory letter of this work addressed to Sir Philip Sidney, who was anything but a Hermetic Magus even if he did take lessons in mathematics now and then from John Dee, warns his reader not to consider these dialogues as "assertive" but rather as dramatic in intent. For in them the speakers "speak with their own voices,

... arguing with whatever fervor and zeal is necessary for them to present their case effectively."[14]

Nothing in this text authorizes us to consider what is said by Isis as an expression of Bruno himself. At most, we can notice a certain benevolence on the part of the "omniscient author," as the literary critics would call him, with respect to the discourse of Isis, who is ably used by Bruno to emphasize a Neoplatonic concept that (as the other philosophical dialogues of these years amply confirm) he absorbed into his natural philosophy. This is expressed in a preceding speech by Saulino: "So then, Nature is God in things."[15] Even here, however, the concept appears considerably modified by the extent to which Bruno abolishes the traditional distinction between God's absolute power and His conditional power—that is, His absolute power to create freely whatever universe He desired, and His power conditioned by the actual, finite universe that he was traditionally supposed to have created.[16] For Bruno, on the contrary (as we saw in the previous chapter), an infinite God continually expresses Himself in an infinite world of infinite things. In the context of this idea, Isis expresses wonder at Momo's distaste when faced by her conviction that the divinity should be searched for in excrement and ashes. Given that Isis continues to disagree with him, Momo changes his tone and replies to her with some interest by recalling precisely that traditional idea of a "ladder of nature" that Theophilus/Bruno had just abolished in the preceding *On the Infinite Universe and Worlds*: "so those wise men [that is, the ancient Egyptians] with magic and divine rites ascended the same ladder of nature toward the height of the divinity as the divinity itself descends in order to communicate with the vilest things of nature."

This passage stimulated a particularly irritated reply from an anonymous Neapolitan reader of Bruno's text who noted in the margin that the idea was "detestable."[17] This reader is usually supposed to have been an ecclesiastic, and it has become almost impossible to read this text without taking into account the way in which his disapproval becomes a part of the chorus of ideas that Bruno's text itself presents for impartial consideration, constituting a Socratic debate made up of conflicting ideas that above all aim at underlining a problem that needs to be resolved. This makes it unwise for the modern reader to intervene in this page, upsetting its dialectical structure in order to posit a Brunian "celebration" of any one of the ideas he is playing with. As he himself had announced in an important page at the beginnning of the *De umbris idearum*, his philosophical position with respect to the traditional schools of thought is defined by the extent to which he takes his distance from all of them, although with differing emphases and distinctions.[18]

The central concept considered by Bruno on this page of the *Expulsion of the Triumphant Beast* can be defined as an idea of magic that respects the processes of nature, and therefore of natural laws, rather than rudely contradicting them. This concept undoubtedly includes a reference to the voice of Isis. It also relates to a theme that is central to Bruno's early works from their beginnings in the drama *Candelaio*, where Bruno (who here defines himself as an "Academic of No Academy") proposes as his comic hero the odious and ironically named Bonifacio, who places all his hopes in the vanity of magic superstitions. Bruno invites his public to laugh at Bonifacio's absurd dreams of magic potions of such power that "they would make rivers flow in directions contrary to their natural bent."[19] Such dreams are likened to his desire to stop the passing of the years, and his pretense of being still young, without realizing that it is only by allowing oneself to be carried willingly by the current of time, "which gives all things and takes all things away," that "the spirit can be enlarged and the intellect magnified."[20] Equally negative are the cunning and often genial tricks of the conjuror Scarumuré, who is quite aware that the comedy he is playing in "is a good one," but could easily become "a much too troublesome tragedy." So that what appears necessary is the quest for a kind of magic that acts in harmony with the processes and metamorphoses of nature—a magic seen as a way of inserting oneself into the secret mechanisms of nature's ways in order to make them work better, rather than to wreak havoc and destruction.[21]

The distinction made by Bruno here appears in line with that proposed in recent years by Wayne Shumaker, in an anthology volume dedicated to Renaissance magic and its relation to the new science. Shumaker claims that the discipline known to Renaissance philosophers as natural magic appears coherent with the development of a new science, while superstitious or demonic magic appears as an impediment.[22] This judgment would seem to be confirmed by Bruno's philosophical dialogue *Of the Cause, Principle, and One*, where we find a therapeutic concept of natural magic being emphasized. Theophilus (a character in the dialogue whom the reader is invited to identify with Bruno himself) observes that as far as medicine is concerned, as well as those chemical cures that are regularly prescribed and approved of, "he does not disapprove of those cures that are considered magical, such as the application of roots, the use of the power possessed by stones, or the murmuring of chants." It is known, he adds, that these things are comforting, and help to lead the patient back to good health instead of inducing torments and death.[23] However, even here it would be a mistake to interpret these comments as a straightforward celebration of magic. Rather, it is important to note the grammatical construction used by Bruno, who does not write that he "approves" of magical therapies, but only (on pragmatical grounds) that "he does not disapprove of them"—which is something rather different.

The conclusion reached by Bruno could be defined *as De natura iuxta propria principia (About Nature, According to Her Own Principles)*, to use the title of the major work of the southern Italian philosopher Bernardino Telesio, whom Bruno much admired.[24] Such principles, according to Bruno, are to be considered as perfectly rational insofar as they represent the perfect seal of the infinite divine mind and not a limited extension of a transcendent divinity. This means that the processes of nature themselves are not subject to miraculous intervention. Nevertheless, the limits of the coordinates of space and time in which the individual human mind is enclosed—too much inside the processes themselves to be able to acquire a perfect rational knowledge of its real situation—open up a new space of shadow in which the traditional concepts of magic survive, even if only as a weak solution, at least in this phase of Bruno's thought. For, in his Italian dialogues, it is possible to forge a link between Bruno's new natural philosophy and the magical tradition only where magic becomes the pliant and subordinate handmaiden of the infinite natural processes of an infinite world.

It would seem to be a relationship of this kind that establishes the connection, emphasized by Bruno in the dedication of his drama *Candelaio* to Signora Morgana B., between the play itself and his first philosphical work published in the same year (1582), the Latin *De umbris idearum (Of the Shadows of the Ideas)*.[25] The *Candelaio* launches a powerful accusation, of both an ethical and an intellectual kind, against superficial magical practices, particularly when they become the instruments of uncontrolled egoism and greed. On the other hand, the *De umbris idearum* appears to find in the concept of protective shade, or the veil of imperfect knowledge that separates the mind from divine certainties, the necessary and ineluctible definition of the human condition. Bruno, in this work too, however, inveighs against the dishonest magicians, who claim to be able to rent the veil of shade with a violent or an inconsiderate gesture, rather than progressing toward the light methodically by following the footsteps traced within the world of nature by the divinity itself. The punishment for such temerity is seen to be an obscure form of blindness, accompanied by pain and anguish, for it is forbidden to interfere with the processes of nature in their regular cycles and rhythms.[26] This theme will be reproposed with particular eloquence in the final dialogue of the *Heroic Frenzies*, where the nine philosophers are blinded by Circe precisely because they had dared, "with excessive ambition," to look directly into "that most intense of all lights which illuminates the world."[27] Nor can they be healed by Circe herself, but only by the nymph of the Thames, who symbolizes precisely the peaceful flow of natural things. For at the beginning of his dialogue titled *Of the Cause, Principle, and One*, written and published in London in 1584, Bruno had written of the Thames that it was the only river in a blood-stained Europe that, during the sixteenth century, continued "to flow peacefully and gaily between its grassy banks."[28]

What is of interest here is not so much the arcadic vision of Elizabethan England as the emphasis placed on the just and regular rhythm of the river of natural life, whose principle nymph gives the philosophers back their sight. It is true that the purifying waters she sprinkles on their eyes were originally given to them by Circe herself. However, Circe was powerless to open the vase that contained them, for her influence was too weak. Now her waters can join with those of the river Thames whose nymph represents a superior principle of knowledge with respect to the inferior magic of Circe. Whereas Circe had condemned the philosophers to "dark blindness" and "weary tasks," before directing them to the place where they could be healed, the nymph of the river Thames offers them a double illumination, of matter and spirit, goodness and truth. This is specifically defined as a stimulus to the work of the intellect that, as the philosophers themselves affirm in their final *Canzone de gl'illuminati* (The Song of the Enlightened), leads to a new and more complete understanding of the "force of those eternal laws" that regulate the infinite universe in its infinite processes of change.[29]

Rather than considering this an attack against every form of magic, it is more proper to notice here that it is in terms of a superior principle of the intellect that we find the major reference to magic in this first phase of Bruno's works. The page concludes the dialogue *Of the Cause, Principle, and One*, the second of Bruno's philosophical dialogues to be written and published in London between 1584 and 1585. It says that "the most profound form of magic is that which is able to deduce the contraries after having found the point of union."[30] A thorough consideration of this statement would require a close reading of the whole of this work, which is usually and rightly considered the most philosophical of Bruno's Italian dialogues, and has been the subject of much learned attention.[31] Here it is only necessary to underline how this concluding page of Bruno's work *Of the Cause, Principle, and One*, situates the magical moment at the beginning of a process of diversity or multiplicity after the intuition of an essential unity of being. There is clearly a reference here to the concept of contraries in the thought of Nicholas of Cusa, although with significant modifications. For Cusanus, every difference is substantially irrational and is situated in the heart of the contractions that represent the fall of the infinite and divine truth toward the shadowy nothingness of the phenomenological world. For Bruno, on the other hand, being cannot ever contract itself into nonbeing, so that every diad, every negative or diversity or contrary, is to be seen in its relationship with the activity of the mind in its perception of a world of finite things inscribed in a substantial and infinite unity of being. The magic that this final page of Bruno's dialogue defines thus finds its origin in the principle of multiplicity that gives rise to the world of the phenomena, but it also resides at the birth of the processes of thought itself. Magic in this sense seems to be intimately related to the powers of the imagination. It creates, beyond the sphere of our will, and according to an intimate necessity of method, the infinite number of diads, or contraries, or differences, through which the mind attempts to rise, by elaborating a series of ever more elaborate syntheses, to an understanding of the ultimate unity of the whole. Hélène Vedrine writes: "Without manipulating the texts, and without any desire to discover an anticipation of a Kantian scheme of the transcendental, it is necessary to insist on the unifying power of the imagination in Bruno's thought. The most powerful expression of the soul, and the mediating term between temporal and eternal things, the imagination for Bruno is the source of all invention."[32]

Among the diads, or contraries, proposed by Bruno at the level of an ontology, the most significant one, at least in the later phase of his work, appears to be related to his concept of the atom as the physical minimum, which he places in a dialectical relationship with the infinite and infinitely powerful maximum. If the physical minimum contracts within itself all the power of the maximum, it is evident that the concept of the atom contains an element of excess with respect to a definition of matter conceived of as completely rational and comprehensible. Precisely this element of excess seems to be in Bruno's mind when he claims in the first book of his Latin work *De triplici minimo (The Triple Minimum)*, published in Frankfurt in 1591, that "the minimum exceeds in energy any corporeal mass to which it has given rise by aggregation."[33] This makes it no surprise if Bruno feels the need to conclude his reasoning on the existence of the atom in the *De triplici minimo*, where the insistence on the number three already indicates a clear element of number mysticism, by calling on the Biblical Old Testament as well as Hermes Trismegistus. These references serve to indicate precisely that element of imperfect logicality that, as Bruno foresees, will accompany any prediction of the behavior of the ultimate components of matter. What is being raised here is a problem that continues to confound the scientists today, in our era of post-quantum physics with its principles of uncertainty, although it would clearly be imprudent to propose Bruno in any simple sense as a precursor of such developments.[34] All that is necessary is to take note of the at least apparent illogicality that, in Bruno's view, necessarily characterizes the behavior of the ultimate components of matter.

With the *De triplici minimo*, we arrive at the final phase of Bruno's philosophical activity, where the relationship between science and magic becomes more complicated. On the one hand, we find in the volumes presented for publication an increasingly scientific character, culminating not only in the work on atomism that opens the Frankfurt trilogy, but above all in the final work, *De immenso et innumerabilibus*, where Bruno's cosmological inquiry reaches its peak. It is true that between these two parts of the trilogy Bruno inserts the *De monade*, which develops a mystical discourse centered on Pythagorean number symbolism, repeatedly returning to the magical themes already present in previous works such as the *Sigillus sigillorum*. However, Bruno assures his reader in the dedicatory letter of the trilogy addressed to Henry Julius, duke of Brunswick, that the *De monade* takes into consideration received opinions rather than his own ideas. What is being presented is thus "what the author has heard others say"—moreover, it is to be considered a work in which he searches for the truth "not without uncertainties." The final work of the trilogy, on the other hand, the *De immenso et innumerabilibus*, presents Bruno's cosmology as truths in which he believes "without a shadow of doubt."[35]

Perhaps it is this distinction, made by Bruno himself in relation to the truth-value of the diverse works that make up the Frankfurt trilogy, that explains the fact that the considerable quantity of manuscript material related to magic, most of which is to be considered as not having found its final form, was never presented by Bruno himself for publication. Especially after the recent publication of this heterogeneous material in a new edition that adds a massive critical apparatus with respect to the first publication of these manuscripts in the nineteenth-century edition of Bruno's complete works, it is possible to see how a situation has developed with respect to Bruno's oeuvre that is not dissimilar to that which has occurred in the case of Isaac Newton.[36] For we now know that the great British scientist himself also left to posterity the unravelling of a problem that consists of a large quantity of mostly unpublished material relating to his studies of traditional subjects such as alchemy and Biblical prophecy, creating a relationship with his advanced research in the field of physics that it is not proving easy to decipher.[37]

Here it is possible to make only some brief comments on this final group of Bruno's magical texts, limiting them to some aspects that relate to the complex relationship between his magic and his science. In the first place, it needs to be emphasized that Bruno repeats here the severe condemnation of superstitious and demonic magic, or the *magia desperatorum* commonly known as "black magic," which was associated with necromancy and at times with deviated forms of religious magic.[38] Bruno does not infer from this condemnation the necessity of denying beings that he continues to call "demons." Nevertheless, as the editors of the recent volume of Bruno's magical works underline, his demons become only one component of an infinitely varied natural universe full of beings that, in known and unknown ways, combine matter and spirit and are therefore not to be considered as in any way "exceptional." This implies that Bruno's magic is concentrated on the possibility of reinforcing the capacities of the human mind, rather than tapping in to powers that lie outside it.[39]

At the same time, the emphasis on the concept of a world soul, which the Platonic and Neoplatonic tradition had already postulated as the basis of those magical arts that were known as "white magic," remains unchanged, even after the atomistic conclusion of the De triplici minimo. The final outcome of this idea, however, appears now to be defined by the concept of "links," or "*vinculi*," which substitutes the more traditional concept of "correspondences" between a crystalline sky above the elemental sphere and the sphere of the base elements itself. In the traditional cosmology, the elemental sphere was thought of as lying under the sphere of the moon, and of being fixed at the center of the universe. One of the canonical Renaissance texts based on this concept was that same De vita coelitis comparanda (For a Comparison between Life in the Heavens and on Earth) by Marsilio Ficino that Bruno had been accused of plagiarizing at Oxford.[40] It is clear, however, that Bruno's concept of "links," which unite the infinite universe in all its parts on the basis of a perfect ontological equality, tends to modify radically the speculative bases underlying Ficino's text, as well as the entire magical and astrological tradition that precedes him. Precisely this traditional context of thought about the magic of correspondences between the sky and the earth continues, nevertheless, at times to echo in Bruno's own magical works of these late years, as they had in a more youthful work such as the Sigillus sigillorum. There are thus pages of these works that tend to contradict some of his own most deeply held and original cosmological theses, so that it seems an oversimplification to claim, as the editors of the recent volume of Bruno's magical works do in their commentary, that "his thoughts on the natural world ... are completely coherent with his thought on magic, which indeed constitutes one of their foundations."[41]

The magical doctrine of the Renaissance was firmly founded on the concept of a hierarchical ladder of being, both material and spiritual, which constituted a vertically orientated connecting link between the natural world and the sphere of the divine. It was difficult, if not impossible, to rethink that doctrine without referring to this axis, which found one of its most essential supports in precisely that Aristotelian–Ptolemaic cosmology that Bruno had spent so many years denying. So it is perhaps not surprising to note that many of his references to the "ladder of nature" in these magical works of his later years tend to lead the reader back into the perspective of the traditional cosmology that Bruno had fought against so valiantly at Oxford, in London, in Paris, and again in

Wittenberg and Frankfurt. To cite a single example, we find a page in the text titled *De magia mathematica*, where, in the context of a long digression on divine names, Bruno appears to be arguing in relation to the traditional Aristotelian *primo mobile*, or "first mover," as well as to equally traditional cosmological entities such as the "sphere of the fixed stars" or the lower planetary spheres of Aristotelian fame.[42] That is to say, these pages, which include no comment on the contradiction they represent with respect to his earlier cosmological thought, appear to deny precisely those cosmological theses that Bruno had argued for so strenuously from his first work on cosmology, *The Ash Wednesday Supper* of 1584, to the final *De immenso* of 1591. It comes naturally to ask oneself, as many commentators have already done, whether these late magical works should not be considered as purely didactic in nature, written with an entirely explicatory aim, or perhaps even personal notes of his reading on the subject. If so, it would be a mistake to attempt to consider them as an essential part of Bruno's own philosophy.

This argument was put forward by Felice Tocco in the nineteenth century and carries with it the weight of his prestige as still today one of Bruno's most acute and perceptive critics, as well as being the editor of the volume in which those so far unpublished manuscripts made their first appearance in print.[43] Nevertheless the many pages of the new volume of *Opere magiche*, including the major work titled *Lampas triginta statuarum (The Lamp of the Thirty Statues)*, appear too dense and complex to justify such a reductive reading of these various and varied texts. The editor of the new volume, Michele Ciliberto, makes an alternative suggestion in his introduction to this volume worthy of note, where he claims that these final magical works of Bruno show him passing from a sphere of pure speculation toward a sphere of practice. In the latter sphere, according to Ciliberto, conceptual contradictions can be justified in a Machiavellian sense, insofar as the concepts called into account are those accepted by the culture of the time.[44] It may be added that, in relation to Bruno's scientific speculation, the doctrines of Renaissance magic appear at times to act as a surrogate of a modern empirical science, for neither the new infinite cosmology nor the new atomistic doctrine of matter could offer Bruno, at that time, many means of intervening in the world of action or of things. For that to happen, new technologies and scientific instruments would have to be developed that for the moment appeared only as dim possibilities on the horizion of an uncertain future. Nor was Bruno alone in his awareness of a hiatus between the new scientific theories that were revolutionizing the idea of nature, and the still limited possibilities of applying them in technological terms. Francis Bacon, to take only one example, would solve the problem only a few decades later by developing an imaginary and utopian vision of the new scientific and technological society in his *New Atlantis*. Bacon's text too, as Marta Fattori underlines, calls the traditional doctrines of magic to account insofar as they connect the *res naturales* (the things of nature) to the *res artificiales* (the things of artifice).[45] Surely this is what Bruno too had in mind when he referred to magic in the *Sigillus sigillorum* as one of the *rectores actuum* (the axes of action).

At the end of the sixteenth century and the beginning of the seventeenth, the operative powers of the new science appear more clearly in words than in deeds, even if a new empirical science can be seen as gradually emerging in the background. Perhaps this is the reason why Bruno's final works on magic take an increasingly interior bent, passing from the ontological to the psychological sphere. That it to say, magic appears in this final phase as increasingly the work of the mind. It is above all the imagination that is called upon in the attempt to penetrate lovingly into the secret links that bind all things together in a series of complex and always varying relationships implying both dominion and submission. For through such penetration, it becomes possible to manipulate intelligently the processes of natural evolution. The necessary instrument in such manipulation is seen to be the word, and it is of interest to note that Maurizio Cambi in a study of Bruno's final works on magic has underlined the increasing importance assumed in them by the art of rhetoric. The intimate link between Renaissance magic and rhetoric had already been emphasized by previous scholars such as D. P. Walker and Cesare Vasoli, and it seems undeniable that Bruno reaches a conclusion of this kind.[46] It was one that he had already proposed in his Triangle of the Graces, to be found in the second work of the Frankfurt trilogy, the *De monade*. For there, language was proposed as the moment of completion of that profound magic that is represented by the opening of the diad that lies at the origins of dialectical thought. The third line of the Triangle of the Graces is for Bruno that of discourse itself: the connecting link, defined by its attempt to arrive at ever more meaningful syntheses of the differences, or contraries, which characterize our imperfect perception of the infinite whole.[47]

Walter Benjamin wrote in his essay *On Language in General and on the Language of Men*:

> every language communicates in its own terms: language is, in the purest sense, the 'medium' of communication. The mediating element, that is the immediacy of every spiritual communication, is the fundamental problem of linguistic theory, and if one wishes to call such immediacy "magic," it can be said that the original problem of language is its magic.[48]

In this sense of the word "magic." it would seem possible to find the immediate connection between spirit, link, and word proposed by Bruno in the *De vinculis in genere (A General Theory of Links)*.[49] This final, unfinished work, probably still in an early draft, illustrates how, as the opening pages of this chapter underlined, Bruno finishes up by proposing a radically new concept of magic with respect to the Hermetic, Neoplatonic, and Kabbalistic traditions, which underlie his thought. For his concept of magic is intimately related to the dialectic of thought itself, within the immanence of an infinite universe. This new infinity of the universe is the context in which thought now attempts to become articulate, no longer dependent on the vain hope of tuning in to privileged messages from a world beyond, transcendent with respect to language itself. At the same time, it is a concept of magic that is fully transitive—that is, that no longer finds its effects only in the subject but in which the subject attempts to control not only an objective natural world but also the emotions of other subjects. The proper sphere in which the immediacy of language operates becomes in the *De vinculis* the sphere of practical and social life. In the social sphere, the emblematic link is seen to be the erotic relationship, considered by Bruno the highest form of communication of the affections. Nevertheless, other forms of love are contemplated in this work, such as that which overwhelms the nine philosophers in the final pages of the *Heroic Frenzies*, where the light that pervades the infinite universe of infinite vicissitudes inspires them to search with patience and method for the connecting links that can lead them to knowledge of those "eternal laws" that regulate the processes of the natural world. The very fact that they have to search for such knowledge means, as Cambi has rightly noticed, that any kind of original, hieroglyphical language that may (as the traditional magical and Hermetic doctrines suggested) have been inscribed into the natural world has since been lost, giving rise to the necessity for a new and arduous research into the particular events of nature's laws and ways.

This train of thought brings us back to what today we would call a methodical form of scientific research into the natural world. In this field, the major work of Bruno's last years may be considered the fragment titled *De rerum principiis et elementis et causis (The Principles and Elements and Causes of Things)*.[50] In this text, which has been included in the volume of the *Opere magiche*, Bruno appears to place himself firmly once again within the infinite universe of his earlier cosmology. Writing of the work of the soul and the intellect, he claims that underlying these principles is an infinite space, capable of containing an infinite substance, in which something can exist.[51] Within what is once again an infinite cosmology, it is interesting to note how the atomism of *De triplici minimo* is modified to place the source of the original energy, which gives rise to all the vicissitudes within the infinite whole, from the "arid" element represented by the atoms themselves to the humid substance that unites and fertilizes them. This could be taken to represent an attempt to subject the atoms to fields of energy that determine their ways of interacting. But above all worthy of note is the passage in which Bruno explains how to move beyond a life subject to the caprices of fortune and fate through a patient search based on observation and investigation into "all particular events, however they are articulated, insofar as they are subject to universal causes." Bruno insists that such a search would not be difficult if only we could succeed in purifying our intellect, which is too often muddied and vague because it is dedicated to vain and frivolous occupations. These are words in which it is already possible to hear the voice of Francis Bacon warning against the dangers of our mental idols, which deflect our attention from a methodical search into those universal laws that regulate the processes of nature.[52] Bruno's *De rerum principiis* leads toward the founding in the seventeenth century of the Accademia dei Lincei in Rome, presided over by the spirit of Galileo Galilei, and the Royal Society in London, presided over by the spirit of Frances Bacon. At the center of the intellectual stage, we now find a study of the particular events of nature in a search to define the universal laws that unite them—a development that will lead to the enactment of a drama in the form of a new science that, for better and worse, has marked the history of the modern world.

NOTES

1. See Gatti (1999).
2. See Yates (1964), 324.
3. My translation. See also Bruno (1977), 90.
4. On Bruno's doctrine of contractions, see Catana (2005).
5. See Bruno (2000a), 310.
6. On this subject, see Granada (1994).
7. See Bruno (2000c), 14.
8. On this subject, see Wyatt (2002).
9. See Bruno (2009), 522–31, and the comment by Nicoletta Tirinnanzi at 909–12.
10. Bruno (2000a), 16. My translation.
11. On this distinction, see Canone (2003).
12. The *Sigillus* is available, with Italian translation and comment, in Bruno (2009).
13. See the relevant entries in Ciliberto (1979).
14. Bruno (2000a), 463.
15. Ibid., 631.
16. See, on this subject, Granada (1994).
17. For this passage in the *Spaccio*, see Bruno (2000a), 632. The annotations of the Neapolitan reader were printed as notes in Bruno (1958).

18. See Bruno (2004), 39.

19. See Bruno (2002), 284.

20. Ibid., 263–64.

21. Ibid., 393.

22. See W. Shumaker, *Natural Magic and Modern Science: Four Treatises (1590–1657)*, Binghampton, NY, Medieval and Renaissance Texts, 1989.

23. See Bruno (2000a), 243–44.

24. For the terms of Bruno's admiration of Telesius, see Aquilecchia (1993a), 293–310.

25. For the link between his play and this first philosophical work, see Bruno (2002), 262–63.

26. For a reading of the *Candelaio* in rigorously ethical terms, see Buono Hodgart (1997). For the idea of shadow as a protective veil held before the inquiring mind, see in particular the fifteenth *Intention* of the *De umbris idearum* in Bruno (2004), 65.

27. See the final dialogue of the *Heroici furori* in Bruno (2000a), 951–60.

28. Ibid., 204.

29. Ibid., 958–59.

30. Ibid., 295.

31. See in particular Dagron (1999) and Mancini (2000), 109–56.

32. My translation. See Vedrine (1990). These comments should be related to the introduction, "Beginning as Negation in the Italian Dialogues of Giordano Bruno," in this volume.

33. Bruno (2000c), 22.

34. For the present-day debate concerning the unpredictability of the behavior of the ultimate components of matter, see *The Ghost in the Atom*, eds. P.C.W. Davies and J. R. Brown, Cambridge, UK, Cambridge University Press, 1986.

35. Bruno (2000c), 234–35.

36. First published in the third volume of Bruno (1879–1891), these manuscript works are now available in Bruno (2000b).

37. The discussion concerning this material originated with the paper by J. E. McGuire and P. M. Rattansi, "Newton and the Pipes of Pan," in *Notes and Records of the Royal Society* XXI, no. 2 (1966): 108–43. See also B.J.T. Dobbs, *The Foundations of Newton's Alchemy*, Cambridge, UK, Cambridge University Press, 1975. The conflicting reactions to these papers is a major subject of discussion in *Rethinking the Scientific Revolution*, ed. Margaret Osler, Cambridge, UK, Cambridge University Press, 2000.

38. See Bruno's *Theses de magia* (79–81), in Bruno (2000b), 383–85.

39. Ibid., 310.

40. For the possibility that Bruno at Oxford made an important reference to the *De vita coelitis comparanda* in order to derive from it the concept of *spiritus*, see chapter 1, "Between Magic and Magnetism: Bruno's Cosmology at Oxford," in this volume.

41. My translation. See Bruno (2000b), 305.

42. Ibid., 47.

43. See Bruno (1889–1891), vol. 3.

44. Leen Spruit also writes: "Magic aims at understanding nature, conceived of as the reign of motion, in order to dominate and manipulate it." My translation. See Spruit (1986), 146–69.

45. See M. Fattori, "'Phantasia' nella classificazione baconiana delle scienze," in *Linguaggio e filosofia nel seicento europeo*, Florence, Olschki, 2000, 37–57; 55.

46. See Cambi (1993). For previous work on this subject, see Walker (1958) and Cesare Vasoli, "La metafora del linguaggio magico rinascimentale," *Lingua nostra* VIII (1977): 8–14.

47. See the fourth chapter of the *De monade* in Bruno (2000c), 286–302.

48. My translation. Benjamin's essay was first published posthumously in the volume of essays and fragments titled *Angelus novus*.

49. In Bruno (2000b), 413–584.

50. Ibid., 585–759.

51. Ibid., 588–89.

52. For Bruno, see ibid., 708–9. Bacon's well-known pages on the four idols of the mind are in the *Novum organon*, aphorisms XXXVIII–XLIV.

15

BRUNO AND METAPHOR

GIORDANO BRUNO WAS BORN ONLY five years after the first publication of Copernicus's *De revolutionibus* in 1543, and only thirty-odd years after Martin Luther's excommunication from the Catholic Church had divided Europe and its culture into two militantly hostile factions. Bruno's lifetime in the second half of the sixteenth century thus covers a vital if often turbulent moment of cultural transition, which would radically affect the history of both science and the humanities. This chapter will primarily be concerned with his thinking about language, and especially with his thoughts about metaphor, thus aligning itself with an interpretative model of early modern culture that establishes "representation," both of thought and of the world itself, as a problem of which historians are increasingly aware.[1] For it is clear that the sixteenth century witnessed what one commentator has called a "Crise des signes" that would radically destabilize not only the way of reading texts but also the reading of the world.[2] By following this path, the paper hopes to show how in some ways Bruno anticipated the coming Enlightenment, while in others he tended rather to indicate alternative routes, some of which would be pursued only at a much later date.[3]

This chapter will take as its starting point a passage in Bruno's first cosmological work, written and published in London in 1584, *The Ash Wednesday Supper*. A still humanist text in its use of dialogue as an appropriate way of facing up to a scientific dilemma, Bruno here celebrates in realist terms the new heliocentric theories of Copernicus that were destined to replace the old geocentric cosmology propounded by Aristotle and Ptolemy and sanctioned by numerous Biblical texts. Later, in the seventeenth century, Galileo would take over many of Bruno's formal solutions, as well as many of his pro-Copernican arguments in his *Dialogue on the Two Major World Systems*, which would get him too into serious trouble with the inquisitors in Rome.[4] Bruno already puts forward in the fourth dialogue of the *Supper* an argument, later to be repeated by Galileo, against the use of Biblical texts in cosmological discussion. Cosmological and Biblical discourse, Bruno maintains, are of two quite separate kinds, and he defines the difference between them in terms of metaphor. In English translation, the passage reads like this:

> When the divines speak as if they found in natural things only the meanings commonly attributed to them, they should not be assumed as authorities, but rather when they speak indifferently, conceding nothing to the vulgar herd. Then their words should be listened to, as should the enthusiasm of poets, who have spoken of the same things in lofty terms. Thus, one should not take as a metaphor what was not intended as a metaphor, and, on the other hand, take as truth what was said as a similitude.[5]

This passage is based on the traditional idea of Biblical discourse as containing four different levels of meaning: the literal level, the metaphorical level, the tropological level, and the anagogical level.[6] Protestant theologians, from Luther to Calvin, had reduced these levels of meaning to two: the literal and the metaphorical level. Indeed, at times Calvin seems to consider the whole of the Bible as essentially metaphorical, insofar as the human mind is, in his opinion, to be considered incapable of contemplating God directly. Bruno seems to be using such ideas to compare metaphor with scientific truth.[7] The Copernican discussion within which this statement occurs makes it quite clear that the Copernican principle of heliocentricity, particularly when expanded to include the infinity of the universe, is considered by Bruno as a cosmological picture of universal truth, and not as a purely instrumental hypothesis to facilitate astronomical calculations. The passage suggests that Frances Yates, in her distinguished and much discussed book on *Bruno and the Hermetic Tradition*, was wrong to consider Bruno's Copernicanism as a Hermetic hieroglyph or diagram—let us say a metaphor—within which, hidden and concealed, lay "potent divine mysteries"—mysteries that, she went on to claim in the same book, make him into a "reactionary" who had nothing to do with the advance of the new science.[8] On the contrary, the Copernican heliocentric principle is, for Bruno, not a metaphor but the truth itself, which has recently been brought to light. Copernicus is, for Bruno, the genius who dragged the heliocentric principle from under the shadows of a centuries-long distorted picture, or false metaphor, of a geocentric universe. It was Copernicus's heliocentric principle that supplied Bruno with the foundations on which to construct what he thought of as a true picture of cosmological infinity. Arguing in favor of an infinite universe in *The Ash Wednesday Supper*, and filling that universe with an infinite number of solar systems in which all the celestial bodies revolve around their central suns, Bruno strongly rejected the objections of his Neoaristotelian critics that such a vision was pure hypothesis, or even fantasy, claiming that he was, on the contrary, talking about "real things" (*ista sunt res, res, res*).[9] Metaphorical expression (here identified with the geocentric universe to be found in so many Biblical texts) and scientific truth (here identified with the Copernican heliocentric principle extended to infinity) seem at this point to be antithetical.

Does this mean that for Bruno metaphor as such is to be banned? Surely not. In many contexts, metaphor seems to define what we may call for Bruno "the humanities" as opposed to natural philosophy or science—that is to say, the universe of words and images through which the mind conducts its search for truth. Like Francis Bacon after him, Bruno had no qualms about "praying metaphors to come to his aid" for heuristic, explanatory, and evaluative purposes.[10] Bruno tends to associate "the humanities" in this sense with above all three groups: the true divines, or those philosophers who attempt to reveal the hidden face of divine truth; the true poets, who are closely associated by Bruno with true divines (this is consistent with his choice of the Biblical *Song of Songs* as one of the greatest texts ever written); and the true painters, whose visual images combine with words to form Bruno's universe of languages. The intimate relationship that Bruno envisages between these three groups is expressed in an early work on the art of memory, where he writes: "Philosophers are in some way painters and poets; poets are painters and philosophers; painters are philosophers and poets. Which is why true poets, true painters and true philosophers search for and admire one another."[11]

Yet if we interpret Bruno's passage in *The Ash Wednesday Supper* in the light of these ideas, we are, I think, obliged to notice that it contains some degree of ambiguity. In what contexts, if any, do we find theologians and poets who renounce the use of metaphor in order to express in some way "directly" the divine truth? Or is his distinction not rather between good and bad metaphors, between metaphors that significantly illuminate the path toward truth, and those, like the Biblical stories, that simply explain difficult concepts in pictorial terms comprehensible to what Bruno rather scathingly calls "the vulgar herd"? Perhaps help might be gleaned on this point by considering Bruno's sources.

Bruno was not only a Renaissance thinker and writer, but his early education had been received at the Dominican monastery in Naples, which had been closely associated in the past with a famous fellow Dominican: Saint Thomas Aquinas. The course in philosophy Bruno followed would have been largely based on Thomistic doctrines inspired by Aristotle. Undoubtedly Bruno was strongly anti-Aristotelian in many respects, particularly as far as his physics, and especially his celestial physics, was concerned. Nevertheless, he always included Aristotle in his many lists of the true philosophers, and it is only natural that Aristotle's concept of the metaphor would have been present in Bruno's mind when he wrote on that subject.[12] It could, in fact, be claimed that Bruno's distinction between two different types of discourse, in which the metaphor assumes different values, derives fairly directly from Aristotle himself. For example, in book II of the *Posterior Analytics*, in a passage where he is concerned with a discourse referring to an external world of objects, and thus with the necessary clarity of definitions, Aristotle writes: "And if one should not argue in metaphors, it is clear too that one should not define either by metaphors or what is said in metaphors, for then one will necessarily argue in metaphors" (97b37). In book VI of the *Topics*, Aristotle makes an even more critical comment, claiming that a metaphorical expression is always obscure, if not actually false (139b34). In his study of *Rhetoric*, however, where he is concerned with a discourse referring to the internal mind and its style of expression, Aristotle's attitude changes to one of praise for the metaphor. "Metaphor gives style clearness, charm, and distinction as nothing else can," he writes in book III, adding that it is not a thing whose use can be taught by one man to another. Metaphors, Aristotle adds, require an acute mind, not only a poetic but also a philosophical mind, capable of perceiving resemblances even in things far apart (1405a–1412a).[13]

This Aristotelian root to Bruno's thoughts on the two types of language is still evident in his final Latin masterpiece, the so-called Frankfurt trilogy published in that town in 1591. In the introductory letter to the trilogy, Bruno formulates a clear distinction between the truth value of the three works of which the trilogy is composed. What he is above all concerned to underline is the "unquestionable certainty" of the final work of the trilogy, the *De immenso*, which he presents as the climax of his cosmological speculation concerning the infinite universe, in which all celestial movements are based on a principle of heliocentricity. In this work, Bruno considers that he has reached an intuition of certain truth, whereas he has no qualms about underlining the relative uncertainty of the other two works of the trilogy, dedicated to the concept of the minimum and to number symbolism.[14] It is in the light of this certainty pertaining to the objective truth of the infinite universe that Bruno makes a series of comments on the need for a new language that, without denying the solemn tones of epic poetry such as he himself is writing here, will be made up, if necessary, of newly invented words, devoid of rhetoric and flourish, capable of describing the world of nature *as it is*. Today we would call that a scientific language. In this passage, Bruno seems to be proceeding toward a Cartesian concept of clear and distinct ideas, already seen as necessary if the truths of nature (thought of as a world of external things) is to be grasped and held firmly in the mind:

> We will be the source of a new (linguistic) usage once we have drawn forth from the deep shadows the famous teachings of the ancient men of wisdom, expressed in their ancient words, to serve as a basis for new things, if need be,

however those teachings may most easily be extracted. We will be inventors of new words. The grammarians are the servants of words, but words serve us. The grammarians should observe the usage we establish ... [15]

On the other hand, Bruno himself would never create a work in which he fully implemented what seem already to be Enlightenment and rationalistic linguistic criteria. In some cases, the apparent divergence between his intentions and his actual practice is strident. For example, in his Italian dialogue, *Lo spaccio della bestia trionfante (The Expulsion of the Triumphant Beast)* Bruno claims that in this work he is speaking simply and literally, naming bread as bread and wine as wine, and giving everything else its proper name. However, he then goes on to develop one of his most complex constructions of myth and fable, envisaging a universal reform of a polluted cosmos in terms of a last-minute reform of the classical astrological images by an aged and rapidly decaying Jupiter, described by Bruno himself as "the subject of our metaphors."[16] If this work is couched in terms of a radical remake of classical Greek mythology, the last work Bruno wrote and published in London, in 1585, the *Heroici furori*, reaches its final ecstatic vision of a now infinite universe through the medium of Petrarchan sonnets rewritten in the light of the imagery of the Biblical *Song of Songs*, which Bruno himself describes as a work in which the images are "clearly and openly treated as metaphors." These are hardly examples of what today we would think of as scientific languages, as Bruno himself seems to underline when, in the passage immediately following the remark on the *Song of Songs* quoted earlier, he asks his reader to believe that his own work is drawn up according to quite different criteria.[17] So it is difficult to know how to construe the fact that Bruno's own sonnets in the *Furori* are themselves composed of complex metaphors—such as the refined conceit of a dialogue between the eyes and the heart; the powerful epistemological images of Actaeon pursuing the moon goddess Diana, only to be consumed in the moment of vision by the hounds of his own thoughts; or the phoenix rising gloriously from the ashes of its funeral pyre. Clearly these too are sonnets composed in a metaphorical mode, and Bruno is obliged to go to great lengths in the prose comment to explicate and deconstruct them into their rational components of argument and reasoning.

In attempting to understand what would appear to be a serious contradiction between linguistic intention and actual practice in Bruno's work, help may be gleaned, in my opinion, by turning to some more modern thoughts on the subject of metaphorical expression. In his seminal study titled *La métaphore vive*, the twentieth-century philosopher Paul Ricoeur also takes as his starting point Aristotle's comments on metaphor that have been quoted earlier. He then goes on to ask himself if the time has not come to give up the opposition between a discourse directed toward the external world, or a scientific discourse of description, and a discourse directed toward the internal world that represents a state of mind and puts everything in hypothetical terms. According to Ricoeur, we need to ask ourselves if it is not the very distinction between "external" and "internal" that has become more and more uncertain, together with that between representation and emotion. To support his point, Ricoeur quotes from Heidegger a statement underlining a difficulty in the concept of language that corresponds to a difficulty in the concept of being.[18] These post-Kantian thoughts on language may be more helpful in understanding Bruno's dilemma than the pre-Cartesian context to which an orthodox historical discourse confines him. It is worth remembering that the Enlightenment placed Bruno in a marginal position, above all recognizing him as an inspiration to philosophical libertinism. Descartes himself thought that there was no need to read his works, writing in his polemical letter to Isaac Beeckman (who was reading Bruno) dated July 13, 1638, that Bruno was like the other philosophical *novatores*, whose many and often contradictory maxims he compared unfavorably with "the certain demonstrations" of geometry.[19] The modern rebirth of Bruno criticism starts with such post-Kantian figures as Schelling and Hegel in Germany, or Samuel Taylor Coleridge, who was both a philosopher and a poet, in England.[20]

It would be easy to further my claim that it can be useful to discuss Bruno's ideas on metaphor in the light of more modern considerations on the subject. For example, the current interest in Vico's anti-Cartesian ideas on metaphor as a foundation of what, in his *New Science*, he calls "poetic logic" is being developed in the light of Charles S. Peirce's definition of metaphor as a type of icon.[21] It is an inquiry that should raise the question of whether the Neapolitan Vico had read the works of his Renaissance countryman, Giordano Bruno. Furthermore, the ideas put forward quite recently by cognitive science, which sees metaphor as a founding element of all language, and of all knowledge scientific and otherwise, could well be related to Bruno's ideas on the art of memory, which he develops as a study of the image-making properties of the creative mind in all branches of its search for knowledge of the infinite whole.[22] From another, if quite closely related, point of view, recent discoveries in neuropsychology, which have led to incomparably more knowledge than we previously had of the workings of our divided brains, are currently being used to further our understanding of such subjects as the nature of dialogue, or of works of art—both subjects on which Bruno himself had much to say—and have been posited also as the bases of a new understanding of scientific knowledge itself.[23] Such recent inquiries remind us that Bruno was already thinking in terms of the connecting powers of the mind, which nowadays we associate with the connecting networks set up by our neurons during the processes of learning and of thought.

Bruno's texts often highlight the capacity of the mind to connect various levels of both discourse and being. One might think, for example, of the passage in the *Heroici furori*, part I, dialogue iii, where Bruno writes: "Beneath sensible images and material objects, he (that is, the frenzied searcher after knowledge) recognizes divine orders and counsels."[24] This remark, with its Neoplatonic overtones, seems to indicate that all knowledge consists of a comparison between different levels of being and is, therefore, always to some extent metaphorical.[25] On the other hand, Bruno always distinguished between what he considered the essential and the accidental truths of nature. The essential truths, as Bruno saw them, are few but absolutely essential, and not subject to metaphorical expression: the infinity of the infinite universe; the heliocentric nature of motion within an infinite space populated by an infinite number of solar systems, and finally, if slightly less certainly, the atomistic composition of an infinite substance. Toward these truths, which Bruno, in the same dialogue of the *Heroici furori* mentioned earlier, describes as representing the "master-plan" of the universe we live in, the individual mind proceeds in myriad ways. If it proceeds more often indirectly than directly, it is because the individual mind is nothing more than a fleeting pinpoint within infinite space and time.

Bruno expressed this last idea with surprising clarity during a session of his trial for heresy in Rome, during which the inquisitors requested him to declare his opinion on the immortality of the soul. Bruno replied that there is no immortality of the individual soul, but only of the infinite, universal framework within which each soul or mind searches for knowledge of an infinite truth. He made his point through the use of a traditional but powerful metaphor, saying to his judges: "it is as if many fragments of a mirror all came together to form an antique mirror. The images animating each single fragment are annihilated, but the glass or the substance remains, as it has always been and always will be."[26] And if Bruno here, in front of the inquisitors, says that he is talking only of the souls of beasts, and not of those of human beings, which continue to live even after separation from the body, it should be remembered that in his works he had frequently advanced the far more audacious claim that there is no substantial difference between the soul of a human being and the soul of an animal—or, indeed, between the souls of animals and those that are to be found in all things: "For all the spirits emerge from the Amphitrite of a single spirit, and to that they return."[27]

Initially, the metaphor of the mirror, or of Amphitrite, may seem little more than orthodox Neoplatonism, with the concept of a world soul surrounded by Biblical overtones reminding the inquisitors (as Bruno clearly intended) of St. Paul's famous dictum in his First Epistle to the Corinthians 13:12 that on this earth we can see only "as in a glass, darkly." However, as always with Bruno, it has to be remembered that everything he says applies for him within the context of a universal infinity, foreign to his sources, that radically transforms the meaning of his images. The mirror becomes a synonym of substance itself, meaning an infinite substance, with the shadow of Spinoza already looming on the horizon.[28] And it is precisely the infinite substance, in my opinion, that—already in Bruno—tends to identify with the divinity itself, and as such to defy metaphorical expression. God is not "like" something else: He simply "is."

It is a point underlined forcibly by Bruno himself when, in part II, dialogue 1, of the *Heroici furori*, he describes the "excellent and magnificent goals" that the heroic mind will forever go on striving for, until it has risen to the point of desiring "divine beauty in itself, without likeness, figure, image or species, if that be possible, and, moreover, if it is able to reach such heights."[29] It is clear from this passage that nonmetaphorical expression is identified by Bruno with the divine truth, or truth (as he calls it) "without likeness." Such purity of truth is seen by Bruno as the ultimate goal of the inquiring mind, but although he often refers to such truth as "divine," both in its beauty and its goodness, Bruno is consistently adamant that his is not a theological but rather a "natural" discourse. The truth he is inquiring into requires the use of mental tools (such as logic, geometry, numbers, the art of memory) to take the measure and probe the evolution of an infinite universe, and the understanding of such truth, as he represents it in the final pages of the *Furori*, is to be seen as the ultimate good "on earth" (*il sommo bene in terra*), and not as a mystical intuition of a transcendental "beyond."[30] Bruno's discourse may frequently make use of a religious terminology, but it moves within the horizon of the new science. At the same time, arriving at the vision of such truths about the natural world has become problematic to the extent of being feasible only in those exceptional circumstances in which the mind is stretched almost to a breaking point, and its language purged, "if that be possible," of all false "likenesses."

It is not my aim, in approaching my conclusion to these few remarks, to make large claims for the importance of Bruno's thought. He was well able to make such claims himself, and recently many of them have been picked up and eloquently illustrated in Ingrid Rowland's new Bruno biography.[31] What I am concerned to point out is rather that Bruno's thought on the nature of the humanities and the sciences, and particularly his thought on the languages within which they are necessarily formulated, may often seem to us familiar today. For we live in a world that has—it would seem definitively—assumed the immense dimensions of both space and time already foreseen by Bruno, just as he foresaw the atomistic frag-

mentation of all bodies within the infinite whole. The crisis of the thinking subject that this new vision implies was solved by Bruno by placing a
special emphasis on the creativity of the individual mind: a neural activity
of imaging, connecting—formulating patterns and ever varying strings of
letters, words, numbers, images—that was, in his time, belied by the increasing emphasis on rules that was beginning to dominate both the arts
and the sciences. Descartes may have conceived of his *Rules to Guide the
Intelligence* as pertaining to the structure of the mind, but he could find
certainty only in pure concepts such as figure or extension that do not
suppose anything that experience has rendered uncertain.[32] The resulting
dualism between mind and matter is about as far as it is possible to imagine from Bruno's view of things.

Bruno knew that the close link he was attempting to forge between the
thinking mind and the infinite amount of matter from which all minds
emerge (today we talk about the "embodied mind") inevitably gave rise
to an idea of all knowledge as fragmented and incomplete. Even mathematics, for Bruno, was knowledge of approximations, in a denial of
the special status of pure mathematical entities of which Kepler (an avid
reader of Bruno) would strongly disapprove.[33] On the other hand, post-
evolutionary philosophers and scientists in the nineteenth century would
rediscover Bruno with enthusiasm. In some manuscript notes on Bruno,
the Italian philosopher Bertrando Spaventa wrote in the middle of the
century: "the same principle which in nature forms and figures things
thinks in the human mind."[34] Later in the century, a disciple of Darwin's
evolutionary theories, John Tyndall, published a book—widely read in its
time, and frequently republished—called *Fragments of Science*.[35] Tyndall
dedicates many pages to Bruno's infinite cosmology and to his theory of
the infinite evolution together of matter and mind. Tyndall's title may
not have been a gratuitous coincidence. He often cites from Bruno's cosmological dialogues, of which the first was *The Ash Wednesday Supper*
(the starting point for this chapter), where Bruno had admitted that his
text hardly added up to a traditional scientific treatise. Sometimes, Bruno
points out, it is poetry; sometimes oratory; sometimes celebration; and at
others vituperation. Only occasionally, he continues, will you find "demonstrations and teaching, in physics and mathematics, in morals and
logic—in short, it can be said that there is no branch of knowledge of
which you will not find some fragment."[36]

The individual minds, then, are seen as passing fragments, destined to
dissolve into the eternal and infinite substance, which they can glimpse
only momentarily in its perfect and incomparable purity. Their attempt
to arrive at this vision is seen by Bruno as heroic: the true hero of the
modern world becomes the intellectual searcher after natural truth. Such
a search involves making an elaborate series of connections between dif-

ferent kinds of being, and will be essentially metaphorical—at least until the infinite object is, even if momentarily, grasped and held in the mind. In this way, knowledge, for Bruno, becomes the object of the search not only of the new scientist, but equally of the true theologian, the true poet, the true painter, and the true philosopher. So, we can conclude by saying that, at the very beginning of the Enlightenment, Bruno vigorously denied that strict division between disciplines and genres that an ever more Neo-aristotelian culture was already busy sanctioning. Not only did he want to see dialogue and collaboration between the different disciplines of the humanities, but he also wanted the humanities and the sciences to come closer together in an effort to share their fragmentary forms of knowledge of an elusively infinite whole.

NOTES

1. The possibility of reading the Renaissance from the point of view of its "language merchants" is proposed by Michael Wyatt in *The Italian Encounter with Tudor England: A Cultural Politics of Translation*, Cambridge, UK, Cambridge University Press, 2005. For the same idea, proposed in a somewhat different context, see "Introduction: History, Culture and Text," in *The New Cultural History*, ed. Lynn Hunt, Berkeley, University of California Press, 1989, 1–22: 16.

2. See the introductory article by Michel Jeanneret in *L'étude de la renaissance: nunc et cras*, eds. Max Engammare et al., Geneva, Librarie Droz, 2003, 12.

3. For a book-length study of Bruno's language and imagery, above all in its Lullian derivation, see Cambi (2002).

4. On Bruno and Galileo, see Aquilecchia (1995a) and Gatti (1997).

5. See Bruno (2002), vol. I, 525.

6. See Henri de Lubac, *Exégèse médiévale: les quattres sens de l'Ecriture*, Paris, Aubier, 1959.

7. See chapter 13 "Bruno's Use of the Bible in His Italian Philosophical Dialogues," in this volume.

8. Yates's contribution to Bruno studies was a rich and complex one that was far from being limited to the Copernican question. Her reading of his Copernicanism, however, has been at the center of a particularly lively debate, including a major critique by Robert Westman. See Yates (1964), 241 and 324, and the reply by Westman (1977).

9. See Bruno (2002), vol. 1, 533.

10. On Bacon's use of metaphor in scientific discourse, see Brian Vickers, "Francis Bacon, Feminist Historiography, and the Dominion of Nature," *Journal of the History of Ideas* 69, no. 1 (January 2008): 117–41, in particular, the section "Misunderstanding Bacon's Metaphors," 122–36.

11. This much quoted passage, whose importance I have already underlined in chapter 5, "Petrarch, Sidney, and Bruno," in this volume, appears in Bruno's explication of his twelfth seal, referring to the images of painters, in the *Explicatio triginta sigillorum*. See note 5 in chapter 5 of this book.

12. Much recent work has been done on Bruno's education in Naples. See, in particular, the section "Nolanus ... Neapolitanus" in Canone ed. (1992), 15–75.

13. For the English translation, see *The Complete Works of Aristotle: The Revised Oxford Translation*, 2 vols., ed. Jonathan Barnes, Princeton, NJ, Princeton University Press, 1984.

14. See the dedicatory letter in Bruno (2000c), 235.

15. Ibid., 14.

16. Bruno (2002), vol. 2, 256.

17. Ibid., 494–97.

18. See Paul Ricoeur, *La métaphore vive,* Paris, Seuil, 1975.

19. In René Descartes, *Tutte le lettere 1619–1650,* ed. Giulia Belgioioso, Milan, Bompiani, 2005, 158.

20. For Bruno's nineteenth-century reputation, see the "Introduction" to Canone (1998a). For Coleridge's reading of Bruno, see chapter 10, "Romanticism: Bruno and Samuel Taylor Coleridge," in this volume.

21. Augusto Ponzio, "Metaphor and Poetic Logic in Vico," *Semiotica* 161–, no. 1/4 (2006): 231–48.

22. On the logical constituents of Bruno's art of memory, see the relevant section in Rossi (2000). On the image-making properties of Bruno's art of memory, see Ciliberto's "Introduction" to Bruno (2009).

23. See Jurij M. Lotman, *La semisfera: l'assimetria e il dialogo nelle strutture pensanti,* ed. Simonetta Salvestroni, Venice, Marsilio, 1985, and John Onians, *Neuroarthistory: From Aristotle and Pliny to Baxandall and Zeki,* New Haven, CT, Yale University Press, 2007. The importance of neuropsychology for an understanding of scientific discovery was posited by Roger Penrose in *The Emperor's New Mind: Concerning Computers, Minds, and the Laws of Physics,* Oxford, UK, Oxford University Press, 1989.

24. Bruno (2002), vol. 2, 558.

25. For this same idea in a modern context, see George Lakoff and Mark Johnson, *Metaphors We Live By,* Chicago, University of Chicago Press, 2003.

26. See Firpo (1993), 302.

27. See the *Cabala del cavallo pegaseo,* dialogue 2, in Bruno (2002), vol. 2, 451.

28. See Deregibus (1981).

29. Bruno (2002), vol. 2, 648–49.

30. Ibid., 747.

31. See Rowland (2008).

32. Descartes's *Regulae ad directionem ingenii (Rules to Guide the Intelligence)* appeared posthumously in a Dutch translation in Amsterdam in 1684, and then in Latin in the same town in 1701.

33. See Johannes Kepler, *Harmonices mundi: The Harmony of the World,* eds. E. J. Aiton et al., Philadelphia, American Philosophical Society, 1997, 88–93.

34. See the Spaventa manuscripts published by Maria Rascaglia in Canone ed. (1998a), 156.

35. John Tyndall, *Fragments of Science,* op. cit.

36. Bruno (2002), vol. 1, 438.

WHY BRUNO'S "A TRANQUIL UNIVERSAL
PHILOSOPHY" FINISHED IN A FIRE

I N ONE OF HIS ITALIAN PHILOSOPHICAL dialogues written and published in London in 1584, *De l'infinito, universo et mondi*, Giordano Bruno described his life's work as an attempt to define a "tranquil universal philosophy": a philosophy that he imagined as a peaceful swim through the infinite ocean of universal being.[1] This was Bruno's third philosophical dialogue written in Italian. In it, he criticizes the fifteenth-century Catholic cardinal, Nicholas Cusanus, who anticipated him in proposing an infinite universe. Cusanus, however, proposed a dualistic universe of Aristotelian origin and with clearly Christian and neo-Thomistic implications, divided between spheres of being of intense light and purity, pregnant with intuitions of a transcendent divinity, and others of intense shade, heavy with materialistic premonitions of mortality. In such a universe, the searcher for truth is likened by Bruno to a swimmer continually tossed between high waves and low troughs, instead of swimming "slowly and gently" through a homogeneous ocean of universal being, as Bruno's own philosophy allows. There are clearly pantheistic tendencies behind Bruno's concept of an infinite homogeneous universe, which tends to identify with the divinity itself. These would later modulate into a new deism, destined to become an important part of the culture of the Enlightenment in the years ahead. In a sixteenth-century Europe, however, caught in the grip of conflict between the aggressive new Protestantism and the militant Catholic Church of the years immediately following the Council of Trent, Bruno's attempt to revive an ancient, infinite cosmology, composed of an infinite number of solar systems—of Pythagorean and Lucretian as well as Copernican origin—turned out to be a failure. Bruno managed neither to live nor to die in peace.

Let us rapidly review the essential facts.[2] Born in Nola in 1548, Bruno entered the Dominican monastery in Naples in 1565 at the age of seventeen, and left it to flee to Rome in 1576 at the age of twenty-eight. In those eleven years in Counter-Reformation Italy, he completed the rigorous course in theology, largely based on Thomist doctrine, and was ordained as a priest in 1573. There were, however, problems with the Dominican authorities, who caught Bruno reading forbidden authors such as Erasmus of Rotterdam, and noted suspect behavior such as his destruction of all his religious icons except the crucifix itself. In Rome, Bruno heard that inquisitorial proceedings had been opened against him, and he started to flee north. In 1579, he left Italy via Venice and spent the next twelve years wandering through northern Europe, sometimes teaching philosophy, and sometimes playing the courtier to an interested prince. Bruno's first works to have survived were published in Paris in 1582. He completed and published his final works to appear with his knowledge and consent in Frankfurt, in 1591.[3]

In those few years of exile, Bruno worked with extraordinary intensity, producing numerous books both in Italian and in Latin that covered a re-markably wide field. He meditated on the new cosmology, becoming not only a convinced Copernican but a realist one at that, while extending Copernicus's still finite universe to infinite dimensions. He subjected to intense review the whole Aristotelian canon, repudiating the cosmology and questioning the logic, preferring to it the pictorial logic of the Cata-lan mystic Raymund Lull. He took into consideration, sometimes critical and sometimes admiring, the whole Neoplatonic tradition, reviewing it from its origins in Pythagoras up to its modern revival in Marsilio Ficino and his Renaissance followers. He noted their admiration for Hermes Trismegistus, with his Egyptian magical and astrological doctrines based on the concept of a world soul. Bruno's own universe was also animated in all its parts, but it was infinite in space and eternal in time. He claimed it as the habitat of an infinite but no longer fully Christian or even tran-scendental God. Above all, Bruno meditated on the ways in which things could be known in such a world. He subjected the orthodox pillars of au-thority to a rigorous scrutiny. In his philosophy, the inquiring individual mind assumes ever more emphatically the center of the intellectual stage. Historically, he saw the Europe of the sixteenth century as locked in the grip of a dark and terrible crisis caused by the wars of religion between deeply divided and angry Christian churches. The golden years of Chris-tendom had passed, he claimed, and violence had prevailed over love. Bruno knew that his own stand for an advanced principle of freedom of thought was likely to finish in punishment and violent death. Desiring peace, he was constantly in trouble, in the Protestant parts of Europe that he visited as well as in the Catholic ones.

Toward the end of 1591, for reasons that have never been fully ex-plained, Bruno made the decision to return to Italy, again via Venice. A nobleman named Giovanni Mocenigo had invited Bruno to attend him in his Venetian palace and to teach him the arts of memory. It was a subject on which Bruno was expert, and had written copiously, but the experi-ment was not a success. What exactly happened in Mocenigo's palace is not known, except that on May 23, 1592, Mocenigo imprisoned Bruno in his palace and denounced him to the Inquisition.[4] The Venetian trial was brief. Bruno attempted to outline his philosophical doctrines to inquisitors whose main concern was to see him kneeling down to ask forgiveness for whatever unorthodox opinions he may have held. In a final hearing of July 30, 1592, Bruno did kneel down and publicly declared his willingness to submit to the opinion of his judges. The trial seemed set to terminate with his release. Then Rome intervened, stressing that Bruno was a citizen not of Venice but of Spanish-dominated Naples, where inquisitorial proceedings had already been initiated against him. Furthermore, the Pope himself, Clement VIII, personally desired that Bruno be extradited and tried again in Rome. Even the relatively autonomous and tolerant Venetian state had difficulty in holding out against such an authoritative request, although it did refuse to concede other inquisitorial prisoners to Rome in these same years—for example, the prestigious professor of a Neoaristotelian philosophy in the University of Padua, Cesare Cremonini.[5] Saverio Ricci has recently claimed convincingly that Bruno's extradition should be seen as part of a conflict in the Roman curia in the early 1590s between what today we would call the hawks and the doves—a conflict that involved Rome's overall political relations with the Venetian republic, as well as crucial issues such as the return to the Catholic fold of the French king, Henry of Navarre. In Bruno's case, the hawks can be said to have won.[6] On February 19, 1593, Bruno arrived in the prisons of the Holy Office in Rome that he would leave only seven years later, when his sentence to death was publicly announced and he was handed over to the secular governor of the city for its execution. On that occasion, Bruno is known to have declared that he would not recant because there was nothing that he needed to recant. He added that his judges feared pronouncing their sentence against him more than he feared receiving it. On February 17, 1600, Bruno was burned alive in Campo dei Fiori as an impenitent heretic—his tongue held in a brace to prevent him from speaking his mind. All his works were immediately placed on the Index of forbidden books.

Bruno's execution for heresy was one of a long series that were taking place throughout Europe in those tragically blood-stained and violent years, not only in the Catholic world but in the newly Protestant one as well. Yet something about that dramatic and complicated trial has given it a particular significance, making it, in the minds of many, into an emblematic event lying at the origins of the modern world. There are, inevitably, conflicting ideas about what it really meant, both then and nowadays, for us. In the remaining part of this epilogue, I shall attempt, first, to outline how much we actually know about it; second, to give an extremely synthetic account of the principal ways in which it has been interpreted, particularly in terms of its intellectual or philosophical implications; and third, to make some concluding remarks of my own.

For many years, indeed centuries, almost nothing was known about Bruno's trial. The principle document that records his execution, with approval, is a letter written by the German Catholic Kaspar Schopp, who had been present at the event.[7] Schopp lists fourteen accusations that had been made against Bruno, and describes him as a dangerous enemy to religion. His letter was partly published in 1621 in Hungary, together with a pamphlet titled *Machiavellizatio* written by the Calvinist extremist Peter Alvinczi—a publication to which I shall return. The trial documents themselves started to emerge only much later, in the second half of the nineteenth century, together with the first modern editions of Bruno's works and the first full-scale biographies. The Venetian trial documents, first published in 1864, became widely known in 1868, when they were reproduced by Bruno's first Italian biographer, Domenico Berti. This biography, republished in a much augmented edition in 1889, was written in the atmosphere of marked anti-clericalism and hostility to the Roman Curia that characterized the early history of the newly united and independent Italian state.[8] The Vatican claimed that the documents relating to the Roman part of the trial had disappeared with the ecclesiastical material that Napoleon took back with him to Paris, much of which was subsequently destroyed. Gradually, however, new documents started to emerge. The text of the announcement that sentenced Bruno to death was first published in a correct transcription in 1921, in the still essential biography by Vincenzo Spampanato. Then, between 1925 and 1927, Monsignor Enrico Carusi published a number of so far unknown documents held by the Holy Office. The real breakthrough came, however, in 1940, when Monsignor Angelo Mercati discovered a summary of Bruno's Roman trial held in the Vatican secret archives, which he published with an extended comment in 1942.[9] Summaries of the more complex trials were drawn up by official clerks to aid the inquisitors at the moment of deciding the verdict and writing out the sentence, and this document is usually referred to as the *Sommario* of Bruno's trial, in spite of some recent doubts about its actual status as such.[10] It is certainly far from filling in all the gaps in our knowledge of what happened during Bruno's trial in Rome. Nevertheless its importance cannot be put in doubt, and Monsignor Mercati was anxious to present his find to the public from the point of view of the Catholic Church, which continued to condemn Bruno together with his secular sympathizers and biographers. Mercati was prepared to pity Bruno's violent end, but insisted on the honesty and legality of the trial. If Bruno, who, in Mercati's opinion, was both depraved and a criminal heretic, had suffered a violent death, that "was not the fault of the inquisitors but of the accused."[11]

With Mercati's discovery, an almost complete overview of Bruno's long trial had at last become possible. In 1949, a major comment on the whole tormented event was published by one of Italy's most prestigious liberal historians, Luigi Firpo.[12] Although fundamentally sympathetic to Bruno's stand for the freedom of thought and expression, Firpo made a bold attempt to move beyond the radical positions "for" or "against" that had so far characterized the gradual publication of the documents. Firpo worked on other inquisitorial trials besides that of Bruno—in particular, those of his fellow philosophers Tommaso Campanella, Francesco Pucci, and Francesco Patrizi. Indeed, Firpo considered the trials for heresy that involved so many of the leading philosophers of this traumatic period in European history as crucial moments for our understanding of the modern world. Firpo worked for forty years to complete the documents of Bruno's trial, preparing a volume dedicated to it that remained unpublished when he died. It eventually appeared in 1993, edited by Diego Quaglione, and is likely to remain for many years the standard account of Bruno's trial. Since then two further trial documents have been discovered and published by Leen Spruit, while a major find has recently been published by Marta Fattori.[13] This is a document of 1621 containing the official Vatican considerations on Peter Alvinczi's Protestant-orientated *Machiavellizatio* requested by the compilers of the Index of forbidden books. It is of particular interest, given that the pamphlet concerned, as we have seen, was published together with Kaspar Schopp's letter approving of the public reading of Bruno's sentence and his execution. Clearly the pamphlet in question was intended as a criticism of Schopp's hostile stance toward Bruno. The Catholic censor understood the meaning of such innuendos. He explicitly mentions the Bruno connection in his report, which condemns the Calvinist-inspired *Machiavellizatio* to the Index while saving and praising the letter by Schopp. The letter is both "useful and fruitful," according to the censor, Camillo Cesare. This—as Professor Fattori points out—constitutes a significant reinforcement of the Inquisition's sentence of 1600 in a meeting of the Index held in 1621. The date is clearly important if it is remembered that in 1616, Copernicus's *De revolutionibus* had been placed on the Index for the first time, and that in 1621 itself, events were already building up that would eventually lead to the trial and condemnation of Galileo. It is clear that in the minds of the Roman Curia, Bruno's trial and execution and the budding Galileo affair were closely connected.[14]

Let us now consider the earliest comments by intellectual historians on Bruno's trial. These were made by his nineteenth-century commentators, who, working before Mercati's discovery of the Roman trial documents, had only the Venetian documents to work on. Their principal concern was to explain Bruno's apparently contradictory behavior in Venice, which consisted of giving remarkably frank answers to the inquisitors about the contents of his philosophy while at the same time repeatedly declaring his willingness to recant. Already in the crucial third session of the trial at Venice, Bruno had admitted that he considered the universe infinite and eternal, populated by infinite worlds, and governed by a universal providence identifiable with nature herself. He confessed to doubts about the incarnation of Christ and about the Trinity, and he declared that he believed in a world soul according to the doctrine of Pythagoras. Interestingly, Bruno calls on the ideas about the creation of St. Thomas Aquinas himself to claim that universal being may be either created or eternal—in both cases, it is to be considered as dependent on a divine cause, so that nothing is ever random or independent.[15] A few days later, we find him on his knees declaring that he would do or think nothing that could dishonor the religion he once served as a monk.

Domenico Berti in his biography of 1889 accompanies the recent publication of these documents with a suggestion on how to read them. He points out how the sixteenth-century discussion of Aristotle's idea of the soul, particularly as it had been conducted at Bologna University by Pietro Pomponazzi, had developed a concept of "double truth" that was important in protecting the Neoaristotelian natural philosophers of Bologna and Padua from interference by the church. This theory reelaborated the claim already made in the Middle Ages that it was possible to argue "philosophically" for theses such as Aristotle's concept of the soul, which Pomponazzi thought was mortal and not immortal, as St. Thomas Aquinas had declared, while at the same time remaining faithful to Christian orthodoxy at a "theological" or "religious" level of truth. Berti calls on this double theory of truth to explain Bruno's behavior at Venice, pointing out how Bruno himself constantly stresses to the inquisitors that, when he is explaining what he himself calls the "impious" aspects of his thought, he is speaking specifically "as a philosopher and according to the principles of a natural light."[16] Later, in the early years of the twentieth century, Giovanni Gentile considered Bruno's strategy at Venice as based on a reading of Machiavelli, who clearly lay behind Pomponazzi's Renaissance theory of double truth.[17] Machiavelli's idea of religion as necessary politically to ensure the moral and social cohesion of community life is what allowed Bruno, the philosopher, to behave as he did, in the opinion of Gentile. Thus, Bruno, by falling on his knees and showing himself willing to recant, never compromised his philosophical conscience at Venice—on the contrary, according to Gentile, his behavior was "a coherent practical demonstration of his philosophical integrity."

With the publication of the summary of the Roman trial in 1942, it became clear that Bruno's strategy of alternating remarkably frank admissions of the Christian unorthodoxy of his philosophy with a willingness to abjure it publicly was never abandoned. Rather, it became particularly prominent in the final stages of the trial. These were characterized by an event that all commentators agree was decisive—that is, the appointment at the beginning of 1597 of the Jesuit Roberto Bellarmino as the official advisor of the Pope on matters relating to the Inquisition, of which he would become formally an inquisitor in 1599. The summary indicates Bellarmino as becoming the dominating figure for the prosecution at Bruno's trial, dating from a crucial hearing, held on January 18, 1599, when the accused was finally handed a list of eight heretical propositions culled directly from his works. He now had to decide if he was prepared to deny them—that is, to deny his essential philosophical creed. Unfortunately this list appears not to have survived, although commentators have succeeded in plausibly reconstructing it. It is probable that it included Bruno's Copernicanism and his doctrine of an infinite universe, while it is certain that it included his idea of the soul as a pilot in a ship. Bruno saw the individual soul (the pilot) as unable to survive the dissolution of the body (the ship), both of them returning on death into the infinite ocean of universal being. This had serious consequences for his moral philosophy, which he had also made clear to the inquisitors. Bruno's philosophy did not contemplate judgment of the individual after death, or the idea of hell, although in the fourth Venetian session of the trial, he had declared his willingness to profess these beliefs "when speaking as a good Catholic."[18] Clearly, however, his doctrine of the soul was a crux marking the distance between his philosophy and Christian theology, and the part of the *Summary* dedicated to the ideas culled from Bruno's books dwells at length on the subject. Bellarmino was quick to see the heretical implications of Bruno's explicit definitions of the individual soul as no more than a fragmentary reflection within the great, universal mirror, or a passing voice mingling with the infinite voices of universal being.[19] Bellarmino rigorously held Bruno down to these ideas when he started vacillating again, attempting to abjure anything and everything except for his doctrine of the soul. From there, events proceeded to their cruel and tragic conclusion.

Luigi Firpo's now classic commentary, the first to take into consideration the whole development of Bruno's trial, is notable for his refusal to mythologize either Bruno's ultimate heroism or Bellarmino's intransigence, which some commentators had not hesitated to brand as "devilish." Firpo was acutely aware of Bellarmino's prestige within the Catholic Church, both in his own times and afterward. He saw Bellarmino as carrying out with particular rigor the duties that his office imposed on him, not without a hope of achieving a genuine conversion.[20] A point that needs to be underlined, however, is that Bellarmino also possessed a remarkable philosophical mind. This allowed him more clearly than the other inquisitors to understand the terms of Bruno's defensive strategy, and to place his finger on precisely that part of his doctrine that Bruno was unable to abjure. For Bruno was indeed a philosopher, and ultimately remained faithful to the central concept of his thought: his anomalous and very personal doctrine of the soul.

In 1964, the British scholar Frances Yates published a book titled *Giordano Bruno and the Hermetic Tradition* that caused an upheaval in the field of Bruno studies, affecting not only the reading of his works but also of his trial. Yates understood the centrality of Bruno's doctrine of a world soul, but she considered it a purely magical and mystical intuition that had nothing to do with his infinite post-Copernican universe or his doctrine of atomism seen as scientific doctrines—that is to say, nothing to do with a cosmology that could be subjected to rational inquiry. Yates consequently proposed a reading of Bruno's works entirely in the light of the so-called *Hermetica*—that is, the series of works attributed to the ancient Egyptian sage known as Hermes Trismegistus who in Bruno's day was associated with an ancient and primordial theology. Seen from Yates's point of view, Bruno was a man with a remarkable spiritual life and creative imagination—subjects on which she made a lasting and valuable contribution. However, she did tend to deprive him of anything that could be called a philosophical logic or reason. Yates's Bruno, therefore, could have no coherent strategy during his trial. Bruno's remarkably frank admissions to the inquisitors about the Christian unorthodoxy of his thought seemed to Yates to signify only some obscure sort of death wish: "megalomania," "state of euphoria," "bordering on insanity," "religious mania," "feverish," "vast and vague hopes" are some of the phrases that can be found in the few pages she dedicates to the trial. Yates only quotes once from the documents, where Bruno refers to Marsilio Ficino (himself a devout Christian) to support a claim that the cross was not originally a Christian symbol but an Egyptian one of astrological origin. From there, she goes on to claim that the trial was "about" magic and astrology, without considering it necessary to investigate how much of the eight-year-long discussion between Bruno and the inquisitors was in fact dedicated to these subjects, which were undoubtedly considered by the Inquisition as "suspect" and "dangerous."[21] More recent studies, however, have underlined that philosophical ideas concerning magic were not always considered by the Inquisition as formally heretical, which was necessary for the death penalty to be applied. Taken alone, they would have been unlikely to have led to Bruno's final condemnation and sentence to death.

It took some decades for the shock of Frances Yates's Hermetic Bruno to be absorbed by the critical tradition. Certainly one of its effects was to banish interest in Bruno's trial as a serious subject of philosophical study for many years. During those final decades of the twentieth century, however, a very notable change took place in the study of the Inquisition itself. An enormous amount of new documentary material appeared, even before the secret archives of the Holy Office were finally made available to public scrutiny. Much more became known about its judicial procedures, with an ever increasing attention paid to the inquisitors themselves as well as to those they accused. Furthermore, the official (if, according to some commentators, only partial) rehabilitation of Galileo in 1992 indicated a significant change on the part of the Catholic Church with respect to its inquisitorial past. When, in the final years of the twentieth century, we find a resurgence of interest in Bruno's trial, once again on the part of commentators interested in its philosophical implications, all these new factors could be called upon to enrich the debate.[22]

The immediate stimulus behind this renewal of attention for Bruno's trial was the four-hundredth anniversary of his death on February 17, 1600, which was the occasion of numerous commemorative events throughout the world. One of the first and most notable developments consisted in a growing awareness that earlier commentators had indeed been right to consider Bruno's trial as closely linked to that of Galileo. In a paper published in 2001, the French scholar Jean Seidengart underlined the particular emphasis to be found throughout the trial on Bruno's doctrine of a plurality of worlds.[23] Considered nowadays as Bruno's most significant contribution to the modern cosmological discussion, it appeared then to the inquisitors as particularly dangerous in its denial of a creative act of God, for although Bruno thought of each of the individual worlds as subject to becoming and perishing, he claimed that the infinite universal space that contained them was eternal, not created in time. Seidengart correctly notices that the specific Copernican question of heliocentricity, on the other hand, became an issue only in the Roman sessions of the trial, indicating that the Curia was increasingly worried by any denial of those many passages in the Bible that presuppose a geocentric earth.[24] Copernicus's book, however, was not yet on the Index, and Copernicanism not a formal heresy. At most, it can be said that with Bruno's trial, a cosmological problem emerges as a premonition of the storm that would accompany the publication of Galileo's *Dialogue of the Two Major World Systems* in the 1630s.

Seidengart's paper was followed in 2002 by a major study in English by Maurice Finocchiaro comparing Bruno's trial, seen as a conflict between religion and philosophy, with the trial of Galileo, seen as a conflict between religion and science.[25] Finocchiaro recognizes that these categories could be considered as simplifications, but he convincingly claims them as useful ones. Undoubtedly Finocchiaro's thorough and detailed analysis of these two traumatic events marking the early modern world constitutes the most complete study yet of Bruno's trial to appear in English, while at the same time establishing beyond reasonable doubt the close link between Bruno's and Galileo's clashes with the religious authorities of their times. Nevertheless, the two trials were not identical, and Finocchiaro is careful also to point to their differences. Bruno, as we have seen, was openly admitting that his philosophy denied many of the fundamental doctrines of the Christian religion, which Galileo never rejected. This is the fundamental issue raised by Bruno's eight-year-long discussion with the inquisitors, and it is the note on which I wish to close this book.

In a useful study of the terms in which Bruno developed his defense, Diego Quaglione, the editor of Firpo's posthumous volume containing the complete trial documents, has recently claimed that Bruno proposed to the inquisitors an academic debate on the fundamental doctrines of Christianity, transposed, by force, into the halls of a tribunal.[26] Bruno prepared himself specifically for this task, receiving from the inquisitorial authorities on December 22, 1593, shortly after his arrival in Rome, a cloak, a hat, and a copy of St. Thomas's *Summa* in octavo.[27] It should never be forgotten that Bruno spent twelve years as a Dominican monk, living in Naples in close proximity to the cell that was St. Thomas's own. Whereas earlier scholars tended to ignore or underplay Bruno's Dominican past, in recent years much work has been done both outside the monastery, by secular scholars such as Eugenio Canone and Ingrid Rowland, and inside by Padre Michele Miele, which has filled in the picture of Bruno's cultural and philosophical preparation imbibed in those crucial early years.[28] We now know much more about what books would have been available to Bruno in the monastic libraries of Naples, who his teachers were and what ground their courses would have covered, as well as what happened to the contemporaries who shared Bruno's religious life. When Bruno challenged the inquisitors to a theological debate, he did it on the basis of an impressive amount of knowledge of the ideas he was calling into question.

One of the most significant outcomes of this recent study of Bruno's early Christian education has been a growing awareness that there were elements of the Christian religion that he never abandoned. For example, the Christian insistence on both the social and the cosmic values of harmony and love, or Christ's preference for a culture of dialogue and peace rather than one of dogma and violence.[29] Yet there can be no doubt that Bruno's mature philosophy transposed these values into an eternal and fully animated cosmos where there was no room for revealed religions of any kind. For Bruno took an explicit stand against any special forms of incarnation or revelation of the divine love, which he saw as linking together every form of life animating an infinite substance, conceived of as in itself divine. Faced by ideas that seem to us to anticipate Spinoza, it is not surprising to find the inquisitors particularly concerned with Bruno's unorthodox stand concerning the Christian incarnation. Maurice Finocchiaro's quantitative breakdown of the trial sessions shows clearly that it was the subject that figured at most length and in most detail throughout the trial. Typical is the fifth session of the Venetian proceedings against him, when, asked whether he had not declared that Christ is merely human like the rest of us, and whether he was prepared to consider this a grave error, Bruno replied that he might have so erred in his premises, but not in his conclusions.[30] This was an elegant way of explaining to the inquisitors that, given that his infinite universe is to be thought of as permeated throughout by a divine love and reason, Bruno too thought Christ divine, but for reasons and in a mode quite different from their own.

Ultimately, the fundamental issue raised by Bruno's trial cannot be formulated by remaining within the network of theological questions asked by the inquisitors in those long years of the 1590s. These were claimed by the inquisitors themselves, and—as we have seen—by many after them, as "legitimate" because the Inquisition was a judicial system that imposed obedience to the theological canons of the Catholic Church by law, in extreme cases on pain of death. Bruno, however, by admitting so candidly his distance from the Catholic theology, was indirectly questioning such a system of law, which imposed on his conscience views different from his own. This was all the more of a problem for the inquisitors because he was not challenging the post-Tridentine system of inquisitorial law in the name of an alternative concept of Christian values, as so many of the newly militant Protestants were doing in the countries of northern Europe that Bruno had visited. Rather, he was doing it in the name of a principle of religious pluralism that derived directly from his cosmology. For Bruno's infinite number of worlds were conceived of as in a state of eternal becoming, or a process of never-ending flux. This concept leads Bruno to the idea of an infinite number of paths available to the mind in its heroic attempt to gain a glimpse of what he called "the inaccessible divine face." It was a remarkably modern principle of pluralism in the religious sphere that had for Bruno the value of a natural law. A crucial passage in his work that confirms this principle can be found in the *Proemiale epistola* of the same one of his Italian dialogues written and published in London in 1584, *Of the Infinite Universe and Worlds*, where Bruno criticized Cusanus. There Bruno writes: "we (that is the philosophers) will become true observers of the history of nature, which is written within us, and makes us into disciplined executors of the divine laws, which are engraved in the center of our hearts."[31]

This quotation suggests that Bruno had read attentively the *Treatise on Law* in the *Summa theologica* of St. Thomas Aquinas, of which we know he received a copy while in prison. It should be noticed that whereas St. Thomas, in the opening sentences of his *Treatise on Law*, refers to an "*extrinsic* principle" moving us to good, or a transcendental God, Bruno refers to an *intrinsic* principle moving us to good, which is God within the infinite universe, and particularly within the individual mind. A more detailed consideration of Bruno's words would have to relate it also to St. Thomas's discussion of his Third Article: "Whether the Reason of Any Man is Competent to Make Laws" where, to the first objection, which says that anyone can make a law for himself, Thomas replies with a quotation from Isidore (*Etym.* v.10) that refers to ethically valid persons as those who agree to participate in the law rather than to make it for themselves, and thus to "show the work of the law written in their hearts." Bruno agrees with St. Thomas's reply rather than the objection. Generally speaking, then, Bruno agreed with St. Thomas that there is a natural law that acquires meaning only when thought of in relation to the divine law.[32] Bruno maintained this idea of natural law even when his ideas about the eternal truths that contain the divine law had assumed a very different character from those of St. Thomas himself. Bruno also agreed with St. Thomas's further claim that each society must have its network of judicial rules, the laws of the peoples rather than of God, to which the good citizen should at times submit even when unconvinced of their absolute justice.[33] Following the dictates of St. Thomas himself, as well as Machiavelli, whom he certainly knew but never named, Bruno could have justified his repeated gestures of submission both in Venice and in Rome. The point of no return came when Bellarmino drew up a list of heresies taken directly from Bruno's texts, giving him to understand that it was these that must be repudiated. There must be an authentic conversion to the Christian faith. At that point, it was no longer sufficient to submit to a specific set of human laws, historically defined. It was no longer sufficient to say: I will submit to a theology in which I do not believe because my society requires it of me, by law. Bruno's last act in the trial was an attempt to appeal above Bellarmino by sending a written defense of his thought to the Pope himself—a document that has not survived. All that is known is that, in the final session of the trial with the Pope present, the memorial was opened but not read.[34] Only then did Bruno declare that he would not retract because there was nothing that he needed to retract.

So, in what terms can we define exactly the final stand taken by Bruno in Rome in those dramatic days that opened the jubilee year of 1600, bringing with them a new century in the Christian era? Help can be found

in attempting this definition by consulting the thought of a modern Thomist such as Jacques Maritain. For the twentieth-century French philosopher, religious liberty clearly represents a problem. No one would wish to
deny the importance of Maritain's commitment to human rights, or the
influence on his thought of American democracy with its principles of pluralism. Nevertheless, for Maritain, such principles have become necessary
only because the modern world has lost sight of the straight and narrow
Christian way. The mixed city of modern liberalism, writes Maritain, in
his *Essay on Liberty*, must necessarily tolerate "les divisions religeuses que
le progrès du temps, et sa malice, ont inscrites dans l'histoire du monde"
[the religious divisions that the progress of time, in its malice, has written
into the history of the world]. Liberty there must be, but, for Maritain, as
far as religious liberty is concerned, it is essentially what he calls "la liberté
de l'erreur" [the liberty to make mistakes].[35] Over three centuries earlier,
however, Bruno had argued that there is no error in liberty, particularly
not religious liberty. His philosophy inquires into the gods of the ancient
world, the gods of the newly discovered new world, the God of Islam and
the God of Israel, as well as the God of the Christian religion both in the
Catholic and the Protestant formulations. That is to say, Bruno, together
with a small number of other sixteenth-century humanists, thought of
religious pluralism as an essentially positive value.[36] His God is a God of
infinite variety. That is why he could claim his philosophy as a tranquil
universal philology: a way of reading the natural world anew according to
its differences. For these differences were, for Bruno, none other than the
multiple traces of a divine presence through which the heroic mind chooses
to pursue the principle of unity, the monad: the metaphysical foundation
of the infinite facets of being to which the philosopher ultimately dedicates his quest. This does not mean that Bruno thought all religious faiths,
or indeed all secular societies, were good—he could, on the contrary, be
harshly critical. For he would always prefer those religions and those societies that valued universal life and harmony. Ultimately, what Bruno was
proposing was a form of philanthropy—a gesture of friendship and peace
rather than of violence and hate.

Bruno's symbol of philanthropy was the dolphin—that most gentle
of creatures that swims through an ocean of infinitely changing waters,
while constantly attempting to reach the light of a sky it can only occasionally glimpse. In a world once again lacerated by religious conflict and
war, Bruno's proposal made to the Roman inquisitors is surely as relevant
today as it was then. For all he was asking of them was that they should
settle their differences by discussion—by dialogue and debate. That was
clearly a dangerous idea in 1600. It is sad to reflect that it can still be a
dangerous idea today.

NOTES

1. See Bruno (2000), 381. See also my comment on this page in Gatti (1999), 118–19.

2. Bruno's life story has recently been narrated in a new biography in English by Rowland (2008). Recent biographies in Italian are Ricci (2000) and Ciliberto (2007).

3. The standard bibliography of Bruno studies is Salvestrini (1958). A recent bibliography that takes account of the many studies that have surrounded the four-hundredth centenary of Bruno's death is Severini (2002).

4. The known documents relating to Bruno's trial are in Firpo (1993).

5. For Cremonini's case, see Leen Spruit, "Cremonini nelle carte del Sant'Uffizio romano," in *Cesare Cremonini, Aspetti del pensiero e scritti,* eds. E. Riondati and A. Poppi, Padua, Accademia Galileiana di Scienze Lettere e Arti, 2000, vol. I, 193–204.

6. See Ricci (2002), 235–66; 260–63.

7. The text of Schoppe's letter is included in the original Latin in Firpo (1993), 348–55.

8. See Berti (1868; 2nd ed., 1889).

9. See Spampanato (1921); Carusi (1925), 121–39; and Mercati (1942).

10. See Beretta (2001), 15–49; 28–29.

11. See Mercati (1942), 52.

12. Firpo's lengthy and thorough analysis of Bruno's trial first appeared in two parts in the *Rivista storica italiana* in 1948–1949 and then as a single volume in Naples (Edizioni Scientifiche Italiane) in 1949. It was continually revised throughout his life and appears in its final form in Firpo (1993), 3–140.

13. See Fattori (2003), 191–200.

14. For a recent discussion of such connections, see Finocchiaro (2002).

15. For the third session of the trial, see Firpo (1993), 167–68. For the much discussed subject of St. Thomas's ideas about creation, see J.B.M. Wissink, ed., *The Eternity of the World in the Thought of St. Thomas Aquinas and His Contemporaries,* Leiden, Brill, 1990—in particular, Aertsen's contribution at 9–19.

16. "I have always defined my ideas philosophically and according to principles of natural reason" (*io sempre ho diffinito filosoficamente et secondo li principii et lume naturale*). See Firpo (1993), 166, and the relevant comment in Berti (1868; 2nd ed., 1889). On the question of the doctrine of double truth, see Tullio Gregory, "Discussioni sulla doppia verità," *Cultura e Scuola* 2 (1962): 99–106.

17. See Gentile (1907), 59. Gentile's assessment of Bruno's behavior during his trial is discussed, with the relevant quotation, in Spampanato (1921), 518.

18. "... *cattolicamente parlando.*" See Firpo (1993), 176.

19. See Ibid., 299–302.

20. See the introduction in Firpo (1993), 91–93.

21. See Yates (1964), 372–93. On the vexing question of what exactly constituted a formal heresy, and the way the inquisitorial court attempted to clarify this vital aspect of its juridical proceedings during Bruno's trial, see Spruit (2002).

22. There is a vast recent bibliography on matters relating to the Inquisition. The most influential study in English is John Tedeschi, *The Prosecution of Her-*

esy. Collected Studies on the Inquisition in Early Modern Italy, Binghamton, NY, Medieval and Renaissance Texts and Studies, 1991. For the Catholic Church's centuries-long process of meditation on the Galileo affair, see Maurice A. Finocchiaro, *Retrying Galileo (1633–1992)*, Berkeley–Los Angeles–London, University of California Press, 2005—in particular, 189–209.

23. See Seidengart (2001), 21–38.

24. See Firpo (1993), 302–3, where the *Sommario* considers the problem: "*Circa motum terrae*," which was not discussed in Venice. Copernicus's name never figures explicitly in the available documents of Bruno's trial.

25. See Finocchiaro (2002), 51–96.

26. See Quaglione (2003), 29–46.

27. It is not clear from the official account of the inquisitors' visit to the jail whether Bruno requested these items or whether they were simply assigned to him. See Firpo (1993), 217.

28. See Canone (1992), Rowland (2002), and Miele (2002).

29. On this subject, see in particular Ciliberto (2002a) and Tirrinanzi (2004).

30. "*Posso aver errato nelli principii ma non già nelle conclusioni.*" See Firpo (1993), 186.

31. See Bruno (2000), 315. For the Thomistic *Treatise*, see St. Thomas Aquinas, *Treatise on Law*, in *Summa teologica of St. Thomas Aquinas*, trans. Fathers of the English Dominican Province, Westminster, MD, Christian Classics, 1981, vol. 2, 993–1119.

32. See on this subject Oscar J. Brown, *Natural Rectitude and Divine Law in Aquinas*, Toronto, Pontifical Institute of Medieval Studies, 1981.

33. On St. Thomas's idea of human law, see the relevant pages in James P. Reilly, *St. Thomas on Law*, Toronto, Pontifical Institute of Medieval Studies, 1988, and John Finnis, *Aquinas: Moral, Political and Legal Theory*, Oxford, UK, Oxford University Press, 1998. In the *Summa theologiae* (I–II, q. 96, A. 4), St. Thomas, referring to Augustine (*De libero arbitrio*), writes that laws that are not so much laws as acts of moral violence (*magis sunt violentiae quam leges*) have no moral authority over us. One may have a moral obligation to forgo one's rights, however, if insistence on one's rights were to lead to public disorder.

34. "Fratris Iordani Bruni de Nola etc., carcerati etc., memoriale directum Sanctissimo fuit apertum, non tamen lectum." See Firpo (1993), 337.

35. See Jacques Maritain, *Une philosophie de la liberté*, in Jacques and Raïssa Maritain, *Oeuvres complètes*, Fribourg, Switzerland, Editions Universitaires, and Paris, Editions Saint-Paul, 1982, vol. V, 325–87; 378–79. The essay was composed in 1933.

36. For a brief overview of the development of the argument for radical forms of religious liberty on the part of other sixteenth-century humanists such as Pico della Mirandola, Guillaume Postel, Sebastian Castellio, and Jean Bodin, see Quentin Skinner, *The Foundations of Modern Political Thought*, Cambridge, UK, Cambridge University Press, 1978, vol. II, *The Age of Reformation*, 244–54.

WORKS CITED
BY AND ON GIORDANO BRUNO

This bibliography contains only those cited works by, or that refer directly to, Giordano Bruno, either in full-length volumes or in part or parts of a work on a larger subject. The bibliography is limited to works in English, Italian, French, and Spanish. Details of all the other cited works are specified in the notes to the individual chapters.

BRUNO BIBLIOGRAPHIES

Salvestrini, Virgilio (1958). *Bibliografia di Giordano Bruno (1582–1950)*, 2nd ed. posthumous ed., ed. Luigi Firpo. Florence, Sansoni.
Severini, Maria Elena (2002). *Bibliografia di Giordano Bruno: 1951–2000.* Rome, Ed. di Storia e Letteratura.
Sturlese, Rita (1987). *Bibliografia, censimento e storia delle antiche stampe di Giordano Bruno.* Florence, Olschki.

PRIMARY SOURCES

Bruno, Giordano (1879–1891). *Opera latine conscripta*, 4 vols., 7 tomes, eds. F. Fiorentino et al. Naples, Morano.
———. (1958). *Dialoghi italiani I (Dialoghi metafisici)* and *Dialoghi italiani II (Dialoghi morali)*, 3rd ed., ed. Giovanni Gentile, rev. Giovanni Aquilecchia. Florence, Sansoni.
———. (1977). *The Ash Wednesday Supper*, eds. and trans. Edward A. Gosselin and Lawrence S. Lerner. New York, Archon Books.
———. (1980). *Opere latine*, ed. and trans. Carlo Monti. Turin, UTET.
———. (1989). *Summa terminorum metaphysicorum*, ed. Eugenio Canone. Rome, Edizioni dell'Ateneo.
———. (1998). *Cause, Principle and Unity*, eds. and trans. Richard J. Blackwell and Robert de Lucca. Cambridge, UK, Cambridge University Press.
———. (1999a). *Opere italiane: ristampa anastatica delle cinquecentine*, 4 vols., ed. Eugenio Canone. Florence, Olschki.
———. (1999b). *Gli eroici furori*, ed. Nicoletta Tirinnanzi. Milan, BUR.
———. (2000a). *Dialoghi filosofici italiani*, eds. Michele Ciliberto et al. Milan, Mondadori.
———. (2000b). *Opere magiche*, eds. Michele Ciliberto et al. Milan, Adelphi.
———. (2000c). *Poemi filosofici latini*, ed. Eugenio Canone. La Spezia, Agora. An Italian translation of these works is in Bruno (1980).
———. (2001). *Corpus iconographicum*, ed. Mino Gabriele. Milan, Adelphi.

Bruno, Giordano. (2002). *Opere italiane*, 2 vols., eds. Giovanni Aquilecchia and Nuccio Ordine. Turin, UTET.

———. (2004). *Opere mnemoniche, tomo primo*, eds. Michele Ciliberto et al. Milan, Adelphi.

———. (2007). *Centoventi articoli sulla natura e sull'universo contro i peripatetici / Centum et vigenti articoli de natura et mundo adversus Peripateticos*, ed. Eugenio Canone. Pisa–Rome, Fabrizio Serra Editore.

———. (2009) *Opere mnemoniche, tomo secondo*, eds. Michele Ciliberto et al. Milan, Adelphi.

Copernicus, Nicholas (1978). *De revolutionibus (On the Revolutions)*, ed. and trans. Jerzy Dobrsycki, comment by Edward Rosen, London, Macmillan.

Firpo, Luigi, and Diego Quaglione, eds. (1993). *Il processo di Giordano Bruno*. Rome, Salerno Editrice.

Gilbert, William (1651). *De mundo nostro sublunari philosophia nova.* Amsterdam, Ludovicum Elzevirium.

Kepler, Johannes (1965). *Conversations with the Sidereal Messenger recently sent to Mankind by Galileo Galilei*, ed. and trans. Edward Rosen. New York, Johnson Reprint Corp.

Schelling, F. W. (1844; 2nd ed. 1859). *Bruno, ossia un discorso sul principio divino e naturale delle cose*, voltato in italiano dalla Marchesa Florenzi Waddington, aggiuntavi una prefazione di T. Mamiani. (Bruno, or a treatise on the divine and natural principle of things, translated into Italian by the Marchioness Florenzi Waddington, with a preface by Terenzio Mamiani. Milan, Molina; 2nd ed. Florence, Le Monnier.

SECONDARY SOURCES

Abbott, George (1604). *The Reasons which Doctour Hill Hath Brought, for the Upholding of Papistry, Which is Falselie Termed the Catholike Religion.* London, Joseph Barnes: 88–89.

Aquilecchia, Giovanni (1991). *Le opere italiane di Giordano Bruno: critica testuale e oltre*. Naples, Bibliopolis.

———. (1993a). *Schede bruniane.* Manziana, Vecchiarelli.

———. (1993b). "Tre schede su Bruno ad Oxford," in *Giornale critico della filosofia italiana* LXXII: 376–93. An English-language version of this paper was published in 1995 with the title "Giordano Bruno at Oxford" in *Giordano Bruno, 1583–1585: The English Experience*, eds. Michele Ciliberto and Nicholas Mann. Florence, Olschki.

———. (1995a). "Giordano Bruno in Inghilterra (1583–1585)," *Bruniana e campanelliana* I, no. 1/2: 21–42.

———. (1995b). "I 'Massimi Sistemi' di Galileo e la 'Cena' di Bruno," *Nuncius: Annali di Storia della Scienza* X, no. 2: 485–96.

———. (1996). "Sonetti bruniani e sonetti elisabettiani (per una comparazione metrico-tematica)," *Filologia antica e moderna* XI: 27–34.

Atanasijevič, Ksenija [1923] (1972). *The Metaphysical and Geometrical Doctrine of Giordano Bruno.* St. Louis, MO, Warren H. Green.

Badaloni, Nicola (1988). *Giordano Bruno: tra cosmologia ed etica*. Bari, De Donato.

———. (1994). "Riflessioni sul tema dell'*individuo* nella concezione metafisica e morale di Giordano Bruno," *Nouvelles de la république des lettres* II: 31–45.

Barbera, Maria Luisa (1980). "La Brunomania," *Giornale critico della filosofia italiana*, 59, vol. I-IV: 103–40.

Barbieri Squarotti, Giorgio (1958). "L'esperienza stilistica di Bruno fra rinascimento e barocco," in *La critica stilistica e il barocco letterario*. Florence, Le Monnier, 154–69.

Bassi, Simonetta (2004). *L'arte di Giordano Bruno: memoria, furore, magia*. Florence, Olschki.

Berggren, Lars (2002). "The Image of Giordano Bruno," in *Giordano Bruno: Philosopher of the Renaissance*. London, Ashgate: 17–49.

Berretta, Francesco (2001). "Giordano Bruno e l'inquisizione romana: considerazioni sul processo," in *Bruniana e campanelliana*, VII, no. 1: 15–49.

Berti, Domenico (1868; 2nd ed. 1889). *Giordano Bruno da Nola: sua vita e dottrina*. Turin, Paravia.

———. (1876). *Copernico e le vicende del sistema copernicano in Italia*. Rome, Paravia.

Besant, Annie (1913). *Giordano Bruno. Theosophy's Apostle in the Sixteenth Century: Lecture Delivered in the Sorbonne at Paris on June 15th, 1911*. Madras (India), Theosophical Publishing House.

Blum, Paul Richard (1998). "Franz Jacob Clemens e la lettura ultramontistica di Bruno," in Canone ed., *Brunus redivivus*.

Bolzoni, Lina (2002). "Images of Literary Memory in the Italian Dialogues: Some Notes on Giordano Bruno and Ludovico Ariosto," in *Giordano Bruno: Philosopher of the Renaissance*, ed. Hilary Gatti. London, Ashgate, 121–41.

Bönker-Vallon, Angelika (1994–1996). "The Mathematical Aspect of Natural Philosophy in Giordano Bruno's Latin Work. An Example of Early Modern Thought," in *Proceedings of the Patristic, Medieval and Renaissance Conference*, Villanova, PA, 19/20: 125–32.

Bossy, John (1991). *Giordano Bruno and the Embassy Affair*. New Haven, CT, Yale University Press.

Braden, Gordon (1999). *Petrarchan Love and the Continental Renaissance*. New Haven, CT–London, Yale University Press.

Buono Hodgart, Amelia (1978). "*Love's Labour's Lost* di William Shakespeare e il *Candelaio* di Giordano Bruno," *Studi secenteschi* XIX: 3–21.

———. (1997). *Giordano Bruno's* Candle-Bearer: *An Enigmatic Renaissance Play*. Lewiston, NY, E. Mellen.

Calcagno, Antonio (1998). *Giordano Bruno and the Logic of Coincidence: Unity and Multiplicity in the Philosophical Thought of Giordano Bruno*. Frankfurt–New York, Lang.

Calogero, Guido (1963). "La professione di fede di Giordano Bruno," *La cultura* I: 64–77.

Cambi, Maurizio (1993). "Il 'De magia' e il recupero della sapienza originaria. Scrittura e voce nelle strategie magiche di Giordano Bruno," *Archivio di storia della cultura* VI: 9–33.

Cambi, Maurizio. (2002). *La macchina del discorso*. Naples, Liguori.

Canone, Eugenio, ed. (1992). *Giordano Bruno: Gli anni napolitani e la 'peregrinatio' europea*. Cassino, l'Università di Cassino.

———., ed. (1998a). *Brunus redivivus. Momenti della fortuna di Giordano Bruno nel XIX secolo*. Pisa–Rome, Istituti Editoriali e Poligrafici Internazionali.

———. (1998b). "Il concetto di 'ingenium' in Bruno," *Bruniana e campanelliana* IV: 11–35.

———. (2003). *Il dorso e il grembo dell'eterno: Percorso della filosofia di Giordano Bruno*. Pisa–Rome, Istituti Editoriali e Poligrafici Internazionali.

Canone, Eugenio, and Ingrid Rowland, eds. (2007). *The Alchémy of Extremes: The Laboratory of the* Eroici furori *of Giordano Bruno*. Pisa–Rome, Istituti Editoriali e Poligrafici Internazionali.

Carusi, Enrico (1925). "Nuovi documenti del processo di Giordano Bruno," in *Giornale critico della filosofia italiana* VI, vol. II: 121–39.

Catana, Leo (2005). *The Concept of Contraction in Giordano Bruno's Philosophy*. London, Ashgate.

Ciliberto, Michele (1979). *Lessico di Giordano Bruno*. Rome, Edizioni di Storia e Letteratura.

———. (1986). *La ruota del tempo: interpretazione di Giordano Bruno*. Rome, Editori Riuniti.

———. (1990). *Giordano Bruno*. Rome–Bari, Laterza.

———. (2002a). "Bruno, Mocenigo e il nuovo mondo: un esercizio di lettura," in *Rinascimento* XLII: 113–41.

———. (2002b). *L'occhio di Atteone: nuovi studi su Giordano Bruno*. Rome, Edizioni di Storia e Letteratura.

———. (2007). *Giordano Bruno: il Teatro della Vita*. Milan, Mondadori.

Ciliberto, Michele, and Nicoletta Tirinnanzi (2002). *Il dialogo recitato: per una nuova edizione del Bruno volgare*. Florence, Olschki.

Clucas, Stephen (2002). "*Simulacra et Signacula*": Memory, Magic and Metaphysics in Brunian Mnemonics," in *Giordano Bruno: Philosopher of the Renaissance*, ed. Hilary Gatti. London, Ashgate: 251–72.

Corsano, Antonio (1940). *Il pensiero di Giordano Bruno nel suo svolgimento storico*. Florence, Sansoni.

Dagron, Tristan (1999). *Unité de l'être e dialetique: l'idée de philosophie naturelle chez Giordano Bruno*. Paris, Droz.

De Bernart, Luciana (1986). *Immaginazione e scienza in Giordano Bruno*. Pisa, ETS.

———. (2002). *Numerus quodammodo infinitus: per un approccio storico-teorico al 'dilemma matematico' nella filosofia di Giordano Bruno*. Rome, Edizioni di Storia e Letteratura.

De Léon Jones, Karen (1997). *Giordano Bruno and the Kabbalah*. New Haven, CT–London, Yale University Press.

Deregibus, Arturo (1981). *Bruno e Spinoza: la realtà dell'infinito e il problema della sua unità*. 2 vols. Turin, Giappichelli.

Ellero, Maria Pia (2005). *Lo specchio della fantasia: retorica, magia e scrittura in Giordano Bruno*. Lucca, M. Pacini Fazzi.

Farley-Hills, David (1992). "The *Argomento* of Bruno's *De gli eroici furori* and Sidney's *Astrophel and Stella*," *Modern Language Review* 87: 1–17.

Fattori, Marta (2003). "*Qua epistola cum nimium utilis, et fructuosa:* un nuovo documento sulla lettera di Gaspare Scioppo," *Nouvelles de la république des lettre*, I/II: 191–200.

Feingold, Mordechai (1984). *Occult and Scientific Mentalities of the Renaissance*. Oxford, UK, Oxford University Press.

———. (2004). "Giordano Bruno Revisited," *Huntingdon Library Quarterly* 3: 329–48.

Finocchiaro, Maurice (2002). "Philosophy versus Religion and Science versus Religion: The Trials of Bruno and Galileo," in *Giordano Bruno: Philosopher of the Renaissance*, ed. Hilary Gatti. London, Ashgate: 51–96.

Frith (Oppenheimer), I. (1887). *Life of Giordano Bruno the Nolan*. London, Nicholas Trübner.

Garin, Eugenio (1976). *Lo zodiaco della vita*. Rome–Bari, Laterza.

———. (1984). Introduction to *Spiritus. IV Colloquio Internazionale Europeo: Atti*, eds. Marta Fattori and Massimo Bianchi. Rome, Edizioni dell'Ateneo, 3–14.

Gatti, Hilary (1989). *The Renaissance Drama of Knowledge: Giordano Bruno in England*. London, Routledge.

———. (1994). "Telesio, Giordano Bruno e Thomas Harriot," in *Accademia cosentina: Atti 1991–2*. Cosenza, Accademia Cosentina. 63–74.

———. (1997). "Giordano Bruno's 'The Ash Wednesday Supper' and Galileo's 'Dialogue of the Two Major World Systems,'" *Bruniana e campanelliana* III, no. 2: 283–300.

———. (1999). *Giordano Bruno and Renaissance Science*. Ithaca, NY, Cornell University Press.

———. (2002). "Bruno and the Protestant Ethic," in *Giordano Bruno: Philosopher of the Renaissance*, ed. Hilary Gatti. London, Ashgate: 145–66.

———. (2008a). "Copernico (sez. Giordano Bruno)," *Bruniana e campanelliana* XIV, no. 2: 511–20.

———. (2008b). "Giordano Bruno and Shakespeare's Idea of a Play," in *Shakespeare e l'Italia, Memoria di Shakespeare*, vol. 6. Rome, Bulzoni: 111–19.

———. (2008c). "Paolo Giovio in Inghilterra: la traduzione inglese di Samuel Daniel," in *Con parola brieve e con figura*, eds. Lina Bolzoni and Silvia Volterrani. Pisa, Edizioni della Normale: 185–98.

Gentile, Giovanni (1907). *Giordano Bruno nella storia della cultura*. Milan, Remo Sandron.

Granada, Miguel A. (1990). "L'interpretazione bruniana di Copernico e la *Narratio prima* di Rheticus," *Rinascimento* 30, 343–65.

———. (1993). "Giordano Bruno e la tradizione filosofica," in *L'interpretazione nei secoli XVI e XVII*, eds. G. Canziani and Y. C. Zarka. Milan, Franco Angeli: 59–82.

———. (1994). "Il rifiuto della distinzione fra 'potentia absoluta' e 'potentia ordinata' di Dio e l'affermazione dell'universo infinito in Giordano Bruno," *Rivista di storia della filosofia* XLIX, 3: 495–532.

Granada, Miguel A. (1999). "'Esser spogliato dall'umana perfezione e giustizia.' Nueva evidencia de la presencia de Averroes en la obra y en el processo de Giordano Bruno," *Bruniana e campanelliana*, V, no. 2: 305–31.

———. (1996). *El debate cosmológico en 1588: Bruno, Brahe, Rothmann, Ursus, Röslin*. Naples, Bibliopolis.

———. (2005). *La reivindicaciòn de la filosofia en Giordano Bruno*. Barcelona, Herder.

Gregory, Andrew (2002). "Macrocosm, Microcosm and the Circulation of the Blood: Bruno and Harvey," in *Giordano Bruno: Philosopher of the Renaissance*. London, Ashgate: 365–80.

Ingegno, Alfonso (1978). *Cosmologia e filosofia nel pensiero di Giordano Bruno*. Florence, La Nuova Italia.

———. (1985). *La sommersa nave della religione. Studio sulla polemica anticristiana del Bruno*. Naples, Bibliopolis.

———. (1987). *Regia pazzia. Bruno lettore di Calvino*. Urbino, Quattro Venti.

Knox, Dilwyn (1999). "Ficino, Copernicus and Bruno on the Motion of the Earth," *Bruniana e campanelliana* V, no. 2: 333–66.

———. (2001). "Bruno's Doctrine of Gravity, Levity and Natural Circular Motion," *Physis* XXXVIII: 171–209.

Koyré, Alexandre (1939–1940). *Etudes Galiléennes*, 2 vols. Paris, Hermann.

———. (1957). *From the Closed World to the Infinite Universe*. Baltimore, MD, John Hopkins University Press.

Kristeller, Paul O. (1964). Section on Bruno in *Eight Philosophers of the Italian Renaissance*. Stanford, California University Press: 127–139.

Kuhn, Thomas (1957). *The Copernican Revolution: Planetary Astronomy in the Development of Western Thought*. Cambridge, MA, University of Harvard Press.

Lerner, Michel-Pierre (1996–1997). *Le monde des spheres*, 2 vols. Paris, Les Belles Lettres.

Limentani, Ludovico (1933). "La lettera (di Bruno) al Vice-Cancelliere dell'Università di Oxford," *Sophia* I: 317 ff.

Lüthy, Christophe H. (1998). "Bruno's Area Democriti and the Origins of Atomistic Imagery," *Bruniana e campanelliana* 4, no. 1: 59–92.

Maggi, Armando (2003). "L'uomo astratto. Filosofia e retorica emblematica negli *Eroici furori*," *Bruniana e campanelliana* IX, no. 2: 319–44.

Maiorana, Giancarlo (1982). "A Voice of Its Own Birth: Bruno and the Foundations of Coleridge's Poetics," *Comparative Literature Studies* 19, no. 3: 296–318.

Mancini, Sandro (2000). *La sfera infinita. Identità e differenza nel pensiero di Giordano Bruno*. Milan, Mimesis.

Massa, Daniel (1973). "Giordano Bruno and the Top-Sail Experiment," *Annals of Science* 30, no. 2: 201–11.

McIntyre, J. Lewis (1903). *Giordano Bruno*. London, Macmillan.

McMullin, Edward (1987). "Bruno and Copernicus," *Isis* 78, no. 1: 55–74.

McNulty, Robert (1960). "Bruno at Oxford," *Renaissance News* XIII: 300–305.

Mendoza, Ramon G. (1995). *The Acentric Labyrinth: Giordano Bruno's Prelude to Contemporary Cosmology*. Shaftesbury, UK, Element Books.

———. (2002). "Metempsychosis and Monism in Bruno's *nova filosofia*," in *Giordano Bruno: Philosopher of the Renaissance*, ed. Hilary Gatti. London, Ashgate: 273–97.

Mercati, Angelo, ed. (1942). *Il sommario del processo di Giordano Bruno*. Rome, Città del Vaticano.

Michel, Paul-Henri (1960). "L'atomisme de Giordano Bruno," in *La science au sezième siècle*. Paris, Hermann: 249–64

———. [1962] (1973). *The Cosmology of Giordano Bruno*. Ithaca, NY, Cornell University Press.

Miele, Michele (2002). "La formazione di Giordano Bruno a S. Domenico Maggiore," in *Giordano Bruno: oltre il mito e le opposte passioni*, eds. Pasquale Giustiniani et al. Naples, Biblioteca Teologica Napolitana: 63–79.

———. (2003). "Giordano Bruno: i documenti napolitani," *Bruniana e campanelliana* IX, no. 1: 159–203.

Monti, Carlo (1994). "Incidenza e significato della tradizione materialistica antica dei poemi latini di Giordano Bruno," *Nouvelles de la république des lettres* II: 75–87.

Nelson, John Charles (1958). *Renaissance Theory of Love: The Context of Giordano Bruno's* Heroici furori. New York–London, Columbia University Press.

Ordine, Nuccio (2003). *La soglia dell'ombra: letteratura, filosofia e pittura in Giordano Bruno*. Venice, Marsilio Editore.

———. (2007). *Contro il vangelo armato: Giordano Bruno, Ronsard e la religione*. Milan, Raffaello Cortina Editore.

Provvidera, Tiziana (2002). "John Charlewood, Printer of Giordano Bruno's Italian Dialogues, and His Book Production," in *Giordano Bruno: Philosopher of the Renaissance*, ed. Hilary Gatti. London, Ashgate: 167–86.

Quaglione, Diego (2003). "L'autodefénce de Giordano Bruno," in *Mondes, Forme et Société selon Giordano Bruno*, eds. Tristan Dagron and Hélène Vedrine. Paris, J. Vrin: 29–46.

Ricci, Saverio (1990a). "Infiniti mondi e mondo nuovo: la conquista dell'America e la critica alla civiltà europea in Bruno," *Giornale critico della filosofia italiana* 10, no. 2: 204–21.

———. (1990b). *La fortuna del pensiero di Giordano Bruno (1600–1750)*. Florence, Le Lettere.

———. (1991). "La recezione del pensiero di Giordano Bruno in Francia e Germania. Da Diderot a Schelling," *Giornale critico della filosofia italiana* LXX: 431–65.

———. (2000). *Giordano Bruno nell'Europa del Cinquecento*. Rome, Salerno Editrice.

———. (2002). "Da Santori a Bellarmino: la politica romana e il processo a Giordano Bruno," in *Giordano Bruno: oltre il mito e le opposte passioni*, eds. Pasquale Giustiniani et al. Naples, Biblioteca Teologica Napolitana: 235–66.

Roche Jr., Thomas P. (1989). *Petrarch and the English Sonnet Sequences*. Brooklyn, NY, AMS Press.

Rossi, Arcangelo (2001). "Bruno, Copernico, Galilei," *Physis* XXXVIII: 283–303.

Rossi, Paolo [1960] (2000). *Clavis universalis: Logic and the Art of Memory*. London, Athlone Press.

Rowland, Ingrid D. (2002). "Giordano Bruno and Neapolitan Neoplatonism," in *Giordano Bruno: Philosopher of the Renaissance*, ed. Hilary Gatti. London, Ashgate: 97–119.

———. (2003a). "Giordano Bruno and Vernacular Poetry," *Bruniana e campanelliana* IX, no. 1: 141–55.

———. (2003b), "Giordano Bruno e Luigi Tansillo," *Bruniana e campanelliana* IX, no. 2: 345–55.

———. (2008). *Giordano Bruno: Philosopher and Heretic*, New York, Farrar, Strauss and Giroux.

Sabbatino, Pasquale (1996). *Giordano Bruno e la 'mutazione' del rinascimento.* Florence, Olschki.

Sacerdoti, Gilberto (1998). *Nuovo cielo, nuova terra. La rivelazione copernicana di* Antionio e Cleopatra *di Shakespeare.* Bologna, Il Mulino.

———. (2002). *Sacrificio e sovranità: teologia e politica nell'Europa di Shakespeare e Bruno.* Turin, Einaudi.

Saiber, Arielle (2005). *Giordano Bruno and the Geometry of Language.* London, Ashgate.

Seidengart, Jean (1992). "La cosmologie infinitiste de Giordano Bruno," in *Infinie des mathématiciens, Infinie des philosophes.* Paris, Belin: 59–82.

———. (2001). "L'infinitisme brunien devant l'Inquisition," in *Cosmologìa, teologìa y religion en la obra y en el proceso de Giordano Bruno.* Barcelona, Universitat de Barcelona: 21–38.

Simoni, Varanini L. (2003). "La *Dissertatio cum Nuncio Sidereo* fra Galileo e Bruno," *Bruniana e campanelliana* IX, no. 1: 207–15.

Singer, Dorothea (1950). *Giordano Bruno: His Life and Thought.* New York, H. Schuman.

Spampanato, Vincenzo (1921). *Vita di Giordano Bruno con documenti editi ed inediti.* Messina, Principato.

Spruit, Leen (1986). "Magia 'socia natura,'" *Centauro* XVII–XVIII: 146–69.

———. (1988). *La dottrina della conoscenza in Giordano Bruno.* Naples, Bibliopolis.

———. (1995). Species Intelligibilis: *From Perception to Knowledge*, vol. 2. Leiden, Brill. 203–13.

———. (2002). "Una rilettura del processo di Giordano Bruno: procedure e aspetti giuridico-formali," in *Giordano Bruno: oltre il mito e le opposte passioni*, eds. Pasquale Giustiniani et al. Naples, Biblioteca Teologica Napolitana. 217–34.

———. (2003). "La psicologia di Bruno nel processo," in *Autobiografia e filosofia: l'esperienza di Giordano Bruno*, ed. Nestore Pirillo. Rome, Edizioni di Storia e Letteratura: 263–85.

Sturlese (Pagnoni), Rita (1985). "Su Bruno e Tycho Brahe," *Rinascimento* XXV, no. 2: 309–33.

———. (1990). "Il *De imaginum, signorum et idearum compositione* di Giordano Bruno e il significato filosofico dell'arte della memoria," *Giornale critico della filosofia italiana* LXIX: 182–203.

———. (1992). "Averroe quantumque arabo et ignorante di lingua greca . . ." in *Giornale critico de la filosofia italiana*, 71, vol. II: 248–75.

———. (1993). "Per una interpretazione del *De umbris idearum* di Giordano Bruno," *Annali della Scuola Normale Superiore di Pisa* XXII: 943–68.

———. (1994). "Le fonti del *Sigillus sigillorum* di Bruno, ossia il confronto con Ficino a Oxford sull'anima umana," *Nouvelles de la république des lettres* 2: 89–167.

Tarantino, Elisabetta (2002). "*Ultima Thule*: Contrasting Empires in Bruno's *Ash Wednesday Supper* and Shakespeare's *Tempest*," in *Giordano Bruno: Philosopher of the Renaissance*, ed. Hilary Gatti. London, Ashgate: 201–25.

———. (2007). "The *Eroici furori* and Shakespeare," in *The Alchemy of Extremes: the Laboratory of the* Eroici furori *of Giordano Bruno*, eds. Eugenio Canone and Ingrid Rowland. Pisa–Rome, Istituti Editoriali e Poligrafici Internazionali: 143–55.

Tessicini, Dario (2001). "Pianeti consorti: la Terra e la Luna nel diagramma eliocentrico di Giordano Bruno," in *Cosmologìa, teologìa y religion en la obra y en el proceso de Giordano Bruno*. Barcelona, Universitat de Barcelona: 159–88.

———. (2007). *I dintorni dell'infinito: Giordano Bruno e l'astronomia dell'infinito*. Pisa–Rome, Fabrizio Serra Editore.

Tirinnanzi, Nicoletta (1996–1997). "Il *Cantico dei Cantici*, nel *De umbris idearum* di Giordano Bruno," in *Letture bruniane I–II*. Pisa–Rome, Istituti Editoriali e Poligrafici Internazionali: 287–306.

———. (2004). "Filosofia, politica e magia nel Rinascimento: l'esperienza di Giordano Bruno," in *Giordano Bruno nolano e cittadino europeo*. Grottaglie, Regione Puglia CRSEC and Scorpione editrice: 31–59.

Tocco, Felice (1889). *Le opere latine di Giordano Bruno esposte e confrontate con le italiane*. Florence, Le Monnier.

———. (1892). "Le fonti più recenti della filosofia del Bruno," *Rendiconti della R. Accademia dei Lincei, Classe di scienze morali ecc.* I: 503–38.

Vasoli, Cesare (1958). "Umanesimo e simboleggia nei primi scritti lulliani e mnemotecniche del Bruno," *Archivio di filosofia* 2–3: 251–304.

Vedrine, Hélène (1967). *La conception de la nature chez Giordano Bruno*. Paris, J. Vrin.

———. (1990). "De la Porta et Bruno sur la Nature e la Magie," in *G. B. De la Porta nell'Europa del suo tempo*, ed. Maurizio Torrini. Naples, Guida Editore: 243–59.

———. (1993). "Alchemie, hérmetisme et philosophie chez Giordano Bruno," in *Alchemie et philosophie à la renaissance*, eds. J. C. Margolin and S. Matton. Paris, J. Vrin: 355–63.

Walker, D. P. (1958). *Spiritual and Demonic Magic from Ficino to Campanella*. London, Warburg Institute Studies.

Westman, Robert (1975). "The Melancthon Circle, Rheticus, and the Wittenberg Interpretation of the Copernican Theory," *Isis* 66, no. 2: 164–93.

———. (1977). "Magical Reform and Astronomical Reform: The Yates Thesis Re-considered," in *Hermeticism and the Scientific Revolution*. Los Angeles, University of California Press: 1–92.

Wildgen, Wolfgang (2001). "La filosofia di Bruno come guida ad una semiotica della scienza moderna," *Physis: Rivista internazionale di storia della scienza* XXXVIII: 473–86.

Wyatt, Michael (2002). "Giordano Bruno's Infinite Worlds in John Florio's World of Words," in *Giordano Bruno: Philosopher of the Renaissance,*" ed. Hilary Gatti. London, Ashgate: 187–99.

Yates, Frances (1934). *John Florio: The Life of an Italian in Shakespeare's England.* Cambridge, UK, Cambridge University Press.

———. (1943). "The Emblematic Conceit in Giordano Bruno's *Degli eroici furori* and in the Elizabethan Sonnet Sequences," *Journal of the Warburg and Courtauld Institutes* 6: 81–101, now in Frances Yates, *Collected Essays* vol. 1. London, Routledge, 1982: 180–209.

———. (1964). *Giordano Bruno and the Hermetic Tradition.* London, Routledge and Kegan Paul.

———. (1966). *The Art of Memory.* London, Routledge and Kegan Paul.

INDEX

Note: *Page references in italics indicate illustrations and captions.*